John Richardson

The Polar Regions

John Richardson

The Polar Regions

ISBN/EAN: 9783337327507

Printed in Europe, USA, Canada, Australia, Japan

Cover: Foto ©berggeist007 / pixelio.de

More available books at **www.hansebooks.com**

THE
POLAR REGIONS

BY

SIR JOHN RICHARDSON, LL.D.

F.R.S. LOND.; HON. F.R.S. EDIN., ETC. ETC.

EDINBURGH:
ADAM AND CHARLES BLACK.
1861.

PREFACE.

The present work is founded on an article written for the last edition of the Encyclopædia Britannica under the same title, and has been expanded by more detailed expositions of the various subjects it embraces, at the request of the Publishers of the Encyclopædia, who thought that its appearance as a substantive work might supply a want.

Its design is to give a connected view of the physical geography and ethnology of the areas comprised within the north and south polar circles, and of the progress of discovery by which our knowledge of these extremities of our globe has been attained. To keep the volume within reasonable limits, the different matters it comprises are necessarily treated in a summary way, but the numerous references to authorities will enable a student to go to the fountains of information.

CONTENTS.

PART FIRST.

	PAGE
INTRODUCTION	1

CHAPTER I.
ANTE-COLUMBIAN PERIOD—B.C. 52–A.D. 1494 . . 14

CHAPTER II.
A.D. 1492–1527 35

CHAPTER III.
VOYAGES TO THE NORTH-EAST FROM ENGLAND—A.D. 1548–1580 53

CHAPTER IV.
DUTCH NORTH-EASTERN VOYAGES—A.D. 1594–1597 . . 64

CHAPTER V.
ENGLISH NORTH-WEST VOYAGES—A.D. 1576–1636 . 76

CHAPTER VI.
AMERICAN CONTINENT, ETC.—A.D. 1668–1790 . . 112

CHAPTER VII.
RUSSIAN VOYAGES ALONG THE SIBERIAN COAST—A.D. 1598–1843 130

CHAPTER VIII.

NINETEENTH CENTURY—ENGLAND—A.D 1817–1845 . . 141

CHAPTER IX.

NINETEENTH CENTURY—(*continued*)—SIR JOHN FRANKLIN . 156

CHAPTER X.

SEARCHING EXPEDITIONS—A.D. 1847–1859 . . 171

CHAPTER XI.

SPITZBERGEN . 203

CHAPTER XII.

CURRENTS OF THE POLAR SEAS . 219

CHAPTER XIII.

ICE . . 237

CHAPTER XIV.

WINDS . . 244

CHAPTER XV.

TEMPERATURE 249

CHAPTER XVI.

VEGETATION 263

CHAPTER XVII.

ZOOLOGY . 274

CHAPTER XVIII.

GEOLOGY 285

CHAPTER XIX.

ESKIMOS 298

CHAPTER XX.

SAMOYEDS 331

CHAPTER XXI.

LAPLANDERS OF YUGRIAN ORIGIN 344

PART SECOND.

CHAPTER I.

A.D. 1576–1840—ANTARCTIC POLAR REGIONS . 351

CHAPTER II.

DISCOVERY OF VICTORIA LAND 367

CHAPTER III.

REMARKS ON THE PHYSICAL GEOGRAPHY OF THE ANTARCTIC REGIONS 374

POSTSCRIPT.

LETTER OF DON PEDRO DE AYALA ON THE VOYAGE OF JOHN CABOT IN 1497 381

INDEX 385

POLAR REGIONS.

PART I.

ARCTIC FRIGID ZONE.

SECTION I.

PROGRESS OF DISCOVERY.

INTRODUCTION.

Ancient ignorance of the Polar Regions—The Phœnicians the first navigators of the Northern Atlantic—Æstrymnades or Cassiterides—Tin—Tarshish—Ireland—America.

In attempting to trace the rise and progress of our present knowledge of the Polar Regions, we naturally turn first to the ancient historians, but we can glean very little exact information from them, their writings containing only obscure notices of countries lying towards the arctic circle, and no account whatever of a corresponding antarctic climate, though some of the old philosophers did express a suspicion of its existence. The Phœnicians alone of the ancients, and their Carthaginian descendants, have a fair claim to the discovery by sea of the western coasts of Europe, and this they achieved in the pursuit of commerce. Tin was the staple commodity sought by their northern voyages, and from Cornwall, then, and still the chief source of that metal, they supplied the world. It is indeed generally supposed, that all the tin in use at the dawn of history came from Cornwall, and if that opinion be correct, the intercourse between the Mediterranean and Britain must have begun at a very early date. In the Book of Numbers tin is specially mentioned among the spoil taken from the Midianites 1452 years before the Christian era. "Only the gold, and the silver, the brass, the iron, the *tin* and the lead, everything that may abide the fire, ye shall make it go through the fire, and it shall be clean."*

The bright white *Kassiteros*† or *Kattiteros*, is named often in the Iliad as a valuable ornament of chariots and armour, of which we have examples in the following passages taken from Cowper's translation:—

> Impenetrable brass, *tin*, silver, gold,
> He cast into the forge.
> *Il.*, xviii. v. 590.

> There, also laden with its fruit, he form'd
> A vineyard all of gold; purple he made
> The clusters, and the vines supported stood
> By poles of silver, set with even rows.
> The trench he colour'd sable and around
> Fenced it with *tin*. *Il.*, xviii. v. 701.

> With five folds
> Vulcan had fortified it: two were brass,
> The two interior *tin*: the midmost gold.
> *Il.*, xx. v. 336.

* Chap. xxxi. v. 22. See also Isaiah i. 25; Ezekiel xxii. 18; xxvii. 12. Mr. Rawlinson supposes that the Phœnicians did not emigrate to the Mediterranean coasts until the thirteenth century before Christ, and if he is correct in giving that date, the Midianites could not have obtained their tin from Cornwall through the Sidonians or Tyrian Phœnicians.—*Translation of Herodotus*, iv. p. 249. Pliny however says, "*India neque æs neque plumbum habet, gemmisque suis ac margaritas hæc permutat.*"

† Pliny says that the *cassiteron* of the Greeks is the metal which he calls *plumbum album*. According to Sir Gardner Wilkinson, tin is termed *Kasdeer* in Arabic, and *Kastira* in Sanscrit. *Stannum* is supposed to have been an alloy of lead, tin, and other metals, combined with silver in the ore, and separated by melting. The word had perhaps a barbaric origin, for the Irish *stán* is as likely to have been the root as the derivative of the Latin epithet. The Icelandic *din*, the Swedish *tenn*, the English and Dutch *tin*, the German *zinn*, and the French *étain* may have come from either the Irish or the Latin. The Welsh *alcan* looks as if it had received an Arabic prefix. In the Truro Museum there is a pig of tin, which, from its peculiar shape, is supposed to be Phœnician, as it differs from the Roman and Norman pigs found also in Cornwall. A figure of this pig is given in Rawlinson's Herodotus, book iii. chap. 116, which see for further details.

INTRODUCTION.

> I will present to him my corselet bright,
> Won from Asteropæus, edg'd around
> With glittering *tin;* a precious gift and rare.
> *Il.,* xxiii. v. 694.

> Radiant with *tin* and gold the chariot ran.
> *Il.,* xxiii. v. 629.

Amber was also an article of Phœnician commerce obtained in the north, and wrought into ornaments by Tyrian workmen.

> A splendid collar, gold with *amber* strung.
> *Od.,* xv. v. 556.

> A necklace of wrought gold, with *amber* rich
> Bestudded, ev'ry bead bright as a sun.
> *Od.,* xviii. v. 358.

As the principal source of amber in the present day, as well as in former times, is the south shore of the Baltic, especially the Gulf of Dantzic; it has been conjectured that the Phœnician ships had penetrated into the Baltic, but so light an article as amber was more likely to have been brought overland from the Baltic, as the furs of the Ural Mountains came further over the continent in the time of Herodotus. Amber is the product also of other coasts in the north of Europe. Pliny tells us that an island lying near the peninsula of *Cartris* (Jutland), which is terminated by the Cimbric cape, was named *Glessaria* by the Roman soldiers because it produced amber (the *Glessum* of the Germans, and *Glær* of the Anglo-Saxons). Considerable pieces of amber are still found occasionally on the Lincolnshire coast by a few men who gain a livelihood by digging up trees from the submarine forest there, to which they get access in the spring-tides.*

---

* "The throwing up of this fossil (amber) during two thousand years on the coast between Memel and Dantzig, leads us to assume that there subsists

Without, however, claiming for the Phœnicians as certain, the honour of having first sailed down the Cattegat, there is little reason for doubting the navigation of the Northern Atlantic by that people, and their pre-eminent skill in nautical affairs, which was indeed readily conceded to them by rival nations.* Pliny, in his list of mythical inventors, says that Hippus, a Tyrian, constructed the *first* "merchant ship" and that "cock-boats" (*cymbæ*) owed their origin to the Phœnicians, who were, moreover, the first people who directed their course on the ocean by the fixed stars. The Carthaginians, he adds, first built "a ship of four banks of oars," and they were also the first who instituted a trade in merchandise, though from father Jupiter came the instinct of buying and selling.

When we consider that the Carthaginians had formed settlements on the western islands of the Mediterranean and on the coasts of Spain,† before the Romans possessed a fleet, and were the discoverers and first colonizers of Madeira, we are not of the number of those who believe that they never ventured out to sea but servilely hugged the shore in long coasting voyages. It is scarcely possible that a people, who, during the many centuries that intervened between their first occupation of Sidon and Tyre, and the destruction of Carthage by the Romans, had enjoyed a monopoly of traffic on the Erythræan, Mediterranean, and Atlantic oceans, should have

at that place a peculiar brown coal formation."—*Erman*, Travels in Siberia, translated by W. D. Cooley, Lond., i. p. 11. Minute grains of amber, or of a substance very similar to it, exist in some of the strata of a (miocene) tertiary formation on the Mackenzie.

* Rawlinson's Herodotus, ii. 414; iv. 46.

† Gades (*Gaddir*) and Utica, according to the Phœnician annals, as quoted by Aristotle, were founded at the same date, about 270 years prior to the building of Carthage, or 1130 B. C.—*Heeren*.

remained stationary in the art of navigation.* It has been alleged that their ships were too small for anything but coasters, though the tonnage of a Carthaginian vessel must have greatly exceeded that of some ketches and fly-boats that braved the Greenland seas in the days of Queen Elizabeth. A pinnace which formed one of Frobisher's first fleet was only of ten tons burthen, and his other two ships were little more than twice as big. Contrast this with the armament of Hanno, composed of sixty ships, capable of carrying thirty thousand colonists of all ages to the western coast of Africa, and averaging five hundred passengers to each ship.†

There is no reason to suppose that the fleet of Himilco, which sailed at the same time to colonize Western Europe, was less efficiently organized, though it may have numbered fewer ships. Himilco's narrative has unfortunately perished, but Festus Avienus, who consulted a record of it deposited in one of the temples of Carthage, states that the Carthaginian admiral navigated the Atlantic four months, planting colonies doubtless on his way. If Strabo be correct when he assigns a Phœnician origin to two-hundred towns in the south of Spain, we cannot but believe that many of the maritime towns of Portugal were also founded by that people. Himilco at length reached the promontory and bay of Æstrymon and the

* See Dr. Redshob, on *Tartessus*, in Notes and Queries, Jan. 1, 1859, 2d. series, vii. p. 3. Strabo says that before the time of Homer the Phœnicians had possession of the best parts of Africa and Spain, iii. p. 104.

† According to Strabo, the Getuli and Libyans destroyed three hundred cities founded by Tyre in western Africa. Professor Owen identifies the "hairy men" which Hanno's sailors slew and skinned, with the gigantic *Gorilla* or Satyr of the country lying under the equator on the west coast of Africa.

As to the size of the ships, Strabo tells us that the merchants of Gades employed very large vessels in their sea voyages, though the poor fishermen of the city had only small boats, which they named "horses," because their prows bore figure-heads of that animal.

islands of the Æstrymnides, which were described as being rich in *tin* and *lead*, and inhabited by a very numerous, spirited, and industrious people, devoted to commerce, and navigating the sea in boats of hide; in which one may recognise the Welsh coracle.* Distant two days' sail from these islands lay the green-turfed holy island† of the Irish race, and near at hand, the Island of Albion. Himilco was contemporary with Aristotle, who applies the epithet of *Keltikon* to *Kassiteros*.

The Æstrymnides have been identified with the Cassiterides or "Tin islands" of the Greeks, and by some moderns with St. Michael's Mount, the Lizard, and the adjacent promontories of Cornwall; by others they are thought rather to be the Scilly Islands. Herodotus (B. C. 450) mentions the Cassiterides, as does also Aristotle (B. C. 340). Polybius (B. C. 160) makes a distinct reference to the British Isles.‡ This knowledge, imperfect as it was, could have been acquired only from the Phœnicians, or from the Phocæan Greeks, who began to frequent Spain in the time of Cyrus (B. C. 566).

Avienus says, moreover, that long before Himilco's time, trading voyages were made from Tartessus to the Æstrymnides, and that *Gaddir* was the Punic appellation of the seaport anciently called *Tartessus*, or, as its Tyrian founders named it, *Tarshish*.‖ When the Phocæans first visited this port, in the sixth century before the Christian era, it had a monarch

---

* There is no proof that the coracle was the only sea-boat of the ancient Britons, though the peculiarity of its construction brought it strongly into notice.

† *Inis-fail* (Hibernice), *Insula fatalis vel sacra*.

‡ Moore supposes that all the British Isles, including Albion, Jerne, Scilly, and the Isle of Man, were called Cassiterides.—*Hist. of Ireland*.

‖ "Ships of Tarshish" came in time to signify vessels built for long voyages, and the ships of Solomon and Hiram (1 Kings x. 22; Chr. ix. 21), destined to bring gold, silver, ivory, apes, and peacocks, were constructed to navigate the

named *Aganthonius*, who is said to have lived to an extreme old age. It was still in existence, but poor and destitute when Festus Avienus saw it about A. D. 370, being, he says, little more than a heap of ruins, though even then an annual feast was held there in honour of the Tyrian Hercules.*

The Romans knew little of the northern and western coasts of Europe until the time of Cæsar, when in the progress of their conquest of Gaul they reached the English Channel by land. The Phœnicians kept the secret of their voyages to the Cassiterides so closely, that one of their mariners is reported to have run his ship on a shoal, that a Roman vessel which followed might not perceive the course he was pursuing, gaining by his patriotism a reward from his own nation. With the same commercial jealousy the Punic seamen magnified the dangers of the Atlantic Ocean, and told marvellous stories of the stagnant and sluggish waters, and of the dense fields of sea-weed which stopped the way of their ships.

The Phœnicians were not only carriers of the metals which were the important articles of their commerce, but they also worked such mines as lay near their maritime depots.—(*Rawl. Herod.* iii. 445.) Avienus says expressly, that not only the common people of Carthage went to the Cassiterides, but also her agriculturists; and we may safely infer that the western Celtic nations learnt the art of reducing metallic ores from

---

Erythræan Sea, as were also Jehoshaphat's, built at Ezion-geber. But when Ezekiel says of Tyre (588 B. C.), "Tarshish was thy merchant by reason of the multitude of all kinds of riches; with *silver, iron, tin* and *lead,* they traded in thy fairs" (xxvii. 12). The Spanish port is clearly indicated, Tarshish being the entrepot for the north-western commerce. In the Hamitic tongue, "*Tarshish*" means, according to Sir Henry Rawlinson, "younger brother"—a suitable appellation of a colony.—*Rawl. Herod.* i. p. 298.

* Called in a Phœnician inscription at Malta, *Adonin Melkarth Baal Tzura,* "Our Lord Melkarth, Baal of Tyre."—*Rawl. Herod.* ii. p. 50.

Phœnician metallurgists. Relics of Carthaginian trade with the British Isles are supposed to exist in the peculiar and very ancient glass beads dug up in Cornwall and Ireland, similar in all respects to those still found in Western Africa.* Cambden says that mining shafts in Cornwall, deserted ages ago, are named in the Cornish language *attal sarazin*, from the belief that they are the work of people that came from Spain or Africa; and the translator of Heeren's *Ideen* adduces as traces of the Carthaginians in Ireland, the existing traditions of a colony of Phenian miners in the county of Wicklow, remarking also on the resemblance which the ancient shafts there bear to the remains of Punic mineral works in Spain. He adds that the brazen instruments occasionally dug up in the Irish bogs, are allied in the same proportions with Carthaginian relics discovered in Italy and Sicily.† The Irish also attribute to the Phenians the introduction of letters into their island.

However shadowy the Irish traditions of the Phenians may

* Rawlinson's Herodotus.
† Moore in his "History of Ireland" mentions the same fact, and states that Phœnician brass implements discovered in an Irish cairn in the year 1848, consisted of 85 or 90 parts of copper alloyed with 15 or 10 of tin. He likewise informs us, that the antique swords found in Ireland are exactly similar to the swords obtained on the field of Cannæ by Sir William Hamilton, and preserved in the British Museum. There were giants on the earth in those days; and the Irish for giant being "Phinn," is a coincidence worth noticing, when speaking of the sons of Phœnix. Phinn M'Coul the Celtic hero had all the attributes of a giant or of a superior caste.

> ——"Far to the west in th' ocean wide
> Beyond the realm of Gaul, a land there lies,
> Sea-girt it lies, where gyants dwelt of old."—MILTON.

Dr. Villanueva in his learned "*Ibernia Phœnicea*" supposes that *Feine* may not signify Phœnician generally, but some principal man, which would be in accordance with the Ossianic fragments. *Fen vel feineh*, he says, denotes the "corner of a building" in Phœnician, and is extended to a leader of the people.—*Ibernia Phœnicia*, Dublin, 1831.

appear to a rigid critic, they may have a real foundation, and it does not require much exercise of the imagination to find a near resemblance between a group of handsome young Milesian girls, gracefully bearing their brown water-pitchers from a cistern in Cork, and a knot of females, of the same ages, carrying jars of exactly the same form, from an Andalusian fountain. Nothing in lineaments, form, or attitudes, militate against both being descendants of one people, probably the *Bastuli* or Phœnician half-breeds.

After the Phocæans had founded Massilia (Marseilles), tin was carried thither across Gaul, and this route would probably be the principal one during the punic wars, when the Carthaginian fleet found full employment in the Mediterranean. Pytheas, a native of Massilia, sailed, as he himself reported, out of the Straits of Gibraltar northwards, and then eastwards, into the Baltic. Strabo, while denying the truthfulness of Pytheas, states many facts on his authority. Among others, he describes *Thule* as being the northernmost of the British Isles, situated on the arctic circle, and having neither sea nor atmosphere, but merely a concretion of the two, like lungs.

Pliny terms this icy region the Cronian Sea, and says that it begins a day's sail beyond Thule, which he speaks of as the last of the islands lying off the Germanic coast, beyond the *Glessariæ* or *Electridas*. Thule, he adds, has no night at the summer solstice and no daylight in the winter. He also mentions by name *Scandinavia*, as an island of the *Sinus Codanus* or Baltic, and calls other islands in the Sound or Cattegat, *Scandia*, *Bergos* and *Nerigon*, the latter word being usually translated Norway. The voyage to Thule is generally made, he tells us, from Nerigon.

In another passage this author, on the authority of

Timæus, names *Raunonia* as an island in the northern ocean on which amber is thrown by the waves, the ocean being that which Hecatæus calls *Amalchium*, and the Cimbri, *Morimarusa*. Its southern shore is the Scythic coast, and in it there is an island of immense size, called *Baltia* by Xenophon of Lampsacus, and *Basilia* by Pytheas. Westward, it extends to the promontory of *Rubeas*, beyond which is the Cronian Sea.* In these passages we have evidence that the Romans, even as late as Pliny's time, had no correct knowledge of the North Sea and Baltic, but that they had obtained the names of many places from the Carthaginians or Massilians.

A mere overland journey from the shores of the Mediterranean to the Baltic, would not have enabled the Massilian or Carthaginian traders to have collected the facts regarding the northern regions which are preserved by Strabo, Pliny, and other writers, and we may infer that the Phœnicians had rounded the Cimbric promontory by sea, though no direct evidence can be adduced in support of such an opinion ; they certainly had the skill and daring required for such voyages, and there is no reason whatever to doubt the existence of a flourishing maritime trade between the western coasts of Europe and the British Isles, long before the Roman rule embraced that remote part of the world.

There are not wanting facts which have led many thoughtful men to believe that Phœnician ships may have been driven across the Atlantic, and thus have preceded the Gothic race in the discovery of a western continent, and that though no trustworthy record of such an accident has been found, yet the classic authors of Greece and Rome make

* *Plinii, Hist. Nat.* iv. cc. 27 et 30.

obscure allusions to the existence of land in the western Atlantic. The drifting of a vessel across the Atlantic is not more surprising than the wreck of a Japanese junk on the shores of Oregon, and the landing of her crew, an event of our own times.

We shall not here repeat the notices of *Atlantis* which are to be found in so many popular works, as nothing certain can be deduced from them ; but as discussions have recently taken place respecting a discovery, or pretended discovery, in America, from which a communication between Carthage and that continent has been inferred, we shall here insert a very brief notice of the facts referred to.

At Grave Creek, near Wheeling, in the valley of the Ohio, there is a conical tumulus having an altitude of sixty-nine feet, and a basal circumference of eight hundred and twenty. This mound was explored upwards of twenty years ago, by the present proprietor of the ground, Mr. Tomlinson. A chamber was found in it, in which lay a human skeleton and various substances, of which a minute account was drawn up and published by the late Dr. Morton. This learned and accurate man, however, made no mention of a small stone with engraved characters which Mr. Tomlinson states he found in the same chamber, and which was actually described soon after the opening of the mound, in 1838. Mr. Squier says that this stone is of sand-stone, of a sort that is very common in the valley of the Ohio. The inscription on it is stated by M. Jomard to be in characters exactly like (*parfaitement conformes*) to those which exist in the bilingual inscription Carthaginian and Berber of Thugga ; that are also cut in other rocks of northern Africa ; and have been probably in use in Libya from time immemorial. This learned geographer argues that in 1838 these characters were unknown in America, and that any one designing to fabricate a fictitious relic, could not have done so without committing some mistake which would instantly have betrayed his imposture. The characters are further said by M. Jomard to be the same as the alphabet of the Touareg, obtained in 1824 at El Ghat by Dr. Oudney, and more recently identified by an officer of the Algerine army with cuttings on the rocks, and numerous inscriptions on the shields, armour, and clothes of several African tribes. This kind of writing is denominated *tsinagh*, is said to have been practised by the Berbers from the most ancient times, and Governor Hanoteau, who has carefully studied the subject, is of opinion that the ancient Libyan idiom mentioned by Herodotus is fundamentally the Berber tongue, which is spoken at this day from one extremity of Africa to the other. It will be observed that between the date of Dr. Oudney's travels and the opening of the tumulus on the Ohio in 1838, there was ample time for the alphabet he made known to have reached the remotest quarters of the United States; and Mr. Squier and other competent ethnologists of that country, who have the best means of judging of the authenticity of the details, have no confidence in the story related by Mr. Tomlinson, on whose authority the whole matter rests.*

The preceding pages, containing notices of the early navi-

* *Voyez* (Jomard) SECONDE NOTE *sur une pierre gravée trouvée dans un ancien tumulus americain et sur l'idiome libyen lue à l'Académie des Inscriptions et Belles Lettres*, 7 *Nov.* 1845; *Ainsi* TROISIEME NOTE (*Bulletin de juillet-aout* 1858, *de la Soc. Geogr. de Paris*, ii. p. 372; Sir J. Alexander, Geogr. Soc. of London; Hanoteau *Soc. de Geogr.*, iv. p. 129, Paris; Smithsonian Contributions to Knowledge, 1856, vol. viii., on the Grave-stone Creek stone, by S. F. Haven, p 28; Jomard, *Bullet. de la Soc. Geogr. de* Paris, 1858, p. 104, et tom. xviii. 1859. *Lettre de M. Squier*, p. 242. *Réponse par M. Jomard*, p. 246). Dr. Latham calls the Berbers *Amazirgh*, and states that they have been supposed

gation of the Atlantic by the pre-eminently maritime people of the ancient world, have been compiled as an introduction to the accounts which are to follow, of voyages into regions still further removed from the centres of civilization by the race upon whom the mantle of the Phœnicians has descended.*

to be the representatives of the tributaries of Carthage. Their language has an affinity in grammar to the Semitic, and the alphabet now in use, which he gives pp. 523 and 566, he believes to be deficient in claims to antiquity.—*Latham, Var. of Man*, 1850, p. 507.

* The authorities consulted in drawing up the introductory chapter are *Ideen über die Völker der Alten Welt von* A. H. L. Heeren, ii. *Das Phönicien*, i. Carthag. This comprehensive work was translated into English in 1832, and to that translation we have had recourse for extracts from Avienus, not in the original German work. The references made by Heeren to Strabo and Pliny have been examined; and also the papers on the tin and amber trades of antiquity by Sir Cornwall Lewis, which contain a condensed statement of the whole subject. These are published in "Notes and Queries, 2d Series, Nos. 110, 115, 118, for 1858, and vol. vi. for July 1858." In Turner's "Anglo-Saxons" there is a dissertation on the Cassiterides and the commentators on Herodotus, more particularly Rawlinson's translation, give much ethnological information respecting the Phœnicians.

The historical work of Sanchoniathon, a Phœnician who lived twelve centuries before the Christian era, was said to have been translated a century after the birth of Christ by Philo of Byblus, a town near Berytus; and some fragments of this pretended translation were preserved by Eusebius, but Philo's work is considered by modern critics to be wholly of his own composition, or, at the best, a translation from a Phœnician writer of much later date than Sanchoniathon.

Dius and Menander compiled from the annals of Tyre a history of which there are no remains. Herodotus was evidently conversant with Tyrian records; Polybius, Pliny, and Strabo, derived some of their statements from Carthaginian authors, and Sallust (B. C. 40) saw and had interpreted to him the *Libri Punici*, then in the possession of Hiempsal the Second, King of Numidia. Rufus Festus Avienus consulted Himlico's narrative as late as the 370th year of the Christian era. It is probable that a great number of precious African manuscripts were destroyed by the burning of the Alexandrian Library.

CHAPTER I.

ANTE-COLUMBIAN PERIOD—B.C. 52—A.D. 1494.

Romans — Alfred — Ohther — Scandinavians — Iceland — Greenland — *Gunningagap* or Baffin's Bay — The Zeni — Szkolni — Colon.

WHEN Cæsar, fifty-two years before the Christian era, planned his expedition against the Britons (*penitus toto orbe divisos*), he excused the meditated aggression by the necessity he was under of cutting off the war supplies which the Gauls received from the island, and he found no difficulty in selecting for its execution ninety-eight ships of burthen from the merchantmen employed in the commerce of the narrow seas, capable of transporting two legions of foot soldiers, and three hundred cavalry, or above 8000 men. The active intercourse between the two sides of the straits of Dover soon enabled the Britons to hear of the hostile designs of the Romans, and to deprecate the invasion by an offer of submission. Cæsar's hopes of conquest were, as is well known, not destined to be fulfilled, and more than a century elapsed before the Romans established themselves in Britain, notwithstanding that they were greatly favoured in their progress by the disunion of the various tribes and nations then occupying the island.

During the Roman rule we hear only of a coasting voyage round the British isles, and the writers of those ages, when mentioning Thule, the frozen north, and the Cronian Sea, did no more than repeat what had been handed down to them from the Phœnicians. When the Romans finally left the Britons to themselves, they abandoned long lines of defensive fortifications, with many well-built towns, and bequeathed a legacy of municipal institutions to the enervated inhabitants; but the art of navigation does not seem to have been cultivated by the well taxed and well governed natives. Its revival was due especially to the Frisians, who were the best seamen of the various Saxon tribes that invaded England, on the departure of the Romans. We hear nothing, however, of an English fleet until Alfred, improving on the Danish and Frisian models, caused "long ships" to be built, with which, aided by Frisian officers and seamen, he gained a decisive victory over the Northmen, who had, during the previous century, carried their raven-flag triumphantly over the northern seas, and who continued for more than a century after Alfred's death to excel the other European nations in deeds of daring on the stormy main.

In the year 860, or eleven years before the accession of Alfred to the throne of Mercia, a Norwegian viking, named Naddodr, discovered Iceland, an island which touches the arctic circle on the north, and answers to the accounts given of Thule by the Phœnicians and Greeks, though most probably no mariner of either of these nations had ever seen it. In the following season it was visited by Gardar, a Swede, and in 874, Ingolfr conducted thither a colony of Norsemen, the ancestors of the existing Icelanders. The Iceland annals state that the Norsemen who first landed on the island discovered traces of some Christian mariners (whom they believed to have been Irish), having preceded them.*

* Saint Brendan, who flourished in the middle of the sixth century, sailed, according to Irish legends, from Kerry westwards to a large island, on which

Alfred, whose acquirements, thirst for knowledge, and princely endowments, have secured him a place in the first rank of English sovereigns, acting on the maxim, *fas est ab hoste doceri*, sought for information of the north from Ohther, or (Audher), a Norwegian. The reply he received to his questions was added by Alfred to his Anglo-Saxon translation of the *Hormista* of Paulus Orosius, and Hakluyt has reproduced it from the version of Geoffrey of Monmouth. The following abridged extract from Hakluyt contains what relates to our subject; and for the better understanding of it, the reader may be told that the Helgoland or Holy Island, which is situated on the coast of Norway, lies on the 66th parallel of latitude, about half a degree from the arctic circle.

"The voyage of Octher, reported about the year 890 unto Alfred, the famous king of England. Octher said that the countrey wherein he dwelt was called Helgoland. Octher tolde his lord, King Alfred, that he dwelt furthest north of any Norman. He sayd that he dwelt towards the north part of the land toward the west coast; and affirmes that the land, notwithstanding it stretcheth marvelous farre towards the north, yet it is all desert and not inhabited, unlesse it be very few places here and there, where certain Finnes dwell on the coast, who live by hunting all the winter, and by fishing in the summer. He said that upon a certeine time he fell into a fantasie, and desire to proove and know how farre the land stretched

he spent seven Easters and then returned to Europe, reaching first some northern islands, probably the Shetlands or Orkneys, but finally attaining his native Ireland with his followers, where he founded many churches.—*Dr. Todd, Nat. Hist. Rev. July* 1860, p. 424. The disciples of St. Columba, in the sixth and seventh centuries, were zealous in propagating a knowledge of the gospel in foreign lands; and Adomnan mentions several voyages into the ocean, made with that view by Cormac.—*Life of St. Columba* by Dr. Smith, pp. 55-74.

northward, and whether there were any habitation of men north beyond the desert. Whereupon he took his voyage directly north along the coast, having upon his steereboord alwayes the desert-land, and upon the leereboord* the maine ocean, and continued his course for the space of three dayes, in which space he was come so far towards the north as commonly the whale hunters use to travell. Whence he proceeded in his course still towards the north so farre as he was able to saile in other three days. At the end whereof he perceived that the coast turned towards the east, or els the sea opened with a maine gulfe into the land, he knew not how farre. Well he wist and remembered that he was faine to stay till he had a westerne winde and somewhat northerly; and thence he sailed plaine east along the coast still so far as he was able in the space of four dayes. At the end of which time he was compelled againe to stay till he had a full northerly winde, forasmuch as the coast bowed thence directly towards the south, so farre as he could travaile in five dayes; and at the fifth daye's end he discovered a mightie river which opened very farre into the land. At the entrie of which river he stayed his course, and in conclusion turned backe againe, for he durst not enter thereinto for feare of the inhabitants of the land, perceiving that on the other side of the river the countrey was thorowly inhabited: which was the first peopled land that he had found since his departure from his owne dwelling: whereas, continually thorowout all his voyage he had evermore on his steereboord, a wildernesse and desert country, except that in some places he saw a few fishers, fowlers, and

* In Anglo-Saxon *Steorbord* and *Bæcbord*. The terms are synonymous with the modern nautical ones, Starboard and Larboard or Port, and mean, the former the helm side, and the other the back or left hand, the helm's-man being seated with his right hand to the rudder and looking forwards.

hunters, which were all Fynnes; and all the way upon his leereboord was the maine ocean. The Biarmes had inhabited and tilled their country indifferent well, notwithstanding he was afrayed to go on shore. But the country of the Terfynnes lay all waste and not inhabited, except by as it were certeine hunters. . . . This he judged, that the Fynnes and Biarmes (Permians) speake but one language. The principall purpose of his traveile this way was to encrease the knowledge and discoverie of these coasts and countreyes for the more commoditie of fishing of horse-whales,* which have in their teeth bones of great price and excellencie; whereof he brought some at his returne unto the king. Their skinnes are also very good to make cables for shippes, and so used. This kind of whale is much lesse in quantitie than other kindes, having not in length above seven elles. And as for the common kind of whales, the place of most and best hunting of them is in his owne countrey; whereof some be forty-eight elles of length, and some fifty, of which sort he affirmed that he was one of sixe, which in the space of three days killed threescore. He was a man of exceeding wealth in the riches of the countrey, having six hundred tame Rane Deere, yet he had but twenty kine and twenty swine, and that little which he tilled he tilled it all with horses.

"He sayd that the countrey of Norway was very long and small. So much of it as either beareth any good pasture, or may be tilled, lieth upon the sea coast, which notwithstanding in some places is very rockie and stonie, and all eastward, all along against the inhabited land, lie wilde and huge hilles and mountaines, which are in some places inhabited by the Fynnes.

* Hakluyt calls these horse-whales "morsses." The Anglo-Saxon word is *hors-hwæl*. (*Ohtheres Reisebericht.* König Aelfred von Dr. Reinhold Pauli, Berlin 1851, where the whole passage is quoted in Anglo-Saxon.) Longfellow has given a pleasant poetic translation of Ohtheres' narrative.

. . . The mountaines be in breadth of such quantitie as a man is able to traveile over in a fortnight, and in some place no more than may be traveiled over in sixe days. Right over against this land, on the other side of the mountaines, somewhat towards the south, lieth Swethland, and against the same, towards the north, lieth Queeneland. The Queenes sometimes passing the mountaines, invade and spoile the Normans; and on the contrary part, the Normans likewise sometimes spoile their countrey. Among the mountaines be many and great lakes, in sundry places, of fresh water, unto which the Queenes use to carrie their boats upon their backs overland, and thereby invade and spoile the countrey of the Normans. These boats be very little and very light."*

As Ohther mentions twice that he had the main ocean on his left hand in his voyage north, we may conclude that he passed to seaward of the Lofodon Isles, and that not being embarrassed by the intricate passages between them and the continent, he was able to advance with a fair wind at the rate of at least fifty or sixty miles a day, which in six days would bring him to the North Cape. The time that remained would scarcely suffice for a voyage to the bottom of the White Sea, and it is probable that he turned back from Varanger Fiord, or the estuary of the river Kola; or at the furthest, from the Gulf of Kandalask. The Biarmes were the people of Finnic extraction, called *Permiaki*† or Permians by the Russians, and in the middle ages the Scandinavians gave the name of

---

* Hakluyt, Principal Navigations, etc., i. 4. A.D. 1599.

† They were the dwarfs of the Scandinavian Edda, who extracted metals from the depths of the earth and practised sorcery and magic. Until a late period, English seamen sailing to the White Sea used to land in Lapland to buy a wind from the natives who traded in their superstitions. *Gand-vick* (Gulf of the Magicians) was a Scandinavian name for the White Sea.

Permia to the country lying between the White Sea (*Bieloé moré*) of modern geographers, and the Ural Mountains.* The Quains (*Qvæn-vick*), Queenes, or Kainulainen, were a Baltic people of the Finnic stock, and from misapprehensions of the meaning of their name, came the notion of their being Amazons. As Ohther feared to land where the people were numerous, he may have supposed that he had reached Permia, while he was yet coasting Finmark. But whatever may have been the extent of his voyage, it is a most remarkable one, as being the first sea voyage made round the North Cape and across the arctic circle that is on record. There is no reason to discredit it, and the courses of other voyages of Ohther related at the same time to Alfred, are consonant with the positions of the ports to which they were made; nor were any of them beyond the nautical powers of a people, who, thirty years previously, had explored the way across the North Sea to Iceland.†

It is of the further exploits of these Northmen, or rather of the colonists established by them in Iceland, that we have now to speak—The oppressions of the Jarls had driven many men of independent mind to seek an asylum in Iceland, and as the climate of that island was not favourable for the cultivation of grain, the support of the increasing population would, until the herds had multiplied by time, be mainly

---

* John Shefferus, in his history of Lapland, identifies Skrithfinnia with Finnish Lapland, and Biarmia with Russian Lapland.

† The beautifully written and well preserved original of Alfred's Orosius, containing the voyages of Ohther and Wulfstan, is preserved in the British Museum. An English translation was published by Daines Barrington in 1773. Dr. J. R. Forster gave a German translation with valuable comments; but the best edition is by the celebrated Anglo-Saxon scholar, Rasmus K. Rask, published at Copenhagen in 1834, with a Danish version. This notice is taken from "Notes upon Russia, edited for the Hakluyt Society by R. H. Major, 1851, p. iv.," where other editions are mentioned.

dependent on success in fishing, whale-hunting, and sealing. While the colony was yet in its infancy, an Icelander named Gunbiörn, the son of Ulf Krake, being out on a fishing excursion, was caught by a storm and driven to a considerable distance westward from Iceland, when he discovered a reef of low rocks or skerries, and further on an extensive land, which he did not explore, but which he called after his own name, distinguishing its remarkable southern snow-clad headland by the designation of *Hvidsærk* (White Shirt). This discovery did not lead to immediate consequences, but about 982 or 983 Erikr Rauthi (Erik the Red) was convicted of manslaughter before the *Thornæs Ting*, or judicial assembly of Iceland, and sentenced to banishment for a term of years. He resolved to pass the time of his compulsory absence in exploring Gunbiörn's land, and having prepared a vessel, sailed with his followers from Sneefieldsjokel, the northern promontory of Faxe Bay, in one of the southern inlets of which the town of Reikiavik has been built. Holding a westerly course, he came in sight of the east coast of Greenland, along which he steered southwards, looking for a habitable spot; and in doing so doubled the *Hvidsærk* of Gunbiörn, called also by the early Icelandic voyagers *Mucklajokel*, and known to modern whalers by the name of Cape Farewell, or *Statenhuk*. It is a very lofty island lying off the southern extremity of Greenland, and being the highest land to the south of the 60th parallel of latitude, and at all seasons capped with snow, it is the first landfall of navigators steering from Europe for Davis Straits; and it is also the point of departure of the whale-ships when bound homewards, whence its appellation of Cape Farewell (*Färväl*), given by English and Danish seamen. The Eskimos of the neighbourhood call it *Omenarsorsoak* or *Kangek-*

Kyerdlek, the latter designation having reference to snow geese.* Erikr having rounded this mountainous island, extended his voyage to a point which he called *Hvarf*, equivalent to the English word Turnagain, and having chosen a wintering place on an island, he named it after himself, *Eriksey*. Captain Graah of the Danish navy, a most competent authority, identifies *Hvarf* with *Cape Egede* on the island of Semersok, lying about $45\frac{1}{4}°$ west of Greenwich. Having spent three years in exploring the western coasts of Greenland, Erikr returned to Iceland and made so favourable a report of the new country, that in 985 or 986 he induced a large body of colonists to sail with him from Iceland in twenty-five ships. Half of the ill-fated fleet perished in the ice, but the remnant reached their destination, and in a few years thereafter all the habitable places of Greenland were occupied. The colony was divided into two districts by an intervening tract of land, named *Ubygd*, "uninhabitable" or "uninhabited." *Bygd* signifies inhabited place, from the Icelandic *byggia* to build, and also to inhabit. The West Bygd reached from latitude 66° down to 62°, and contained, when in the height of its prosperity, ninety farms and four churches. South of it lay a desert Ubygd, of seventy geographical miles, terminated by the East Bygd, consisting of one hundred and ninety farms, and having two towns, Gardar and Alba, one cathedral, and eleven churches. The justiciary or Lagmand resided at Brattahlid, the site of Erikr's farm, which became, at a later date, a royal demesne. The aspect of this Bygd was south-west.

* In the ancient sailing directions for proceeding to Greenland from Bergen without touching at Iceland, the mariner is told to steer directly west, to pass twelve Icelandic sea miles south of Reikianes in Iceland, and to sight *Hvidsærk*, on the day after which they will come to *Hvarf*, and between *Hvarf* and *Hvidsærk* is *Herjolfsnes*.

The occupation of Greenland speedily led to the discovery of America. One of the colonists, conducted by Erikr to Greenland, was Herjulfr Bardson, a descendant of Ingulfr, the first settler in Iceland. This man's son Bjarni was absent on a voyage to Norway at the time of Erikr's expedition; but returned in the same season of 985 or 986 to Eyras in Iceland, when being informed of his father's departure, he immediately resolved to follow him, that they might spend their yule-tide together, as they had always been accustomed to do. He therefore made the proposal to his crew, and obtaining their consent put to sea again, but encountering thick stormy weather, was driven far to the southward of his proper course. On the sky clearing, Bjarni was within sight of a woody part of the American coast, supposed to be Nantucket Island, south of Boston, which, not agreeing at all with the description he had received of Greenland, he directed the prow of his ship to be turned northwards; and after passing several of the projecting headlands of Newfoundland and Labrador, but without landing on any, or even naming them, he finally came in sight of White Sark, and fortunately meeting a boat there, was directed to *Herjulfrsnes*, his father's new abode. The fiord which enters the land at this promontory, is understood to be Friederichstal, in latitude 60°, where in 1828 the Moravian mission, established only four years previously, numbered four hundred members. Hot springs exist on the island of Ounartok in its vicinity.

To understand rightly the accounts of the voyages of the Norsemen of that age, it is necessary to have an estimate of the length of a day's sail, and this Captain Graah furnishes us with from averages taken out of sailing directions given in the Landnama bok, and other very ancient documents. In them the voyage from Stadt in Norway to Horn on the east of Iceland, was stated to be seven days' sail; and from Snæfieldnes to Hvarf in Greenland four days, giving with other averages from ninety-seven to one hundred and seven miles each day. A row-boat Captain Graah found, was expected to average twenty-four miles a-day, much about what it would do at the present time when the rowers are working day after day.*

At the close of the tenth century, Erikr's son Leifr having visited Norway, was, by command of King Olaf Triggveson, instructed in the principles of Christianity, and on his return to Greenland was accompanied by a priest who baptised Erikr and his followers. Malte Brun, on the authority of Schlegel and Beckman, says, that *Peter's pence* were sent from Greenland to Rome in the form of '*dents de boiardo*,' or morse-tusks.

In the year 1000, Erikr having purchased a ship, in which Bjarni Herjulfrson had traded to Norway, prepared an expedition for acquiring a more perfect knowledge of the lands which Bjarni had seen in the west, but in consequence of having fallen as he was embarking (which he considered to be an evil omen), he transferred the command to his son Leifr. The events of this voyage, and of the others which led to the attempted colonization of America, and its frustration by the hostility of the natives, having reference to lands lying far south of the arctic circle, do not come properly within the scope of this historical summary of Arctic discoveries, but it may be briefly stated that Erikr wintered in a place where he found wild vines, and thence named the country Vinland. This country has since been recognised by

* Graah's Greenland, Eng. Tr., p. 161.

its relative position and its productions to be Rhode Island, and "Leifr's buthr," or his winter habitation, to have been situated on the banks of Taunton River.* His *Straumfiorthr*, and *Kialarnes* are identified with Buzzard's Bay and Cape Cod, by the descriptions given of the coast in the annals referred to. *Littla Helluland* and *Helluland it mikla*, the lesser and greater Slatylands, are on the same authority of Rafn, considered to be Newfoundland and Labrador, while by *Markland* (the woody country) Nova Scotia is thought to have been designated—Markland's gulf, being consequently the Gulf of St. Lawrence. The first child of European extraction born in America was Snorre Thorfinnson, who saw the light in *Thorfinn's Buthr*, near the existing town of Taunton.

The intercourse between Greenland and America was from time to time renewed, down to the year 1347 by voyages undertaken chiefly to procure wood for building purposes, but no second attempt at colonization was made. Among the wood brought from America, special mention is made of a piece of *Mösur*, probably bird's-eye or curled maple, which was sold in Norway at a high price ; richly carved bowls of wood, formerly named mazers or masers, being greatly prized.

The entire narrative of the events here merely glanced at, is told in authentic Icelandic annals, and is of deep interest, but we must turn to less attractive indications of voyages made within the arctic circle by the Norwegian Greenlanders.

* The celebrated Dighton Rock, on which certain characters and figures are inscribed, as has been supposed, by the Scandinavians, has had considerable weight in fixing the exact locality of *Leifr's buthr*, but as early as 1789 Washington considered the inscriptions to be of Indian origin, and Schoolcraft, taking to his aid an experienced Indian chief, has confirmed this opinion; the Indian pronouncing the legend to have reference to a battle, but rejecting some of the characters as interpolations destitute of meaning.

We learn from the annals, that whale-hunting, sealing, and fishing, were carried on during the summer far to the northward of the inhabited West Bygd, and that the general appellation of these distant haunts was *Northursæta*. One locality especially resorted to by the sealers, was named *Greipar*, because of the resemblance which the fiords that there indent the coast, had to the intervals, being the fingers of a man's hand, which *grcipar* signifies. Rafn assigns latitude 67° as the site of *Greipar;* and *Bjarney*, or Bear Island of the old colonists, he identifies with the Disco so well known to whalers and Arctic voyagers.

Gardar, the episcopal seat, was situated at the bottom of Erikrfiord, and its cathedral was dedicated to St. Nicholas. At this place an expedition was organized, in 1266, by the priests, for the purpose of visiting the northern summer haunts of the sealers, and exploring the country beyond, named by them *Furthurstranda*, and reported to be a country where the cold was so extreme as to render any habitation there impossible. The sea to the westward in that high latitude the Norsemen called *Gunningagap*, five or six centuries before it obtained the designation of Baffin's Bay. The episcopal discovery ship proceeded northwards, and, passing Greipar, came to an inlet on which, from its curvature, the name of *Kroksfiorthr* was bestowed. There a strong southerly gale and thick weather setting in, the ship was driven northwards, the crew knew not how far; but on the weather clearing up, the mariners found themselves in an archipelago amid much ice, with the sun as high above the horizon at midnight as they had been accustomed to see it at Gardar when in the north-east quarter of the sky on the same day of the year. This is too vague to enable us to fix either the direction of their voyage or the extent to which they penetrated beyond the arctic circle.

A memorial of the ancient visits to the *Northursæta* was discovered by an Eskimo Greenlander named Pelinut, on *Kingitorsoak*, one of the Women's Islands, lying in lat. 72° 55' N., long. 56° 5' W. It is a stone with a Runic inscription dated 1135, and stating that on Saturday before "the day of victory" of that year, Erling Sighvatsson, Bjarni Thortharson and Eindrid Oddson cleared the ground and raised these marks.*

The *Kongskuggsiö* (*Speculum Regale*), supposed to have been written in the twelfth or thirteenth century, mentions that the interior of Greenland is covered with ice, but that the habitable banks of the fiords abound in good pasturage; and that the colonists subsisted by raising cattle and sheep, as well as by the chase of the rein-deer, walrus, and seal, the land being adverse to the production of grain. The country was governed by Icelandic laws and the first of its series of bishops was Arnold, who was elected in 1121 at the instance of Leifr's grandson Sokke, the last being Endride Andreason, who was consecrated in 1406.† Before the latter date the colony had begun to decline. In 1348-9‡ a black pestilence

---

* The stone having been presented to Captain Graah, was by him deposited in the Royal Museum of Copenhagen. The day of victory was an ancient festival kept on the 21st of April, and being so early in the season, implies that the men named had either wintered there or had travelled northwards on the ice.

† Professor Finn Magnusen and several learned Icelanders deduce their pedigree from a couple who were married in Greenland in 1409 by Bishop Endride. The sculptor Thorwaldsen could also trace his descent from these ancient stocks.

‡ This epidemic, known as the blackdeath or great mortality, is stated by Hecker to have commenced in China in 1333, and fifteen years later to have spread through Europe. In some places of England the visitation swept off nine-tenths of the inhabitants; and it was severe in London in 1348. From England the plague was carried by a ship to Bergen, in Norway. Sailors found no safety in

had committed wide ravages among the people of the north; and in 1379 the Skrællings (or Eskimos), the aborigines of the country, had invaded the West Bygd, killed eighteen Icelandic Greenlanders, and carried away two boys captive. As soon as intelligence of this disaster reached the East Bygd, Ivar Bere or Bardsen (called Boty ages afterwards by Barentzoon), a principal man or lay-superintendent of the bishop's court, was despatched with a levy of East Bygd people to the assistance of their countrymen. On arriving at the scene of the massacre he found no man, neither Christian nor heathen, but only sheep running wild, of which he brought away as many as he could; and thus miserably terminated the colonization of the west Bygd.

The East Bygd dragged on its existence twenty or thirty years longer, and the exact date of its extinction has not been so definitely recorded. In the beginning of the fifteenth century the Semiramis of the north, Margaret, Queen of Denmark and Norway, suspecting that she had been defrauded by the traders of tribute due from Greenland, imprisoned the merchants who were accustomed to voyage thither, and deterred others from going, in consequence of which the settlement languished away. The final blow was, according to a pastoral letter of Pope Nicholas the Fifth, given by a hostile fleet (suspected by Captain Graah to have been English), which, in 1418, laid waste the country and carried all the vigorous inhabitants into captivity, the dwellers in remote parishes only escaping. By a treaty made in 1433 between King Erik of Norway and our Henry the Sixth, such captives as had been

putting to sea, and vessels drove about the ocean or drifted on shore, whose crews had perished to the last man. This plague carried off twenty-five millions of Europeans. It was preceded and followed by the dancing mania, or St. Vitus' dance.

disposed of in England were liberated; but the neglect of the mother country continuing, the Greenland colonists either retreated to Iceland, or perished under the repeated assaults of the Skrællingar. Vestiges of the ancient colonization remain in ten different places within the limits of the West Bygd; and ruins of old edifices are still more numerous and in better preservation in the East Bygd, where the roofless walls of six or seven churches remain standing, the most perfect being that of *Kakortok* in the inlet of *Igaliko*, about ten miles distant from Juliana's Hope. Here, according to Eskimo tradition, the last of the colonists was slain by the Skrællingar. In 1830 a gravestone was dug up at this church, on which there was an inscription in Runic characters to the memory of Vigdisa, the daughter of Magnus, but no date. The name of Vigdis was borne by several descendants of Thorfinn the Red.

Accidental circumstances may have kept up in the north a knowledge of Greenland, which was lost to the world in general for want of a periodical press. In the church of Barra an Eskimo kayak was suspended, having been driven thither by winds and currents. In 1682 a Greenlander was seen in his kayak off the island of Eda by several people who did not succeed in bringing him ashore, and two years afterwards another appeared off Westray.*

The condensed notices which we have given of the first European colonization of Greenland have been abstracted mainly from Rafn's *Antiquitates Americanæ*, which contains latin versions of the Icelandic annals with comments. For the identification of the ancient names of places with the modern ones, and for various facts, Captain Graah's authority

* Account of the Islands of Orkney by James Wallace, A.D. 1700.

has been relied on.* The authenticity of the Icelandic manuscripts seems to be fully established, and the facts they record are mingled with fewer extravagances and mythic interpolations or monkish fictions than those of the contemporary nations. Though the discovery of America by the Northmen is mentioned by Adam of Bremen in his Ecclesiastical History, written in 1073-6, and the sailing directions for vessels proceeding from Norway to Greenland by the aforesaid Ivar Bardsen, by Biörn Jonsen, and of the Landnama Bok, were more or less extensively known to the northern seamen, the importance of the discovery seems to have been so completely overlooked, that the exploits of the Icelanders had been forgotten when the genius of Columbus awoke in Europe the spirit of maritime enterprise. Then the attention of the learned was turned to the Icelandic annals, and in process of time various versions and extracts were given to the world;† but the want of maps and the absence of astronomical observations rendered the geographical results of the early voyages scarcely intelligible to the moderns.

The adventures in the northern seas of the two brothers Zeni, M. Nicolo, the knight, and M. Antonio, in the year 1380, are evidently to be placed in the category of fictions, to which some appearance of truth has been imparted by the introduc-

* Expedition to the East Coast of Greenland, by Captain W. A. Graah, in 1828, tr. by G. G. Macdougall, London, 1837. The Icelandic *Kongskuggsiö* and *Landnamabok* were begun to be compiled very early in the history of that Island. Translations into English were published by Beamish in 1841, by Blackwell in his edition of Mallet, in 1847, and a summary is given in the Geographical Journal for 1858.

† Adam of Bremen's *Hist. Eccl. Hamb. et Brem.* was printed in 1579; the *Theatr. Orbis* of Ortelius in 1601; Mylius *De Antiq. Ling. Belg.* in 1611; Grotius *De Orig. Gent. Amer.* in 1642; Olaus Magnus, *De Hist. Gent. Septentr. in* 1550; Torfæus, *Hist. Vinland* in 1705.

tion of a few names and facts. These could readily be gathered by the Venetians in their commercial voyages to England and Scandinavia; and Captain C. C. Zahrtmann (in a paper* which fully exposes the falsehoods of the narrative), gives it as his opinion that it was compiled by Nicolo, a descendant of the Zeni, from accounts current in Italy in the middle of the sixteenth century, after an interest for the fate of the last Christian colony had been awakened at Rome.† The romance of the Zeni might have been dismissed with these observations, but as the mariners of Queen Elizabeth's time, and geographers down to a recent date, have speculated on the situation of the places said to have been visited by the Zeni, it may be well to add that *Frislanda* of the Zeni, is in Captain Zahrtman's opinion the Feröe Islands.‡ Nicolo Zeno, *il cavaliere*, is stated to have made a voyage to the north after the defeat of his countrymen in the battle of Chioggia, by the Genoese, and being overtaken in the Flemish seas by a storm, was wrecked on *Frislanda*, an island under the sovereignty of the king of Norway. *Eslanda* is without doubt Iceland, and *Engrovelanda*, Greenland. To this latter country the narrative transfers the volcano of Heckla, and states, moreover, that there was near it a monastery of preaching friars, and a church dedicated to St. Thomas. Hot water brought from thermal springs in pipes,

* Royal Geographical Society of London, v. p. 102.

† Letter of Pope Nicholas the Fifth to the bishops of Skalholt and Holum, discovered by Professor Mallet in the papal archives.

‡ Columbus, in his *Tratado de los zinco zonas habitabiles*, mentions Frislanda at a date prior to the *publication* of the narrative of the Zeni. Heberstein evidently makes *Novaya Zemlya* and *Engroneland* to be the same, since he says that opposite Petschora and the mouths of the Obi there is reported to be a region called Engroneland, which remains unknown, because of the difficulty of navigating the icy sea which surrounds it.—*Ramusio* II., fol. 182, and Suppl. fol. 66., ed. 1583. See in the same vol., fol. 230, the narrative of the Zeni.

supplied warmth to the monastery in winter. The imagination of the writer here turns the Geysers of Iceland, or the hot springs of Ournatok, in Greenland, to economic uses. But there is no evidence of a church dedicated to St. Thomas having ever existed in either locality. The Zeni papers, however, describe correctly the Skrælling or Eskimo Kayak. "The fishers' boats resemble a weavers shuttle, and are formed of skins of fishes (seal-skins) extended on bones of the same. In these they shut themselves up close, and let the sea and wind toss them about without any fear of breaking or drowning."*

Estotilanda is said to lie about a thousand miles westward of *Frislanda*, and may well be part of the American continent. Southwards of this lies a country named *Drogio;* and *Icaria* was another land, said to have been discovered on a voyage from Frislanda to Estotilanda. The accounts of the inhabitants of these several countries, and of their social condition, are such as to shew, beyond doubt, the fictitious character of the main narrative on which, nevertheless, some imperfect intelligence of the colonization of Greenland has been engrafted, as the passages referred to indicate. The narrative of the Zeni was not published till 1558, when it appeared at Venice in a small volume accompanied by a chart, the original of which is said to have been an old portolano in the Zeno archives.† It was reprinted in 1559 in the second volume of

---

* Cluvierus (A.D. 1661) divides Canada into *Estotiland, Terra Cortorealis*, and *Terra Laboratoris*, and into the adjacent islands of vast size, *Golesme, Beauparis, Monte de Lions*, and *Terra Nova*, which last is the same, he says, with *Terra de Bacalas*. South of Canada lies *Nova Francia*, of which *Norumbega* is a part.

† Hudson, the navigator, by G. M. Asher, LL.D. Printed for Hakl. Society, p. 165.

the *Viaggi* of Ramusius, but without the map. This map is considered by Dr. Asher to be of Scandinavian origin, to be very correct for the time in its outline of the coast of Greenland, and to have served for the basis of Hondius' delineation of that country. The error in its position of the south part of Greenland, by the detachment from it of Frizeland, led to mistakes by Frobisher and Davis, who took copies of it with them, as it was considered to be authentic by the geographers of that time.

The Danes, though they suffered the Greenland colonies to die out, did not permit their memory altogether to perish, and in 1476 Szkolni or John of Kolnus,* a Polish pilot in the service of Christian the Second King of Denmark, carried out a number of emigrant Scandinavians. He is said to have landed in *Græsland* (Grassland), after visiting Greenland, but there is no account of his having founded a colony, or of any of his further proceedings. Michael Lok's map, published in Jones's edition of "Hakluyt's Divers Voyages," places the discovery of Scolvus, written *Grætland*, to the west of Greenland, between latitudes 72° and 76°; and Dr. Asher observes, that Gilbert having placed *Grocland*, as the word has been otherwise spelled, further to the south, led to the fiction of a Dane named *Anskoeld* having been the discoverer of Hudson's Bay.† Gomara says John Scolvo the pilot visited Labrador with the men of Norway.‡

Cristoforo Colon, now universally known by his latinized name of Columbus, had as early as the year 1474 formed the plan of the voyage, by accomplishing which, thirty years

* Variously named Scolvus, Scolmus, Sclolvus, and Scalve by cosmographers.
† Asher, lib. cit. xcviii. The earlier editions of Brockhaus' Conversations Lexikon contain the Anskœld myth.
‡ Select. lett. of Columbus by R. H. Major, p. xxx., printed for Hakl. Soc.

later, he changed the geography of the world. In a letter dated 1477, quoted by his son, he says, that "he sailed a hundred leagues beyond the island of Thule, the southern part of which is distant from the equinoctial line seventy-three degrees, and not sixty-three as some assert; neither does it lie within the line which includes the west of Ptolemy, but is much more westerly. To this island which is as large as England, the English, especially those from Bristol, go with their merchandise. At the time that I was there the sea was not frozen, but the tides were so great as to rise and fall twenty-six fathoms. It is true that the Thule, of which Ptolemy makes mention, lies where he says it does, and by the moderns it is called Frislanda."* As it was not till 1480 that the astrolabe was improved by Martin Behaim and his assistants, so as to become serviceable at sea in ascertaining the latitude by the altitude of the sun, we are not to expect a near approach to the true geographical positions in a seaman's narrative of earlier date; and even in maps constructed a century after the time of which Columbus speaks, the whole of Iceland is placed to the north of the arctic circle instead of to the south of it. The island to which the Bristolians traded can be no other than Iceland, but tides, which rise 156 feet, are not to be found even in the Bay of Fundy, and this passage requires explanation. The West Bygd of Greenland corresponds more nearly in latitude with the Thule of Columbus, but in 1474 that colony possessed neither trade nor inhabitants. The information that Columbus obtained in Iceland must have strengthened, or perhaps originated, his desires for western discovery.

* Major, lib. cit. p. xlv.

CHAPTER II.

A. D. 1492-1527.

Marco Polo—Columbus—The Gabotti or Cabots, John and Sebastian—Labrador—Newfoundland—Fabian's Chronicle—Butrigarius—Sir Humphrey Gilbert—Sebastian Cabot's maps—Cortoreale—Robert Thorne.

DURING the latter third part of the thirteenth century, when the prosperity of Venice was in its zenith, Marco Polo, following the steps of his uncles and other merchants, travelled across Asia to Khan-balik or Pekin, the seat of the Tartar conqueror of China, Kublai-khan. His narratives made Europe acquainted with the advanced civil condition of China, and approximately with the position of Cathay; but the project of reaching the fabulously rich lands of the extreme east by sea, does not seem to have presented itself at that time to the minds of northern navigators, and maritime enterprise lay dormant until the middle of the fifteenth century, when Prince Henry of Portugal gave the impulse by which his countrymen went forth to trace the western coast of Africa down to the *Cabo Tormentoso*, the throne of the Genius of Tempests of Camoëns. In 1492 Columbus, the most noble of the many worthy seamen Genoa had produced, made his glorious discovery of the Western Indies, by which he gave a new world to ungrateful Spain; and six years afterwards Vasco di Gama, doubling the storm-beaten extremity of Africa (to be thenceforth termed the Cape of Good Hope), reached India by the eastern route.

These splendid achievements of the peninsular mariners were not unheeded in the north. The English merchants longed to have a share of the commerce of the two Indies; and as the Pope had assigned the eastern route to the Portuguese, and the western one to the Spaniards, the mariners of Bristol thought that a way to new fields of commercial enterprise might be found by steering to the north-west. How far a knowledge of the doings of the Norsemen in Greenland may have been influential in originating this notion has not been ascertained, but it is difficult to believe that the Bristolians, who traded to Iceland in the time of Edward the Fourth, had not heard of Engrönland, whither so many of the Icelanders had gone, and of the western Helluland, Markland and Vinland, discovered by the emigrants.

It happened fortunately for the interests of geography, that during the reign of Edward the Fourth, a Venetian merchant, Signor Giovanni Gabotto (John Cabot) had settled in Bristol and had prospered largely by the commerce then carried on there, that to Norway and Iceland doubtless included. This man had three sons Lodovico, Sebastian, and Sancio who were associated with him in his maritime enterprises. Whether these children were born to him before or after he left Venice, is a matter of uncertainty; and though both Italy and England claim to have given birth to Sebastian, the most celebrated of the family, the honour seems to belong properly to Venice, for according to Sebastian himself, as quoted by Galeacius Butrigarius, the Pope's Legate, and Pietro Martire Anghiera, he, though merely in boyhood when his father settled in England, was old enough to have acquired some knowledge of the classics and of the sphere; moreover, in letters patent granted when Sebastian was twenty-seven years of age, John Cabot is styled a Venetian citizen.*

As the Cabots are generally held to be discoverers of North America up to the arctic circle by English writers, who overlook the previous doings of the *Norrœnu* Greenlanders, we may be excused for devoting a little space to the various and in some points discordant accounts of their voyages. The foundations of the several reports are Sebastian's recollections, uttered in conversation many years after the events, his written documents having perished. It is not therefore surprising that the exact year of the first voyage should be uncertain, and that it should not be easy to apportion the amount of credit, as discoverers, due to the father and son respectively.

A French writer states that M. d'Avezac has fixed, by authentic documents, the year 1494 as that of Cabot's first arrival on the western continent; and moreover alleges that the Basques claim, by their traditions, to have discovered and established a cod-fishery on the great bank of Newfoundland as early as the middle of the fifteenth century, having been led thither in pursuit of whales. Unless, however, this claim can be supported by some substantial evidence, it must fall to the ground.†

* Hakluyt in "Divers Voyages," says that Sebastian was an Englishman, having been born at Bristol about the year 1467. See, "Of the north-east Frostie Seas by the Ambassador of Moscovy, addressed to G. Butrigarius—collected by Richard Eden," 1555. Pub. Hakl. Soc. 1852, ii. p. 192.

† *Revue des deux Mondes*, An. 1859. The documents on which M. d'Avezac relies, are not named in the paper to which we refer, but he probably took for his authority the Mappemonde of Sebastian Cabot, to be mentioned in a succeeding page.

Hakluyt has printed an authentic document relating to the commencement of the voyages of the Cabots, being letters patent, granted on the 5th of March, by Henry the Seventh, in the eleventh year of his reign, to John Cabot and his three sons Louis, Sebastian, and Sancio, authorizing them to sail with five ships, under the banners of England, to the east, the west, and the north, at their own cost, and to set up the said banners in every town, city, castle and island of them newly found, and which they can subdue and occupy. Hakluyt takes 1495 to be the eleventh year of Henry's reign, mentioned in this paper; but Mr. Jones has pointed out that as the year was then computed from the 25th of March, 1496 is the true date by our present reckoning. Little depends on this correction, since the voyage made under the authority of the letters patent did not take place until 1497, as appears by the only precise information respecting it, that we have access to.

This is to be found in an extract taken by Hakluyt from a map of Sebastian Cabot, cut by Clement Adams, which states that in the year of our Lord 1497, John Cabot, a Venetian, and his son Sebastian, discovered a country which no one before that time had ventured to approach, on the 24th of June, about five in the morning. He (John) called the land *Terra primum visa*, and the island opposite to it he named St. John. Then follows a short description of the inhabitants of the country and its productions, the abundance of the fish called (*vulgi sermone*), *Baccalaos* being noticed.* The map with this inscription was engraved, Purchas says, in the year 1549, at which time Sebastian was yet alive, holding the office of Grand Pilot of England, and standing in high favour with Edward the Sixth—Hakluyt had opportunities of consulting copies to

* See onwards at page 47 for a material part of the Latin inscription.

be seen in her Majesty Queen Elizabeth's privy gallery, and in many ancient merchants' houses. Mr. Hugh Murray in his work on North America, says, that the map in the privy gallery was, he understood, destroyed by fire.

In translating the Latin inscription on Cabot's map, Hakluyt has interpolated "with an English fleet from Bristol," and in fact the Mathew of Bristol is said to have been the first of the fleet which reached the North American continent, not the island of Newfoundland, as generally supposed, but according to arguments adduced by Mr. Biddle, the coast of Labrador.* From a letter to the English Ambassador at Madrid,† written by Master Robert Thorne, long a resident in Spain, we learn further that his father, "a merchante of Bristow," and Hugh Eliot were adventurers in that fleet, and discoverers, he says, of the Newfoundlands, and "if the mariners would then have been ruled, and followed their pilot's minde, the lands of the West Indies, from whence all the gold commeth, had been ours. For all is one coast, as by the carde now appeareth." In this last sentence Thorne refers to a map made by himself, and sent in 1527 to Dr. Ley, the Ambassador to Spain from Henry the Eighth, a copy of which exists in the reprint of "Divers Voyages," published by the Hakluyt Society. In it a deep inlet is shewn on the American shore, about the 44th parallel of latitude, and another about the 54th, beyond which the coast line stretching directly northwards is designated "*Nova terra laboratorum dicta ab Anglis primum inventa,*" or the land of Labrador newly discovered by the English. The delineation of the coast extends to the Straits of Magellan, and Terra del Fuego is expanded into a large southern continent. Hugh Eliot or Elyot's name appears again in a patent of dis-

---

* Memoir of Sebastian Cabot. † Hakluyt, i. p. 214.

covery, granted by Henry VII. to him and three others, at a later date.

John Cabot followed up his discovery of 1497 by obtaining a licence under the sign-manual, dated on the 3rd of February 1498, permitting him to equip six English ships of 200 tons burthen, and to lead the same to "the *land and isles of late founde by the seid John*, in oure name, paying for theym as we should for our owen cause paye, and noon otherwise." John Cabot, to whom this licence is exclusively given, did not, as far as we can now learn, go to sea himself in the fleet fitted out under its authority, but he either deputed his son Sebastian to the command, or as a passage quoted below may lead us to conjecture, having died before the fleet sailed, Sebastian took charge of it by right of inheritance.

A contemporary entry in a chronicle kept by one Robert Fabian, mentions the departure of the fleet of 1498 in the following terms :—" In the thirteenth yeere of K. Henry the 7. (*by meenes of* one John Cabot a Venetian, which made himselfe very experte and cunning in knowledge of the circuit of the world and ilands of the same, as by a sea card and other demonstrations reasonable he shewed) the king caused to man and victuall a ship at Bristow, to search for an Island, which he said hee knew well was rich, and replenished with great commodities : which shippe thus manned and victualled at the king's cost, divers merchants of London ventured in her small stocks, being in her as chiefe patron, the said Venetian. And in company of the said ship, sailed also out of Bristow three or foure small ships fraught with sleight and grosse merchandizes, as course cloth, caps, laces, points, and other trifles. And so departed from Bristow in the beginning of May, of whom in this maior's time returned. no tidings." A later entry says—"This yeere (the fourteenth of his raigne), also were brought unto the king three men taken in the Newfound Island, that before I spake of, in William Purchas' time being maior.*

It is to this voyage doubtless that reference is made by Pietro Martire Anghiera, a member of the Council of the Indies to the Catholic King, when he says that Sebastian Gabotto, a very prudent man and accomplished seaman, his very familiar friend, told him, that on the death of his father, finding himself to be very wealthy, he fitted out two ships at his own cost, and steering between the north and north-west (*tra il vento di Maestro e Tramontano*), went on till he reached the 55th parallel of latitude, where being surrounded by icebergs, even in the month of July, and his ships in great peril, though he had in a manner continual daylight, he was forced to turn and direct his course for some distance to the south and then westwards, according to the inclination of the coast. He traced the land, which he named *Baccalaos*, going southwards until he reached the same parallel of latitude as the straits of Gibraltar, in the longitude of Cuba.†

Galeacius Butrigarius, the Pope's legate in Spain, who also conversed with Sebastian on his discoveries, reports his words to have been,—"And understanding, by reason of the

---

* Hakluyt, ii. p. 9. Fabian's Chronicle was in the hands of John Stow when Hakluyt made these extracts.

† Sebastian evidently uses the name *Baccalaos* as a commercial designation of the *Gadus morrhua* or cod-fish. But the word being of Basque origin (*Bacaleos* or *Bacallos*), has given rise, as mentioned above, to an assertion of the prior discovery of Newfoundland by the Basques, who are said to have been led thither in the pursuit of whales, about the middle of the fifteenth century. The French name for the cod-fish *Cabillaud*, is supposed to have come from the Basque word by a transposition of syllables.—*Revue des deux Mondes*, 1859. Hakluyt, iii. p. 8. Ramusio, iii. p. 35, D.

sphere, that if I should saile by way of the north-west, I should by a shorter tract come unto India, I thereupon caused the King to be advertised of my device, who immediately commanded carvels to be furnished with all things appertayning to the voyage, which was *as farre as I remember in the yeere* 1496, in the beginning of summer, I began therefore to saile toward the north-west, not thinking to find any other land than that of Cathay, and from thence to turne toward India, but after certain dayes I found the lande ranne towards the north, which was to mee a great displeasure. Neverthelesse, sayling along by the coast to see if I could find any gulfe that turned, I found the land still continent to the fifty-sixth degree under our pole. And seeing that there the coast turned toward the east, despairing to find the passage, I turned back againe, and sailed downe by the coast of that land toward the equinoctial line, ever with intent to finde the said passage to India, and came to that part of this firme land which is now called Florida, where, my victuals failing, I departed from thence and returned to England, where I found great tumults among the people, and preparation for warres in Scotland; by reason whereof there was no more consideration had of this voyage."*

There are discrepancies in these several accounts, owing either to Sebastian's forgetfulness of dates, or to errors in the reporters of his conversations. Thus, our navigator is made to say by Butrigarius that his father died at the time of Columbus's great discovery of America becoming known in

---

* Hakluyt, iii. p. 6, translated from Ramusio, iii. Richard Eden, as quoted by Hakluyt (i. p. 498), mentions a voyage made in the eighth year of the reign of Henry the Eighth to the West Indies and Brazil by Sir Thomas Pert and Sebastian Cabot, which was defeated by want of courage in the former.

England, which it must have been at the tardiest, in 1493, whereas the licence of 1498 shews that John Cabot was alive in the beginning of that year, as does also the Chronicle of Robert Fabian. With respect to the preparation for war with Scotland, James the Fourth, carrying Perkin Warbeck with him, invaded England in 1496, a truce was concluded between the two countries in 1497, and Perkin himself was made prisoner in that year—dates not easily reconcilable with Sebastian's reasons for the voyages being broken off after 1498. There is no distinct evidence of more than two voyages having been made to North America by the Cabots, father and son, yet the accounts said to have been derived from the latter of the highest latitude that he reached are very contradictory. In one of the immediately preceding extracts, it is stated to be 55°, in the other, 56°. Francis Lopez de Gomara says, that *Sebastian Cabote* with two ships and three hundred men, took the way towards Island from beyond the Cape of Labrador until he found himself in 58 degrees and better. Then feeling the cold he turned towards the west, refreshing himself at Baccalaos; afterwards he sailed along the coast unto 38 degrees, and from thence shaped his course to returne unto England. (Hakluyt.)

Ramusio in his *Discorso sopra la terra ferma dette del Lavorador et de los Bacchalaos*, affirms that Sebastian searched the land up to the 67th degree; and in his general preface to the same volume, he informs the reader that, "*Il Signor Sebastian Gabotto nostro*" wrote to him many years past that, having sailed a long time west and by north *(ponente e quarta di Maestro)*, behind the islands of *Nova Francia* up to 67° 30′ north latitude, he found the sea open, and would have gone to *Cataio Orientale*, if the malignity of the shipmaster had not forced him back. Perhaps the high latitude Ramusio mentions, may have been an error occasioned by his fancy that the voyage of Steven Burrough to the Sea of Kara was made by Sebastian himself,* to whom in fact he attributed it.

Sir Humfrey Gilbert, in his *Discourse on the North-west Passage*, uses nearly the words of Ramusio, stating Sebastian's highest latitude to have been 67½ degrees, and referring, like Hakluyt, to the map in the Queen's private gallery at Whitehall, but adding the above statement of the latitude to the extract from Adams's map.† Sir George Peckham, in a treatise on the same subject, mentions 63 degrees as the northern limit of the discoveries of the Cabots. While Mr. Richard Willes,‡ in his *Argument to prove a North-west passage*, says, " Graunt the West Indies not to continue unto the pole, graunt there be a passage between these two lands, let the gulfe lie neerer us then commonly we finde it, set namely between 61 and 64 degrees, as Gemma Frisius, in his mappes and globes imagineth it, and so left by our countryman Sebastian Cabot, in his *table which the Earle of Bedford hath at Cheinies* :"§ again, " Cabota was not only a skilful seaman, but a long traveller, and such a one that entered personally that straight, sent by King Henry the Seventh to make this aforesaid discoverie, as in his owne discourse of navigation you may reade in his carde, drawen with his owne hand, that the mouth of the North-westerne straight lyeth neere the 318 meridian, between 61 and 64 degrees in the elevation, continuing the same bredth about 10 degrees west, where it openeth southerly more and more until it come

---

* Romusio, Viaggi. ii. p. 211.   † Hakluyt, iii. p. 16.   ‡ Hakluyt, iii. p. 25.
§ John Baron Russell of Cheyneys, Bucks, was advanced to an earldom in 1549.

under the tropicke of Cancer, and so runneth into *Mar del Zur*, at the least 18 degrees more in bredth there, then it was when it first began."*

Sebastian's "owne mappes and discourses, drawn and written by himself, were (according to Hakluyt) in the custodie of the worshipful master, William Worthington," and were accessible to geographers in the time of Queen Elizabeth, but are no longer extant.

The great interest which attaches to the voyages of the Cabots, as being the beginning of the attempts of England to make the north-west passage, has led us to give the preceding quotations from most of the early notices of any credit. Though Ramusio corresponded with Sebastian, he does not seem to have been accurately informed of the dates and objects of his northern voyages, since, though evidently inclined to award a full measure of praise to his countryman, of whom he speaks proudly and even affectionately, it is to the Portuguese Gaspar Cortoreale that he gives the merit of being the *first, as far as he knew*, who attempted a north-west passage to the Spice Islands, yet he tells us that Sebastian would have gone to China but for the mutiny of his men.

Until a very recent date, geographers had to rely on the statements of the authors we have quoted, when debating the northern limits of Cabot's voyage and the probability of his having entered Hudson's Strait, but a discovery has been made on the continent within a few years, of a great planisphere of Sebastian Cabot, bearing the date of 1544, and now preserved in the Imperial Library of Paris. We have not seen this very interesting document, but Dr. Asher went to Paris to examine it, and he states, that it is attached to a roller, is very large,

* Hakluyt. See also De Laet *Novus Orbis*, p. 31.

and on both sides of the engraving there are, *pasted on*, explanations, which on the one side are in the Latin language, on the other in Spanish. The Latin letter-press has been reprinted, and fills more than twenty pages. It is in one of these explanatory affixes that the date of 1544 occurs.* Dr. Asher does not mention any reason for believing that these explanatory notices are the work of Sebastian Cabot himself, except their correspondence with a notice of Chytræus, which will be immediately referred to. Admitting that Sebastian furnished the explanations, we have shewn in the preceding pages that this eminent navigator, if correctly reported by his friends and correspondents, varied considerably at different times in his dates and latitudes, most likely from defect of memory as age gained upon him; and we may therefore find discrepancies more or less important in the marginal explanations pasted on different copies of his planisphere but forming no part of the original engraving.

Dr. Asher refers, as we have said, to the *Itinerum Deliciæ* of Chytræus, who visited Oxford in 1566, and there copied a series of inscriptions corresponding, except in date, which is 1549, with the Latin inscriptions on the Paris map, and in some minor points. Clement Adams is accused by Dr. Asher of having altered Sebastian's original text by bombastic additions, of changing to 1497 the date of the voyage, which is recorded as 1494 in the printed affixes to the Paris planisphere, and even of falsifying the outlines of the chart after Sebastian's death. These are serious charges, but the question of their justice may remain in abeyance until Dr. Asher gives his reasons for making them, at length, in his promised treatise on

---

* Henry Hudson the Navigator, edited for the Hakluyt Society by G. M. Asher, LL.D., London 1860, p. 260.

the Cabots. In the mean time we believe, on the strength of the contemporary authorities already quoted, that the correct date of John Cabot's discovery of the *Terra primum visa*, was 1497.

The subjoined copy of the geographical part of Adams' inscription will shew that it is in his own words, and is to be taken as being in accordance with the information he had received; and that he mentions John Cabot and his son in the third person. We know not why more credit should be given to the Paris map than to Adams.*

M. Jomard, in his *Monumens d'ancien Geographie*, has published three out of four sheets of Cabot's great planisphere (without letter-press as yet), which may be consulted in the British Museum.† On the sheet numbered provisionally 66, 67, the American coast, under the appellation of *Costa del hues norweste*, is extended to the arctic circle, which it just cuts, the *ysla demene* is on the 64th parallel of latitude, the *rio*

* "Anno Domini 1497, Joannes Cabotus, Venetus et Sebastianus illius filius eam terram fecerunt perviam, quam nullus prius adire ausus fuit, die 24 Junii, circiter horam quintam bene manè. Hanc autem appellavit terram primum visam, *credo* quod ex mare in eam partem primum oculus injecerat. Nam que ex adverso sita est insula, eam appellavit insulam *Divi Joannis*, hâc opinor ratione, quod aperta fuit eo die qui est sacer Divo Joanni Baptistæ." . . . "Imprimis autem magna est copia eorum piscium, quos vulgi sermone vocant *Bacallaos*." The rest describes the native inhabitants and ferine productions of the land and sea (Hakluyt, iii. p. 6). The entire Latin inscription, as given by Hakluyt, is reprinted in the *Novus Orbis* of Joannes de Laet, published by the Elzevirs at Leyden in 1633, p. 31, wherein it is stated that many copies of Sebastian's map (*non paucæ*) existed at that date in England. De Laet's beautifully executed map of Newfoundland is very useful for identifying the places named in the rude maps of the preceding century. Clement Adams is called by M. Richard Eden in his Decades, a learned *young man*, schoolmaster of the Queen's henchmen. This passage occurs in Eden's account of Chancelor's voyage of 1553.

† Mappemonde de Sebastien Cabot, Pilote Major de Charles Quint, de la premier moitie de xvi siecle, format double.

neuado on the 59th, and the *ysla de los aves* and *San Brandon* on the 55½°. *Islanda* is wholly north of the arctic circle, and reaches to 73½°, or more than six degrees higher than its true position. With such an error in the geography of an island comparatively well known, we may be allowed to conjecture that the highest inlet seen by the Cabots was that now known as the Strait of Bellisle, but which is named on Sebastian's planisphere, *Rio nevado*, and which was called by the Portuguese *Golfo de Castello*. In the face of Sebastian's declaration, that at the limit of his voyage he turned back, because he found the coast trending continually to the east, we cannot believe that he entered Hudson's Strait and followed it to the westward through ten degrees of longitude, as Willes, from an inspection of the chart at Chenies, was led to assert. It is more likely that Sebastian's planisphere, constructed half a century after his voyage, embodied what he knew of the Portuguese discoveries, derived from Ribero's chart of 1529 or other sources; and if in any voyage subsequent to 1498, Sebastian did enter an inland sea, it must have been the Gulf of St. Lawrence, behind the islands of Nova Francia, as he stated in his letter to Ramusio.

The atlas of Baptista Agnese contains a map bearing date 22d October 1544, and therefore contemporary with the engraving of Sebastian's planisphere, whereon the American coast is represented as deeply indented by very numerous inlets, not explored to the bottom. On its northern portion the following names occur in order, proceeding from the south,—*Terra che discobrio Steuen Gomes*—*Zalfi*—*Terra di Bertones*—*Cabo raso*—*y. d. Baccalaos*—*y. de los aues*—*y. del fogo*—*y. de la fortuna*, placed in the mouth of an inlet, beyond which the Labrador coast is traced for some degrees without names. The places from Cape Race, northwards, belong all to projecting parts of Newfoundland, mostly represented as islands.*

Gasper Cortoreale is said to have been an able navigator, of a determined and enterprising character, who had been educated in the household of the King of Portugal. He sailed in the year 1500 with two caravels and discovered *Terra Verde* (Labrador) from *Rio Nevado*, a river or strait encumbered with ice, in latitude 60° (Hudson's Straits) to *Rio Lorenzo* and its gulf called *Quadrato* (Gulf of St. Lawrence), which turns at the end of *Los Bacchalaos*. In his voyage down the coast he named *Porto di Molvas* (Cod-harbour) in latitude 56°, a large island of demons, succeeded on the south by another called *Terra Nuova*; then by a smaller one termed *Los Bacchalaos*; and by a detached islet denominated *Cabo de Ras* (Cape Race). All these, with many other small islands and straits, are represented on a rude map in Ramusio's third volume, which was first published in the year 1556. That map, therefore, may be considered as containing Ramusio's notion of Newfoundland and the adjoining coasts at that date; but the headlands are represented as they would appear to a ship a good way from land, making like islands, and the map is otherwise so incorrect, that the trouble of endeavouring to identify all the places by their modern names would be ill repaid, though this may be approximately done by referring to De Laet's map, whereon many of the names are reproduced. Mr. P. Fraser Tytler, in

* This map, though on a small scale, is very neatly executed, and is very far superior as a delineation of the coast to the rude wood-cuts of Ramusio. It does not contain lines of latitude and longitude, and the north point is towards the right-hand corner of the sheet. There is a second map of North America in the same atlas, which does not go so far north. The Atlas is preserved in Trinity College, Dublin.

his *Progress of Discovery in the Northern parts of America*, says that *Terra Verde* is* named *Terra Cortorealis* in a Roman map dated 1508. Cortoreale's voyage was closely-followed by that of Estevan Gomes, but as the latter is not said to have approached the Arctic regions we shall not make further mention of him here, though both contributed to introduce the Portuguese to the Newfoundland cod-fishery.

Rondelet, whose work appeared in 1554, speaks of the fishery of the cod-fish by the Bretons and Normans in the sea of Nova Francia as an established thing at that date, and Bellon, another French icthyologist, identifies the American species of cod with the *stock-fish* which was brought to Germany from the coast of Norway. The Basques also were early in the field, and Samuel Champlain, quoting from Niflet and Antoine Magin, says, that the Bretons and Normans established a fishery on the great bank of Newfoundland in 1504; and Jean Denys of Honfleur is said to have constructed a chart of the Gulf of St. Lawrence in 1506.† Hakluyt also has recorded the private voyages of several Englishmen to the American continent in the beginning of the sixteenth century, but as these were not directed to the high latitudes, and did little or nothing in preparing the way for Arctic enterprise, we must pass them by.

In 1527 Master Robert Thorne, then residing in Seville, addressed what he termed "a Declaration of the Indies" to

---

* This short notice of the voyage of Gaspar Cortoreale is extracted from Ramusio (iii. p. 417). Mr. Tytler says, that the most authentic details of the voyage are to be found in a letter written by the Venetian Ambassador at Lisbon, to Pietro Pasquiligi, only eleven days after Cortoreale's return to Portugal. The pretended voyage of Anus Cortoreale is omitted in the text, as being unsupported by authentic records.

† Charlevoix Histoire de la Nouvelle France, i. p. 4.

King Henry the Eighth, exhorting that monarch to send a naval armament to the north, which he states to be the only way of discovery that remained to be tried, other princes having preoccupied the southern, western, and eastern routes. This writer's father has been mentioned in a preceding page as one of the first adventurers that left England in search of new lands, and the son claims to have a hereditary interest in the subject. A second letter, giving his reasons in detail, and accompanied by a map which we have already referred to, is addressed to Dr. Edward Ley (or Lee), then ambassador to the Emperor, and afterwards Archbishop of York. In this letter he gives a summary view of the discoveries made by the Spaniards, Portuguese and others, up to that time. He adds to his notice of the discovery by the English of the "Newe founde lands." "Nowe then, if from the sayde newe founde landes the sea be navigable, there is no doubte, but sayling northwarde and passing the pole, descending to the equinoctiall lyne, we shall hitte these islandes (the Spice Islands), and it shoulde bee the much more shorter way than either the Spaniards or the Portingales haue." . . . "I judge there is no lande (un) inhabitable nor sea innavigable. So that if I had facultie to my will, it would bee the first thing that I would understande, euen to attempt if our seas northwarde be navigable to the pole or no." This is the first clear enunciation of the desirableness of attempting a voyage across the polar regions, and Master Thorne's arguments had doubtless considerable weight at the time when they were made public and for long afterwards.*

* Thorne's letters are given at length, together with a *fac simile* of his map in "Divers Voyages," edited for the Hakluyt Society by John Winter Jones, 1850, p. 27 *et infra*.

Hakluyt informs us that Thorne's exhortation *took present effect* with King Henry the Eighth, who sent forth in the month of May of that same year "two faire ships well manned and victualed, having in them divers cunning men to seeke strange regions." These ships, the Mary of Guildford and Sampson, effected nothing. The former touched at Newfoundland and came home, the latter was lost in a storm.*

The voyages of the Cabots did not open to the English nation a short way to the Spice Islands or to the fabulous golden regions of the East, but they laid the foundations on which, in the course of a few generations, a lucrative fishery and many prosperous colonies were raised, and through which the Anglo-Saxon race were eventually diffused, as masters, over the northern half of the new world.

The knowledge of America, possessed by the best informed geographers of the period that elapsed between the voyages of the Cabots and those of Frobisher, will be best understood by consulting the ancient maps republished by the Visconte de Santarem in his *Trois Essais sur l'Histoire de la Cosmographie.* (British Museum.)

* Hakluyt, iii, p. 54.

CHAPTER III.

VOYAGES TO THE NORTH-EAST FROM ENGLAND.—A.D. 1548-1580.

Sebastian Cabot—Sir Hugh Willoughby—Novaya Zemlya—Arzina—Richard Chancelor—Stephen Burrough—Kolguev—Petchora—Novaya Zemlya—Vaigats—Sea of Kara—Kara gate, or Burrough's Strait—Pet and Jackman—The Ob—Novaya Zemlya—Petchora—Ygorsky Schar, Southern Vaigats or Pets Strait; or Straits of Nassau—Russian accounts of Samoeid land.—Mathew's land—Kostin Schar — Cape Taimyr — Sievero Vostochnoi nos, or Cheliuskin.

THE unfortunate voyage alluded to at the end of the last chapter, is the only effort made by England in the reign of Henry the Eighth to further northern discovery; but in the auspicious, though too brief reign of his son Edward the Sixth, the spirit of discovery revived, and the merchants complaining of the decay of trade, while the Portugals and Spaniards were bringing home riches yearly from both the Indies,* their views were now directed to the north-east, and to Russia, of which country an account by Sigismund Baron Heberstein was published at Vienna in 1549.† The Hans Towns had hitherto monopolized the Russian trade, under pretence of a league established among themselves (*Hans Law*, Purchas calls it) for repressing the overflowings of the supposed teeming

* Hakluyt, i. p. 243.
† *Rerum Muscovitarium Commentarii. Vindobona.* An Italian translation was brought out at Venice in 1550, and reprinted in the *viaggi di Ramusio* in 1583. See an English translation for the Hakluyt Society by R. H. Major.

northern populations, and securing the southern kingdoms from damage. The English merchants determined to brave the Hans Law, and Clement Adams says that Sebastian Cabot was a prime mover in the business, and, as Mr. Richard Eden testifies, had long beforehand the secret of a voyage toward Cathay in his mind. He had returned to London in 1548 from the service of Spain, either of his own free-will or recalled by the young king, who in 1549 created him Grand Pilot of England, with a pension of £166 : 13 : 4, for the good and acceptable services done and to be done.* As "governour of the mysterie and discouerie of regions, dominions, islands and places unknowne," he drew up an excellent code of instructions for the guidance of the officers and mariners in the company's employment, and carefully superintended the outfit of the three ships that were "prepared and furnished out for the search and discouerie of the northerne part of the world, to open a way and passage to our men for trauaile to newe and unknowen kingdomes."

This enterprise was undertaken in the year 1553; Sir Hugh Willoughby, an able military officer, was appointed Captain-General, and embarked in the Admiral, named the Bona Esperanza, of 120 tons, William Gefferson, master; Richard Chancelor, pilot-major, commanded the Edward Bonauenture, of 160 tons; and the Bona Confidentia of 90 tons, had Cornelius Durfoorth for master. Each ship was furnished with a pinnace and a boat, and provided with eighteen months' provisions and everything that was necessary. The expense was defrayed by a subscription of £25 sterling from each of the members of the company, producing the gross sum of £6000.

* Hakluyt, iii. p. 10.

All being ready, the squadron unmoored from Ratcliffe on the 20th day of May 1553, and passing by Greenwich, where the young king then lay sick, the mariners, apparelled in watchet or skie-coloured cloth, rowed amaine; "the common people flockt together, standing very thicke upon the shoare, the Privie Counsel, they lookt out at the windowes of the court, and the rest of the courtiers ranne up to the tops of the towers: the shippes hereupon discharge their ordnance, and shoot off their pieces after the manner of warre, and of the sea." Very slow progress was made in the voyage down the Thames, and Orfordness was not left behind until the 23d of June. On the 14th day of July the fleet approached Ægoland and Halgoland, Norwegian islands lying on the 66th parallel, and distant by the reckoning 250 leagues from Orfordness. The last named island was Ohther's abode, as mentioned in a former chapter, and a party having landed there in a pinnace, saw about thirty small houses, but the inhabitants had fled.

Off Seynam or Senjän, an island on the Norwegian coast, whose north end is in latitude $69\frac{1}{2}°$, the fleet encountered a storm in which the Admiral and the Bona Confidentia were separated from the Edward Bonauenture, driven far out to sea, and prevented from embarking a pilot to take them to Wardhuus, in Finmark. Fourteen days after leaving Senjän they sighted "Willoughbie Land, in 72 degrees," along which they plied northward for three days, but "the Confidence being troubled with bilge water, and *stocked*" (stagged, hogged), they bore round to seek a harbour. Willoughby's Land is, according to Admiral Beechey, whose opinion has been generally received, that part of the coast of Novaya Zemlya which is named the Goose Coast by the Russian Admiral Lütke, of which the extreme points are the *Syevernuy Muis*, and *Yuzhnuy Guisnuy Muis*, or North and South Goose Capes. A harbour was found in Lapland, on the west side of the entrance into the White Sea, opposite the promontory of *Kanin nos* at Warzina, called by Hakluyt Arzina, near unto Kegor.

There, at no great distance from the Dwina, where relief could have been obtained, the captain-general, officers, and crews of both ships were miserably frozen to death, as some Russian fishermen ascertained in the following spring. The journals and other papers that were recovered mention the discovery of Willoughby's Land, and that the captain-general, believing from the frost, snow, and hail in the middle of September, that further navigation was inexpedient, determined to winter in that desolate haven. He had sent out parties to explore the country in several directions, for the distance of three or four days' journey, but no inhabitants were met with, though bears, great deer, foxes, gluttons and diverse other strange beasts were seen. How long the miserable men sustained the severity of the weather is not known, but a will, found on board the Admiral, proved that Sir Hugh Willoughby and most of that ship's company were alive in January 1554. Had they been skilled in hunting and in clothing themselves,* so as to guard against the severity of the weather, and taken the precaution, moreover, of laying in at the beginning of the winter a stock of mossy turf, and such dwarf *empetra, vaccineæ* and *andromedæ*, as the country produced for fuel; and above

* Voyage towards the North Pole, edited by Capt. F. W. Beechey, 1843, p. 227. See also Three Voyages, edited by Dr. Beke for the Hakluyt Society, 1853, p. 5. In a commission from the Muscovy Company, bearing date 1568, instructions are given to *search* whether that part of *Noua Zembla* (against Vaigats) doe ioyne with the *land of Sir Hugh Willoughby*, disconered '53, and is in 73 degrees.—*Hakluyt*, i. p. 382.

all, had they secured a few of the very many seals and great fishes (whales) which they saw in abundance in the sea around them, they might have preserved their lives, and even passed a comfortable winter. Chancelor, during the time the fleet lay at Harwich, had overhauled his provisions and found much of them corrupted and putrid, which afterwards gave him much anxiety. Matters were not likely to be better on board the other two ships, and want of warmth, with a low and bad diet, would speedily render the crews victims of scurvy. The English agent at Moscow, on being apprised of the sad event, sent men to conduct the ships containing the goods and dead men back to England; but the ships being leaky, sunk by the way, and carried the living navigators to the bottom along with the dead.

The Edward Bonauenture was more skilfully or more fortunately managed by Richard Chancelor, captain and pilot-major. On losing sight of the Admiral, off Seynam, Chancelor made for the appointed rendezvous at Wardhuus, in latitude $70\frac{1}{4}$ N., and after waiting seven days in vain for the arrival of the Admiral, he again sailed, and eventually reached St. Nicholas in the White Sea. From thence he proceeded overland to Moscow, delivered his credentials to the Czar, Ivan Vasilovitch, and obtained from him many privileges for the English merchants. In 1554 Chancelor returned to England; shortly afterwards Cabot's Company received a charter of incorporation from Queen Mary, and in the eighth year of Queen Elizabeth, an act of parliament was passed, in which the company is styled "The Fellowship of English Merchants for the Discovery of New Trades." It was, however, generally termed the Muscovy or Russia Company. This success, and the long and advantageous alliance with Russia which it led to, is another of the benefits resulting from the suggestions and influence of Sebastian Cabot, and England owes gratitude to Venice, from whence her grand pilot came.

In 1556 the Muscovy Company fitted out the Serchthrift pinnace for discovery towards the river Ob, and further search for a north-east passage. The command of this small vessel was given to Steuen or Stephen Burrough, who had been master of Chancelor's ship, and he was accompanied, as he had been in the Edward Bonauenture, by his brother William, afterwards comptroller of Queen Elizabeth's navy. The family name of these seamen is variously spelt by Hakluyt, Borough, or Burro. "On the 27th of April, being Monday, the Right Worshipful Sebastian Cabota came aboord our pinnesse at Grauesende, accompanied with divers gentlemen and gentlewomen, who, after they had viewed our pinnesse, and tasted of such cheere as we could make them aboord, they went on shore, giving to our mariners right liberall rewards, and the good olde gentleman Master Cabota, gaue to the poore most liberal almes, wishing them to pray for the good fortune and prosperous successe of the Serchthrift, our pinnesse. And then at the signe of the Christopher, hee and his friends banketed and made mee and them that were in the company great cheere; and for very joy that he had to see the forwardness of our intended discovery, he entred into the dance himselfe, among the rest of the young and lusty company: which being ended, hee and his friends departed most gently, commending us to the governance of almighty God."* Sebastian Cabot was then in the eighty-eighth year of his age.

By the end of May the Serchthrift had reached the highest

* Hakluyt, i. p. 274.

point of Norway which Burrough says was named by him on the previous voyage of Chancelor. On the 7th the Edward, Captain Chancelor, which had hitherto accompanied the Serchthrift, being bound for St. Nicholas, steered to the southward while Burrough went to the river Kola. There they met a Russian lodia, whose master informed them that the river Petchora was distant seven or eight days' sail. Availing himself of the great courtesy of this Russian shipmaster, whose name was Gabriel, Burrough kept company with his lodia (*lodji*), and with others bound for the fishery of salmons and morses at Petchora. On the 8th of July he reached *Kanin-nos*, the eastern promontory of the Gulf of Archangel, near which, riding at anchor in ten fathoms of water, he obtained a good plenty of haddocks and cods. Thirty leagues further, Gabriel conducted the English pinnace to the harbour of Morgiouets, where there was plenty of sea-fowl with abundance of driftwood, but no trees growing. Here Burrough was presented with three wild geese and a barnacle by a young Samojed. On the 14th the Serchthrift passed to the south of the island of Kolguev or Kolgoi, and next day went in over the dangerous bar of Petchora, on which there was one fathom of water. Burrough ascertained the latitude to be 69° 10′, and the variation of the compass $3\frac{1}{2}°$ from the north to the west. The rise of the tide at full moon is four feet. On the 21st July, after leaving Petchora, monstrous heaps of ice were seen, and at first mistaken for land; and soon afterwards, before the mariners were aware, the Serchthrift was enclosed within it, "which was a fearefull sight to see." Getting clear of this danger, an easterly course was pursued in about 70 degrees of latitude, but the pinnace was daily hampered by ice. On the 25th a monstrous whale came so near the ship that a sword might have been thrust into his side, on which all the company shouted, for they feared that their ship would be overthrown; with the cry however the monster, making a terrible noise in the water (spouting) departed, and the fearful mariners were quietly delivered of him. On the same day certain islands were seen, one of which Burrough named St. James, and made its latitude to be 70° 42′ N., which, according to Admiral Lütke is ten miles too much. From a Russian master of a lodji, Burrough learnt that he was off the southern extremity of Novaya Zemlya, and on the 31st the Serchthrift came to anchor among the islands of Vaigats. On one of the islands Burrough saw a heap of Samojed idols, of very rude manufacture, with bloody eyes and mouths. This spot is Bolvánovsky Nos (Image Cape), at the north-eastern end of Vaigats, which, according to Admiral Lütke, was visited in 1824 and found to be in precisely the same state as it was when discovered by the Serchthrift.* On the 5th of August Burrough saw "a terrible heape of ice approach neere," and therefore thought good with all speed to depart from thence. On the 22d, despairing of discovering any more to the eastward that year, he returned westwards; on the 29th he passed to the north of the island Kolguev, and on the 31st doubled Kanin-nos. On the 11th of September the Serchthrift was brought to anchor in the harbour of Kholmogorui (Colmogro) on the Dwina, where she wintered. Burrough intended to resume his voyage to the Ob next spring, but being dispatched to Wardhuus to search for some English ships, the voyage to the Ob was not performed. "The passage by which Burrough thus sailed between Novaya Zemlya and Vaigats into the sea of Kara, is called by the Russians *Karskoi Vorota*

* Dr. Beke, lib. cit. page x.

(the Kara Gate), and as he was the first navigator who is recorded to have been there, he must be regarded as the discoverer of that Strait."*

Though the Muscovy Company were much occupied with their inland commerce through Russia to Persia, they renewed from time to time their attempts to find a passage eastward along the northern coasts of Europe and Asia. With this in view, they instructed their agents in Russia to collect information respecting the mouths of the Ob and other large rivers that flow into the Arctic sea; and they sent out at least two sea-expeditions. Of one of these, only the instructions are known dated 1568, and addressed to Bassendine, Woodcocke and Browne. The other sailed in 1580 under command of Arthur Pet and Charles Jackman, two able and persevering seamen. They were instructed when they came to Vaigats to pass eastwards along the coast of Samoeda, keeping it always in sight till they came to the mouth of the river Ob, and so to pass eastwards to the dominions of the Emperor of Cathay. Jackman's vessel was very frail, sailed ill, and had a crew of only five men and a boy. Pet pushed on ahead leaving Jackman to follow him to the rendezvous at Vaigatz, and having made Novaya Zemlya in latitude 71° 38′ N., about the south Goose Cape, he ran to the southward, keeping Novaya Zemlya on his larboard hand until he reached Vaigatz, but being unable to approach it on account of the ice, he thought that it was a continuation of Novaya Zemlya, and standing off shore missed the northern or Burrough's Strait which he was seeking for. Continuing his

* Dr. Beke, l. c. Steven Burrough's journal appears in the second volume of Ramusio's "Viaggi," and is there erroneously attributed to Sebastiano Cabota.

course to the southward with a flowing sheet, he ran into the Bay of Petchora, then resuming an easterly course he got sight of the south end of Vaigatz, and on the 19th of July entered the strait between that island and the main land of the Samojeds. This southern passage of Vaigatz or Pet's Strait is called by the Russians *Vyorsky Schar*, and by the Dutch the Straits of Nassau. (Dr. Beke.) On the 25th, while endeavouring to find a way eastward through the ice, Pet was overtaken by Jackman, when they agreed to seek the land again and to confer further at Vaigatz. By warping from one piece of ice to another they reached a clear sea on the 15th of August and gave God the praise. On the 16th they turned to the westward, and on the 26th of December Pet reached Ratcliffe, but Charles Jackman carried the William into a Norwegian port, where he wintered. On the opening of the navigation next year he left his port of refuge in company with a Danish ship and was never heard of after that time. Arthur Pet was one of Chancelor's crew in the Bonaventure.

Purchas* has preserved two documents written in 1584 by Anthony Marsh, a chief factor of the Muscovy Company, in which he quotes a letter from four Russians containing the following paragraph: "Heretofore your people have been at the said river of Ob's mouth with a ship, and there was made shipwreck, and your people were slain by the Samojeds, who thought that they came to rob and subdue them. The trees that grow by the river are firs and *yell*, a kind of soft light fir." Dr. Beke, after giving the two documents at length, observes, that we learn from them two very remarkable facts. The first is, that previously to the year 1584, an English vessel had crossed the sea of Kara, and reached the mouth of

* Pilgrimes, iii. p. 804.

the Ob. The second is, that at that time the best marine route from the White Sea and the Petchora was by the isles of Vaigatz and Novaya Zemlya, and by the land of Matpheone, being a record of the discovery of the entrance into the sea of Kara by the Matochkin Schar. In this strait the Russian pilot Rosmuistor wintered in 1768, and through it penetrated into the sea of Kara in the following season, but could not proceed far for the ice. The navigation of the sea of Kara has always been obstructed by ice, and on that account the eastern side of Novaya Zemlya has not yet been fully surveyed. Mathew's Strait (which is the translation of *Matochkin Schar*, or *Mattuschan Yar*, is in latitude $73\frac{1}{2}°$ N.) and separates Novaya Zemlya proper from Mathew's Land, north of which, and separated from it by a narrow winding channel, lies Lütke's Land, the largest of the Novaya Zemlya islands. Mathew's Strait lies nearly on the same parallel of latitude with *Olenii nos*, the eastern promontory of the estuary of the Ob.

The Novaya Zemlya islands, including Vaigats, which may be considered as the fourth and southernmost of the chain, stretch in a crescentic curve, concave to the east from a little below the seventieth parallel northwards through seven degrees of latitude. The east side of the sea of Kara has a general north-east direction from Kara Bay, in latitude 69, to Cape Taimyr and Cape Cheliuskin of Middendorf or the *Sievero Vostochnoi nos* (Sacred Promontory) of Baron Wrangel. Several large rivers fall into the sea from the land of the Samojeds, of which the Ob and Yenisei have long estuaries. The Piasina, and Taimyr or Legata, are more northern, the latter entering the bottom of Taimyr Bay.

CHAPTER IV.

DUTCH NORTH-EASTERN VOYAGES.—A.D. 1594-1597.

William Barentzoon—Novaya Zemlya—Cornelison Nai—Sea of Kara—Barentzoon and Rijp—Bear Island—Circumnavigation of Spitzbergen—Doubling of the north end of Novaya Zemlya—Barentzoon's miserable winter and death—escape of the survivors.

THE Netherlanders had loooked on the progress of the English in Russia with no small commercial jealousy, and after employing in vain John de Walle, merchant-ambassador at the court of the Czar, to shake the credit of the Muscovy Company, determined to compete with them in the search for the north-east passage. The merchants of Middleburgh in Zeelandt were the first to move in this business, and in conjunction with the Syndic of West Friesland, resident in Enkhuysen, fitted out the Swane of Ter Veere, under the command of Cornelis Corneliszoon Nai, who had served for some years as a pilot in the Russian trade; and the Mercurius of Enkhuysen, commanded by Brant Ysbrantzoon, otherwise Brant Tetgales, also an experienced seaman. Amsterdam likewise, desiring to participate in the enterprise, fitted out a vessel named, like that of Enkhuysen, the Mercurius, which was wisely entrusted to William Barentzoon, by contraction Barentz, (meaning the son of Barent or Bernard,) a most skilful mariner and a burgher of Amsterdam. He had prepared himself for voyaging in the north-sea by a study of the Icelandic records, and Purchas has preserved a translation of part of

Ivar Bardsen's or Boty's account of Greenland, made out of High Dutch into Low Dutch by William Barentzoon, and out of Low Dutch by Mr. Stybre in 1608, for one Henric Hudson.* Purchas has also printed a paper of Barentzoon's on the tides of Kara.

The three ships having assembled at the Texel on the 4th of June 1594, Cornelison Nai was appointed commodore, and an agreement made that they should keep company as far as Kildin, in Lapland. On the 29th of the month Barentzoon parted from the other two ship-masters, on his separate voyage to Novaya Zemlya, and a few days afterwards Nai and Ysbrantzoon sailed for Vaigats. In this first voyage Barentzoon coasted the west side of the Novaya Zemlya Islands from Langenès, on Mathew's Land, to the Islands of Orange, the most northerly points of the range. His men being exhausted by constant labour among the ice, he turned back on the 1st of August, and on the 15th reached Matvyeea Ostrov and Dolgoi Ostrov, to the eastward of Pet's Strait, and south of the 70th parallel of latitude.

Here he rejoined the other two ships, whose masters reported that they had entered the Kara Sea by the Yugorsky Shar (Pet's Strait), which they renamed the Strait of Nassau,† and sailing eastward to the longitude of the Ob, had been not far they thought from Cape Taimyr. Linschoten, who was supercargo of Ysbrantzoon's ship, and wrote the history of the

---

* Ivar Bardsen, a Greenlander by birth, and proctor of the episcopal city of Gardar in the beginning of the fifteenth century, is mentioned in a preceding chapter. His description of Greenland was preserved by Erik Wakkendorph, Archbishop of Drontheim, and translated by Torfæus in his *Grœnlandia*. The imperfect copy used by Barentzoon was found in the Ferüe Islands.

† In Blome's English edition of M. Sanson's map, printed in 1670, this strait is named Gorgossoio Schar.

voyage, is thought by Gerrit de Veer to have made too favourable a report of the extent of the voyage across the Kara. On the 14th of September all the three ships regained the Dutch coast, and shortly afterwards reached their respective ports in safety.

A second fleet of seven ships from Zeelandt, Enkhuysen, Rotterdam, and Amsterdam, did not get farther than Pet's Strait and made no discovery.

The third Dutch voyage was performed by Amsterdam ships only, the other ports declining further expenditure. Barentzoon was chief pilot, and Jacob Van Heemskerck and Ian Corneliszoon Rijp were captains. They sailed early in the summer of 1596. Barentzoon desired to reach Cape Taimyr by rounding the north end of Novaya Zemlya, in 77 degrees of latitude, but Rijp wished to avoid the eastern land, and Barentzoon giving way to his authority, or to the urgency of his persuasions, they steered to the northward on a more westerly meridian. On the 9th of June 1596 they saw a high steep island in latitude $74\frac{1}{2}°$, to which they gave the name of Bear Island. Seven years later, Stephen Bennet, a ship-master in the service of the Muscovy Company, called the same rock Cherie, after Sir Francis Cherie, a member of the Company. Continuing their voyage northwards and eastwards, the mariners saw land again in 79° 49′ N. and supposed that they had reached a part of Greenland, but they had in fact arrived on the east side of the Spitzbergen group of islands, and were the discoverers of that archipelago. Running westwards, along the south coast of the north-east island, they entered the Vaigats or Hinlopen Strait, and passing round the north ends of New Frizeland and West Spitzbergen, in the 80th parallel of latitude, they returned to Bear

Island on the 1st of June. Here the two vessels separated. Rijp again sailed north to Bird Cape, on the west side of Spitzbergen, and from thence home.

Barentzoon, on the other hand, held to the eastward until he reached Novaya Zemlya, in latitude 73¼°, whence he coasted the western shore, northwards, until he passed the Islands of Orange, the limit of his first voyage; and having fairly rounded the north-east extremity of the land, and traced its eastern coast some way, was arrested by the ice, and shut up in Ice Haven. "In the evening of the 26th of August," says Gerrit de Veer, "we got to the west side of Ice Hauen, where we were forced, in great cold, pouerty, misery and griefe, to stay all that winter." On the 11th of September the poor mariners took counsel among themselves, and after debating the matter, determined to build a house upon the land, to keep therein as well as they could, and so commit themselves to the tuition of God. "And to that end we went further into the land, to find a convenient place, and yet we had not much stuffe, in regard that there grew no trees, nor any other thing in that country to build the house withall. But at last we found an unexpected comfort in our need, which was, that we found certain trees, roots and all, which had been driven upon the shoare, either from Tartaria, Muscouia, or elsewhere, for there was none growing upon that land." On the 15th of September, a bear having come to where the crew were working at the wood, put its head into the harness-tub, to take out a piece of salt-beef, and was shot in the act of doing so. After this, in the fine days, they dragged the wood on sledges to the building site they had chosen, and in so doing, ran considerable risk from the bears; but in stormy weather they kept close under hatches on board, being unable to endure the severity of the cold, when it blew hard. On the 12th of October the house was so far finished, that half the crew slept in it, but endured great cold, not having clothes enough, and because they could keep no fire, on account of the chimney not being made, whereby it smoked exceedingly. From this time they were engaged, when the weather permitted, in landing provisions, breaking up parts of the ship to get deals, and in other necessary preparations for the winter. On the 20th of October, a party going on board to get spruce beer, found it frozen, with the barrels burst and the iron hoops broken. This day, in calm sunshine, they saw the sea open. On the 24th, the whole crew took up their abode in the house, part of them having up to that night slept on board. On the 4th of November the sun ceased to rise above the horizon, and at this time the bears began to depart, and the arctic foxes to come about them. The foxes which they took in the dark season gave them an occasional and seasonable supply of fresh food. They endeavoured to warm themselves in their sleeping berths by putting hot stones under their feet, for the cold and the smoke were alike unsupportable. The heat was so great a comfort that they endeavoured to make it continue by stopping up the doors, chimney, and all the avenues of fresh air, but (as was sure to happen) when they had succeeded, "we were taken with a great swounding and dazeling in our heads, so that some of us that were strongest, first opened the chimney, and then the doores, but he that opened the doore fell downe in a swound, with much groaning, upon the snow. On casting vinegar in his face, he recovered and rose up. And when the doores were open we all recovered our healthes again, by reason of the cold aire." They continued in fine weather to go abroad and set springes for foxes, but were often frostbitten in the face and ears. By an observation of Bellatrix, a star in the left shoulder of Orion, the latitude of the house was ascertained to be 75° 43' N.*

On the 28th of December, one of the men made a hole at the door and went out, but staid not long, on account of the hard weather. He found the snow higher than the house. On the 5th of January 1597, the festival of twelfth night was kept, and the gunner was made king of Novaya Zemlya, which, says Gerrit de Veer, "is at least 800 miles long, and lyeth between two seas."

On the 22d, some of the men going abroad perceived that daylight began to appear, and said that the sun would soon be visible, but William Barentzoon replied that it was yet two weeks too soon. On the 24th in fair clear weather, Gerrit de Veer, Jacob Heemskerck, and another going to the seaside, saw, contrary to their expectation, the edge of the sun, and hurried home to tell William Barentz and the rest the joyful news; "but William Barentz, being a wise and well experienced pilot, would not believe it, esteeming it to be about fourteen daies too soone." After two days of misty weather, the sun was again seen on the 27th. From some astronomical data, given by Gerrit de Veer, Dr. Beke infers that it was on the 25th, not the 24th, that the sun was first seen, and even with this correction, the extraordinary refraction of 3° 49' must be allowed for. The daylight, by this time, had so increased, that the men were able to refresh themselves by playing at the ball. With the return of daylight, the bears came again about the house, and some being shot, afforded a very seasonable supply of grease, so that they were able to burn lamps and pass the time away in reading.

* Beke, l. c. p. 131.

The bears, at this time, ransacked the ship, and drawing from under the snow the cook's cupboard, which was empty, carried it on shore.

In May 1597 it was determined that, if the ship were not clear of ice by the end of the month, the crew should depart in the schuyt and boat, which were accordingly made ready to put to sea. For this purpose, on the 29th of May "ten of us went unto the scute to bring it to the house to dresse it and make it ready to sayle, but we found it deep hidden under the snow, and were faine with great paine and labour to dig it out, but when we thought to draw it to the house we coulde not, because we were too weake, wherewith we became wholely out of heart, doubting that we should not be able to goe forwarde with our labour; but the maister, encouraging us, bad us strive to do more than we were able, saying that both our lives and our welfare consisted therein, and that if we could not get the scute from thence and make it ready, then, he said, we must dwell there as burghers of Novaya Zemlya, and make our graves in that place. But there wanted no good will, but only strength, which made us to let the scute lye, which was no small greefe unto us, and trouble to thinke of. But after noone we took hearte again, and determined to tourne the boate that lay by the house with her keele upwards, and to amend it, that it might be the fitter to carry us over the sea." One of the bears killed in the end of May had in its stomach a piece of rein-deer with the skin, shewing that these animals frequent Novaya Zemlya in the spring, if indeed some do not remain there all the winter. If the poor men could have procured an adequate supply of so sane an aliment as rein-deer's flesh, they might have kept off scurvy, which had by this time seriously impaired their strength, the sudden and unexpected debility when called upon for exertion being one of its most unequivocal signs. Had they been able to go out on the ice and procure seals, their flesh, dark and unsightly as it is, would have been an excellent resource, and even the polar bears might have been eaten with impunity and advantage, but the narrative makes no mention of their having recourse to the flesh of those they killed until the 31st of May. Then they unfortunately took the only noxious part of the animal, the liver, "and drest and eate it; the taste liked us well, but it made us all sicke, specially three that were exceeding sicke, and we verily thought that we should have lost them, for all their skins came of from the foote to the head; but yet they recouered againe, for the which we gave God heartie thankes, for if as then we had lost these three men, it was a hundred to one that we should neuer have gotten from thence, because we should haue had too few men to draw and lift at our neede."

On the 14th of June, having made all the preparation in their power, cut a road through the ice and snow to the ship, for the more easy transporting the most valuable of their merchandise, and launched their two boats, they set sail, after having written and signed a letter of protest, stating that they had abandoned the ship which was still fast in the ice to save their lives. Eleven of the crew placed their signatures to the document, but four others either could not write, or were too ill to do so. On the first day, though occasionally hampered by ice, they sailed twenty miles, to Island Cape. Next day passing Hooft-hoek (Angle-head) and Flushing Point they reached Point Desire, being a distance of fifty-two miles; and on the 16th they went thirty-two miles further to the Islands of Orange, thereby emerging from the Sea of Kara. Having obtained three birds here, they drest them for the use of the sick. "And being there, both our scutes lying hard by each other, the Maister called to William Barentz to know how he did, and William Barentz made answeare and said—Well! God be thanked, and I hope before we come to Wardhuus to be able to goe. Then he spake to me and said, Gerrit, are we about the Ice Point? If we be, then, I pray you lift me up, for I must view it once againe, at which time we had sailed from the Islands of Orange to the Ice Point, about twenty miles."

From this date they kept sailing southwards down the west coast of Novaya Zemlya as the weather and streams of ice permitted. On the 17th of June their boats were so squeezed by large masses of ice, that they were forced to land their sick and the cargo on a floe, which was done with much difficulty and hazard, and the boats were hauled up and repaired.

"On the 20th Claes Adrianson began to be extreme sicke, whereby we perceived that he would not live long, and the boateson came into our scute and told us in what case he was, and that he could not long continue alive; whereupon William Barentz spake and said, I think I shall not live long after him; and yet we did not iudge William Barentz to be so sicke, for we sat talking one with the other, and spake of many things, and William Barentz read in my card which I had made touching our voiage; at last he laid away the card and spake into me, saying, Gerrit give me some drinke; and he had no sooner drunke but he was taken with so sodain a qualme, that he turned his eies in his head and died presently, and we had no time to call the Maister out of the other scute to speak unto him; and so he died before Claes Adrianson.

The death of William Barentz put vs in no small discomfort, as being the chiefe guide and onely pilot on whom we reposed ourselves next under God." On the 27th of June the boats doubled Cape Nassau; this is Admiral Lütke's north-extreme in 1828, and all that part of Novaya Zemlya which is to the east of it has been justly named Barentz's Land, he and his companions having been the sole explorers of it down to the present time.

One of the boats foundered, by pressure of the ice, on the 1st of July, but most of the merchandise, though damaged by the salt water, was saved with much exertion, the sick landed, and the boat recovered and repaired. The hardships the crew endured, however, were fatal to another of the sick men, John Franson, nephew to that Claes Adrianson who died on the same day with William Barentz.

On the 27th of the month, the boats passed a place named in the narrative *Constinsark*, and by later writers Coasting Search. Dr. Beke has clearly shewn that this is the Kostin Shar of the Russians, a strait leading round an island named Medusharsky, previously visited by Oliver Brunel, already mentioned, and subsequently by Henry Hudson and others.* Soon afterwards they fell in with two Russian *lodij*, from whose crews they had a friendly reception, and got some small supplies, but being unable to understand their language, they obtained no precise directions as to their future course. In shaping this they missed William Barentzoon sadly, and made a great circuit into the bay of Petchora, instead of crossing direct to Kanin nos, as they had hoped to do. They met, however, other Russian vessels from which they obtained occasional supplies of provisions, and eventually reached

* "Three Voyages," etc., pp. 30, 202, 222.

Kanin nos on the 18th of August. Then they made arrangements for crossing the White Sea by dividing their candles and other stores between the boats. This voyage of 160 miles they performed safely in their small and crazy boats, and on the 20th they were near the Lapland coast. Having reached the island of Kildin on the 25th of August, they were received kindly by some Russians, who told them of vessels lying up the river at Kola, and by their means the Master Heemskerck hired a Laplander to guide one of his men overland to Kola, with a letter setting forth the destitute condition of his party. Having despatched their messenger, the rest, after lightening the boats by the removal of the goods, drew them up on the beach. "Which done we went to the Russians and warmed us, and there dressed such food as we had; and then again we began to make two meales a day, when we perceiued that we should euery day find more people, and we dranke of their drinke which they call *quas*, which was made of broken pieces of bread, and it tasted well, for in a long time we had drunke nothing else but water. Some of our men went inland and there found blew-berries and bramble-berries, which we plucked and eate, and they did us much good, for we found that they healed us of our loosenesse"—(scorbutic diarrhœa).

On the 29th the Laplander was seen returning without the man, "whereat we wondered and were somewhat in doubt; but when he came unto vs, he shewed vs a letter that was written unto our Maister, the contents thereof being, that he that had written the letter wondered much at our arrival in that place, and that long since he verily thought that we had beene all cast away, how that he was exceeding glad of our arrival, and would presently come vnto us with victuals and all other necessaries to succour vs withall. We being in no small admiration who it might be that shewed vs so great favour and friendship, could not imagine what he was, for it appeared by the letter that he knew vs well. And although the letter was subscribed *by me John Cornelison Rijp*, yet we could not be persuaded that it was the same John Cornelison who the yeere before had set out in the other ship with vs and left vs about Beare Island." It was, however, the same kindly Ian Cornelizoon, who being in Lapland on a trading voyage, next day came to them in a Russian *jol* (yacol or jolly-boat), bringing Roswicke beere, wine, aqua vitæ, bread, flesh, bacon, salmon, sugar and other things, rejoiced with them for their unexpected safety, gave God great thanks for His mercy, and finally carried the rescued mariners to their native country. The open scutes in which they had sailed 1600 miles in a stormy and ice encumbered sea, were, by permission of the Boyard, deposited in the merchants' house at Kola, for a remembrance of their extraordinary voyage.

The more eastern seas and coasts of arctic Europe and Asia, having been explored at much later dates by Russian subjects, will be mentioned in a future chapter, but the efforts of England for the discovery of a north-west passage, subsequent to Cabot's time, claim the precedence chronologically.

CHAPTER V.

ENGLISH NORTH-WEST VOYAGES.—A. D. 1576-1636.

Sir Martin Frobisher's, First, Second, and Third voyage to Meta Incognita—Desolation—Queen Elizabeth's Foreland—Frobisher's Straits—Terra septentrionalis—Gold ore—Mistaken, or Hudson's Straits—Greenland—Davis—Desolation—Davis' Straits—Hudson's Straits—Cumberland Inlet—Labrador—Baffin's Bay, latitude $72\frac{1}{4}°$—Women's Islands—Sanderson's Hope—Maldonado—Weymouth—John Knight—Henry Hudson—Land of East Greenland in 82° N.!!—Dr. Scoresby—Hudson's River—Hudson's Bay—Hudson sent adrift by his mutinous crew—Sir Thomas Button's, Hope's check'd—Nelson River—*Ut ultra*—Prince Henry's instructions—Baffin and Bylot—Southampton Island—Baffin's Bay—Horn Sound—Wolstenholme Sound—Whale Sound—Sir Thomas Smith's Sound—Carey's Islands—Alderman Jones' Sound—Sir James Lancaster's Sound—Jens Munck—Luke Foxe—Hudson's Bay—*Ut ultra*—Sir Thomas Roe's welcome—New Wales—Foxe's farthest—Tenudiakbeek.

THE project of a north-west passage, though afterwards a favourite enterprise in England, and resumed from time to time, was suffered to rest for nearly eighty years after the failure of Sebastian Cabot's attempt in 1498. Sir Martin Frobisher, an educated man, "thorowly furnished of the knowledge of the sphere, and all other skilles appertayning to the arte of navigation," had for fifteen years been endeavouring to move his friends and the London merchants to fit out an expedition for north-west discovery, but though he had the patronage of Dudley Earl of Warwicke, and the active assistance of one Michael Lok, who helped him with money and credit, he could not overcome the opposition of the Muscovy

Company, until 1574, when a mandate of the Lord Treasurer compelled that company to grant a license for the voyage. Three several expeditions in successive years were the results of Frobisher's agitation, and in point of date, they take the precedence by eighteen or twenty years, of that of Barentzoon, mentioned in the last chapter.

The first expedition, projected on a small scale, consisted of two barks, the Gabriel and Michael, of between twenty and twenty-five tons a-piece, and of a pinnace which measured ten tons, manned, in the aggregate, by thirty-five men, and victualled for twelve months. Sir Martin Frobisher, captain and pilot, embarked in the Gabriel, and Christopher Hall, the master of that ship, wrote a journal of the voyage as he did of the two following ones, being master of the Ayde in the second voyage, and chief pilot of the fleet in the third one. George Best, Frobisher's lieutenant in the first two voyages, and captain of the Anne Francis in the third, also wrote accounts in considerable detail of all the three voyages. Dionise Settle kept a journal of the second voyage, and Thomas Ellis of the third. All these narratives, viz., two of each voyage, were published by Hakluyt, and there are discrepancies in them with regard to some dates and names of places. Neither Settle nor Ellis troubled themselves with recording astronomical observations; the latitudes were taken, Best tells us, with the staffe rather than with the astrolabe, because the divisions of the latter were too small to give the minutes. The real errors in the geographical positions are generally small, compared to what might have been expected with such intruments.

The small vessels furnished for the first voyage dropped down from Deptford to Greenwich on the 7th of June 1576, and following the precedent of Sir Hugh Willoughby's expedition, Hall says, " We set saile, all three of vs, and bare downe by the Court, where we shotte off our ordinance and made the best show we could; Her Majestie (Queen Elizabeth) beholding the same, commended it, and bade vs farewell with shaking her hand at vs out of the window. Afterward she sent a gentleman aboord of vs, who declared that her Majestie had good liking of our doings, and thanked vs for it, and also willed our Captaine to come the next day to the Court to take his leave of her."

On the 11th of July land was discovered, in latitude 61°, rising like pinnacles of steeples, and all covered with snow, to which the ships were unable to approach because of the ice. Our navigators supposed this to be Frizeland, but, in fact, the land so named is the south part of Greenland, and they were then off Cape Desolation or *Torsukatek*, at the north-west extremity of the deserted colony of *East Bygd*. Charts of this period place Greenland to the north, in the position of Baffin's Bay, and Frizeland is represented as an island having the form of the real southern extremity of Greenland. Off Desolation the pinnace foundered, carrying down her crew of four men, and the Michael, Owen Gryffyn master, " mistrusting the matter," returned to England and reported that Frobisher was cast away. But that " worthy captaine, notwithstanding these discomforts, although his mast was sprung, and his toppe-mast blowen overboord with extreame foule weather, continued his course towards the north-west, knowing that the sea must needs haue an ending, and that some land should haue a beginning that way; and determined therefore, at the least, to bring true proofe what land and sea the same might be so farre to the northwestwards, beyond any man that hath heretofore discovered." On the 28th of July 1576, Frobisher saw a headland which he named Queen Elizabeth's Foreland, but a landing could not be effected till the 10th of August, when Hall rowed in the boat to a small island and found the flood-tide setting south-west. Next day the latitude at noon was 63° 8′ N., and this day the Gabriel entered the strait. This is from Hall's journal—Best differs in the dates, and says as follows: "And the 20th of July he (the worthy captain) had sight of a high land, which he called Queene Elizabeth's Foreland, after Her Majestie's name. And sailing more northerly alongst that coast, he descried another foreland with a great gut, bay, or passage dividing as it were, two maine lands or continents asunder." . . . "Coveting still to continue his course northwards, he was alwayes by contrary winds dereined ouerthwart these straights." . . . "Wherefore he determined to make proofe of this place, to see how farre that gut had continuance, and whether he might carry himselfe thorow the same into some open sea on the backeside, whereof he conceived no small hope, and so entered the same, the one and twentieth day of July, and passed on above fifty leagues therein, as he reported, having upon either hand a great maine or continent. And the land upon his right hand, as he sailed westward, he judged to be the continent of Asia, and there to be divided from the firme of America, which lieth upon the left hand over against the same."

Of these dates, Hall's appears to be the most trustworthy, as they are given day by day in his journal, while Best's narrative of the three voyages, bears marks of having been composed after their conclusion. And one day is evidently too short a time for Frobisher to have made a decided effort to go northwards before he resolved on entering the strait.

Frobisher's Straits have not been explored by a competent surveyor since the days of their discoverer, and for long their position was held to be so uncertain, that every map-maker took the liberty of placing them according to his fancy, several transferring them to the south end of Greenland. They are now set down in the latest Admiralty circumpolar charts within the limits of the errors arising from defects in the original observations for latitude. A map of Sir Humfrey Gilbert's, in the British Museum, gives a very rude draft of the straits, placing them on the American side.*

The entrance of the strait is described by Best in his account of the second voyage, as follows—" About noone we made the North Foreland (Cape Labrador of Hall), otherwise called Halles Island; also a small island bearing the name of the said Hall, whence the ore was taken. . . . This North Foreland is thought to be divided from the continent of the northerland by a little sound, which maketh it an island, and is thought to be little less than the Isle of Wight, and is the first entrance of the straights upon the norther side, and standeth in the latitude of 62° 50'." "Queene Elizabeth's Foreland, being the entrance of the streits of the southerland, standeth in the latitude of 62° 30' N., northwards of Newfoundland, and upon the same continent, for anything that is yet known to the contrary. . . . The narrowest place of the straights, from land to land, between Jackman's Sound and the Countesse of Warwick's Sound, which is reckoned scarcely thirty leagues within the straights from Queene's Cape, was judged nine leagues over at least." Gabriel's Island is ten leagues from the mouth of the straits, and Prior's Sound lies

* Sir Humfrey Gilbert's Voyage, April 1576, 4to, by Henry Middleton, printed for Richard Jhones.

ten leagues further in. Beyond that is Thomas Williams' Island, and bearing north-west from the latter, at the distance of ten leagues, is Burcher's Island, which is the limit of Frobisher's passage westward, in the straits. Trumpet Island is situated between Williams' and Gabriel's Islands; Mount Warwick stands on the south side of the straits, and the Countess of Warwick's Sound on the north side.

At Burcher's Island a party of Skrællings or Eskimos was seen, with whom Frobisher had sundry conferences, and some of them came aboard his ship, and bartered skins for looking glasses and other toys. This giving the crew undue confidence, five of them went inland, contrary to the captain's orders, and were never seen again. After this the natives became wary, but at length Frobisher entrapped one and took him captive, in revenge for the loss of his men, that he supposed to have been intercepted on their way back to the boat. This man died after reaching England. If Frobisher's men were merely detained among the Eskimos, or voluntarily designed to remain among them, this was the worst measure that could have been resorted to for their safety, as the natives would, without fail, wreak their vengeance upon the captives.

Frobisher, however, was not less humane than other navigators of that age, few of them scrupling to carry off the natives of the new world to make a show of in Europe. As a seaman, Frobisher seems to have had few superiors. The following account of his behaviour in the storm off Desolation, mentioned above, is taken from an extract, given by Mr. Rundall, of a manuscript of Michael Lok's, preserved in the British Museum.*—"In the rage of an extreme storme the vessel was cast flat on her syde, and, being open in the waste

* MSS. Cotton; Otho, E. 8-47.

was fylled with water, so as she lay still for sunk, and would neither weare nor steare with any helpe of the helme; and neuer have rysen agayn, but the meruelous work of God's great mercy to help them all. In this distress, when all the men in the ship had lost their courage, the captayn, like himselfe, with valiant courage, stood vp, and passed alongst the ship's side, in the chayne-wales, lying on her flat syde, and caught holde on the weather leche of the fore-saile; but in the weather-coyling of the ship the fore-yarde brake, and the water yssued from both sydes, though withall, without anything fleeting over."

On his return to England from his first voyage, Sir Martin Frobisher "was highly commended of all men for his great and notable attempt, but specially famous for the great hope he brought of the passage to Cataya," and her majesty condescended to name the broken lands, bounding the straits, *Meta Incognita*. A tract on the north side is termed, in Sir Humfrey Gilbert's chart, *Terra Septentrionalis*. This prospect, however, of a passage to India, was less attractive to the London merchants than the bait of immediate gain. A piece of black stone, brought home by one of the company, being submitted to one Baptista Agnello, he (by coaxing nature, as he privately admitted, to Michael Lok), obtained therefrom a grain of gold; thereupon money was speedily raised to defray the outfit of a second expedition, not for discovery, but to bring home the supposed golden ore, and this in the face of the report of the master of the Tower, and of two skilful assayists, who declared it to be but a marquesite, containing none of the precious metal.

The instructions given to Frobisher for his conduct on this second voyage, were to search for the ore, and defer discovery to a future time; but though it must have cost him a heart-pang to forego a project that he had cherished so long, he was faithful to the orders he received, and loaded his three ships with the worthless stuff. On this voyage the mariners erected, with sounding of trumpets and other ceremonies, a cairn of stones on Mount Warwick. Various interviews and skirmishes with the natives took place, and in Yorke's Sound, some remnants of clothing were found belonging to the men they had lost the year before. Several of the poor natives were slain in these conflicts, many more wounded, and a man and two women captured. One of the women being old and ugly, was thought to be a devil or a witch, and was therefore set at liberty; but the other, who was young, with a child at her back, was kept.

Fifteen ships were fitted out in 1578 for the third voyage, to bring home ore, but one of them, the bark Dennis, foundered in a great storm off the Queen's Foreland. In the meanwhile a swift current coming from the north-east carried the remainder of the fleet southwards towards Frobisher's Mistaken Straits, now known by the name of Hudson's Straits. We have already mentioned that a claim has been set up in behalf of Sebastian Cabot as discoverer of this strait, and there is greater reason to believe that it is actually the opening which was called by Cortoreale *Rio Nevado*, which name of Nevado has been transferred to some mountainous islands on its north side, that even in summer are covered with snow. Part of the fleet following the General, "entered within the said doubtful and supposed straights, having alwayes a faire continent upon the starboorde, and a continuance still of an open sea before them; and had it not been for the charge and care he had of the fleete and fraughted ships, the General both would and could have gone through to the South Sea, and dissolved the long doubt of the passage which we seek to find to the rich countrey of Cataya; . . . And where in other places we were much troubled with yce, as in the entrance of the same, so after we had sayled fifty or sixty leagues therein, we had no lets of yce."

Respecting the opinion here expressed by Master Best, there is no doubt that had Frobisher been on a voyage of discovery and not on a mercantile enterprise, he would have pushed onwards and entered Hudson's Bay. But his sole object was to get into the straits named after himself, where he could "provide the fleete of their lading," and which he had overshot in thick weather, by mistaking Resolution Island for the north Foreland. Though the chief pilot, Christopher Hall, had openly declared that they had never been in that strait before, Frobisher, to keep up the spirits of his followers, held out that they were in the right way; and after many days, having doubtless ascertained his true position by astronomical observation, for "it pleased God," says Best, "to give us a clear of sunne and light for a short time, . . . he perceived a great sound to goe thorow into Frobisher's Straights." By this channel Frobisher recovered his intended port, and the Gabriel being sent round, proved the Queen's Foreland to be an island. Many years afterwards, La Peyrouse, going to attack the Hudson's Bay posts, entered Frobisher's Straits in mistake, but found his way into Hudson's Straits, probably by the same channel through which Frobisher passed in the opposite direction.

On the 1st of August, after many dangers past and the dispersion of the fleet by a storm, most of the ships were assembled at the Countesse of Warwick's Island, and every captain was commanded to bring ashore all such gentlemen, soldiers, and miners as they had under their charge, with such provision as they had of victuals, tents, and other necessaries, for the speedy lading of the ships with the contents of the mine. It was designed that one hundred men should winter there in a fort to be built, but portions of the house were lost in the storm or were stowed in the missing ships, so this part of the scheme was abandoned. Whereupon they buried the timber provided for the intended fort, and sowed some pease to prove the fruitfulness of the soil against the next year. Master Wolfall preached a godly sermon and administered the communion to many of the company on a spot called Winter's Fornace; as the said preacher did at sundry other times and places, because the whole company could not conveniently come together at once. On the last day of August the whole fleet departed homewards.

Best tells us nearly as much of *Meta Incognita*, its natural productions and inhabitants, as we know in the present day. "It is now found," he says, "that Queene Elizabeth's Cape, being situated in latitude $61\frac{1}{2}°$, which was before supposed to be part of the firme land of America, and also all the rest of the south side of Frobisher's Straites, are all several islands and broken land, and so will likewise all the north side of the said straites fall out to be; and some of our companions being entered above sixty leagues within the *Mistaken Straites*, in the third voyage mentioned, thought certainly that they had descryed the firme land of America towards the south, which I think will fall out to be. These broken lands and islands, being very many in number, do seem to make there an *archipelagus*, which as they all differ in greatnesse, form and fashion, one from another, so are they in goodnesse colour, and soyl, much unlike. They are all high lands, mountanous, and in most parts covered with snow all the summer long." . . . "The people are great enchanters, and use many charmes of witchcraft." . . . "They use to traffike and exchange their commodities with some other people, of whom they have such things as their miserable countrey and ignorance of art to make, denied them to haue, as barres of yron, heads of yron for their darts, needles made foure square, certain buttons of copper, which they vse to weare upon their foreheads for ornament, as our ladies of the court of England doe vse great pearle."* Best goes on to give a very fair and full description of the habits of the Eskimos. The iron articles he saw were either remnants of their intercourse with the ancient Icelandic colonies in Greenland, or articles that had ascended so far north by coast traffic in the course of the seventy or eighty years that had elapsed since the discovery of Newfoundland and the neighbouring parts of Labrador. The Eskimos have a natural aptitude for barter, and articles pass in that way rapidly from tribe to tribe.

On each of Frobisher's three voyages, "Frizeland" (as part of Greenland continued to be named), was seen, when outward bound; and this is just the course the whalers of the present day are forced to pursue. They find that the sea opens earlier in the season on the Greenland coast, and that it is only when the whale-fishing is far advanced that they can penetrate the ice on the west-side of Davis' Straits. On the third voyage, after sailing for a time along the Greenland coast, " a very hie and cragged land, almost cleane covered with snow," our mariners landed on a place somewhat void of ice. There they

* Hakluyt, ii.

saw certain tents made of skins, and boats (*kayacks*) much like those of Meta Incognita; but along with the usual Eskimo furniture, there was found a *box of nails*, whence it was conjectured that the natives had traffic with other nations. Ellis, who mentions this fact, does not describe the nails, and we are left to conjecture whether they had been extracted from some ancient colonial buildings or from the drift timbers of some wreck.

Frobisher was the first, except perhaps Skolni, who landed on Greenland after the destruction of the Scandinavian settlements, whose former existence on that coast he seems not to have known, and believing that he had made a discovery, he took possession of the country in the name of Queen Elizabeth, calling it "West England," and naming a conspicuous high cliff "Charing Crosse." The worthlessness of the ore procured on the last two voyages having disgusted the merchant adventurers, Frobisher's career as a discoverer terminated; and luckless Michael Lok, being unable to redeem his suretiship, was shut up in the Fleet Prison, a catastrophe which involved himself and fifteen children in ruin.

A passenger on board the Busse Emmanuel of Bridgewater, one of the ships of Frobisher's third expedition, reported that on the homeward voyage that ship had coasted for three days a large island lying to the south-east of Frizeland, which was observed to be fertile and well-wooded. This land was no doubt the southern extremity of Greenland, or it may be a congeries of icebergs, and the supposed forests merely ocular delusions. The island, however, found a place in charts, though it was never seen again; and even after whalers had often traversed its supposed site, it was thought to have sunk in the sea and was then noted as a shoal, and called the "Sunken land of Busse." Sir John Ross sounded for this bank but found it not. The report of Thomas Wiars the passenger, as quoted by Hakluyt, is silent about woods.* It is curious that another island in the same vicinity, between Iceland and Greenland, was supposed to have perished in a different way. In a map of the world by Roysch, dated 1508, the situation of the island is marked, and a note tells us that it was totally destroyed by fire in the year 1450.†

After ten years had passed subsequent to Frobisher's third voyage, the merchants of London took fresh courage, and again subscribed their moneys for another trial for the north-west passage. This time the charge of the enterprise was entrusted to Master John Davis, "a man well-grounded in the arte of navigation." This truly able navigator made, like Frobisher, three consecutive voyages to the north-west. In the first, the Sunshine and Moonshine were the vessels fitted out for the occasion; and of the crews, numbering conjointly forty-two persons, four were musicians. Davis was "captaine and chiefe pilot of this exployt." In the second voyage the Sunshine was again engaged with two others; and in the third voyage, the same vessel once more made one of three.

Davis followed the practice of sighting the Greenland coast on his outward voyages, and of landing thereon when weather and place were suitable. The name of Frizeland, however, was still retained, and Davis named a part of the coast to the north of West England, the "Land of Desolation." One fjord, situated in latitude $64\frac{1}{4}°$ N., he named Gilbert's Sound, and there he found much drift timber, and had many

---

* Hakluyt, iii. p. 44.

† Universalior cognita orbis. Tabula ex recentibus confecta observationibus. (Roysch) An. 1508. Preserved in the British Museum.

interviews with the natives, who were very tractable. It is here that the settlements of the Moravian Brethren, named Godhaab and Nye Hernhut, have been established.

Leaving Gilbert's Sound, Davis stood to the westward and northward for five days, and on the 6th of August 1585, discovered land in latitude 66° 40′ altogether free from the " pester of yce, and ankered in a very faire rode, under a brave mount, the cliffes whereof are orient as golde." This conspicuous hill was named " Mount Raleigh," the anchorage obtained the appellation of " Totnes Rode," and the sound which compasses the mount, that of " Exeter Sound;" the north foreland was called " Dier's Cape," and the southern one, which projects further, " Cape Walsingham."* These places are on the verge of the arctic circle, and Davis is the first English navigator who attained so high a latitude on the American shore. On coasting the land to the southward, Davis found it rounding off to the westward, and came to the Cape of God's Mercy, its southern extreme. There fogs obliged him to keep close to the north shore, and when the weather cleared, he found that he had shot into a very fair passage, in which, when he had sailed sixty leagues, he arrived at certain islands, having open passages on both sides. These islands were named on the third voyage, in honour of the Earl of Cumberland; and Davis then ascertained that the inlet he

---

* Sir John Ross bears testimony to the accuracy of Davis, and states the geographical position of these places to be as follows:—

Mount Raleigh	lat. 66° 14′ N.;	long. 61° 30′ W.	
Exeter Bay	66° 30′	61° 00′	
Dier's Cape............	66° 42′	61° 06′	
Cape Walsingham ...	66° 00′	60° 50′	
Warwick's Foreland	62° nearly.		(Davis).
Resolution Island ...	61° 21′	66° 55′	(Parry).

had discovered, was separated from Frobisher's Straits merely by a chain of islands. He did not, however, recognise the straits of his precursor as a previous discovery, and gave them therefore a new name, that of "Lumlie's Inlet." The tide is noted as rising four fathoms vertically, among the Cumberland Islands, and it is stated that a south-west by west moon maketh a full sea. On the 29th of July, having coasted the south shore, which trends south-west by south down to 64° north latitude, he got clear of the straits; after which he crossed the mouth of an inlet twenty leagues broad, situated between 62 and 63 degrees, and abounding in noisy races, currents and overfalls. This is evidently Frobisher's Strait; and on the 31st, Davis, after giving his latitude at noon as 62°, says, "this afternoon we were close to a foreland or great cape which is the southerly limit of the gulf passed on the 30th, and the beginning of another very great inlet." This last inlet is Hudson's Strait, and Davis is one of three, if not four, of its discoverers who preceded the famous seaman whose name is now indissolubly associated with it. The northern cape of Hudson's Strait was named by Davis Warwick's Foreland, a synonym of Queen Elizabeth's Foreland of Frobisher; Resolution Island, adjoining the foreland, is the actual north-east point of the strait, and the tides about it are very strong, as Davis mentions. The south cape of Hudson's Straits was named by Davis in honour of "the worshipful Mr. John Chidley of Chidley, in the countie of Devon, Esquire."

Two delineations of Resolution Island, taken from Davis' survey, are, according to Dr. Asher, still in existence; the one is an engraved planisphere, inserted into a copy of Hakluyt in the British Museum; the other is on the globe constructed by Molyneux, quoted in Davis' summary account of his voyages, and preserved in the library of the Middle Temple.*

Cumberland Strait has of late years been explored by whalers, and Captain Penny has named its northern arm Hogarth Sound. There he has repeatedly wintered, and carried on a very successful seal and whale fishery. The only account of the country since Best wrote, that we have met with, is published in the *Missions-Blatt* for 1859 by Brother Warman, who passed the winter of 1857-8 in the sound. Cumberland Inlet, he states, penetrates Cumberland Island in an almost due northerly direction, with a slight inclination to the westward. It extends from about the 65th to the 68th parallel of north latitude; its southern angle being in the same longitude with Cape Chudleigh, the most northerly point of Labrador. Its eastern coast rises abruptly from the sea, attaining in some points an altitude of 3000 feet. The western coast is flatter, dotted with islands, and more inhabited. The face of the country consists almost exclusively of barren granite rocks; moss, scantily intermixed with grass, occurs in the hollows where moisture collects, while in sheltered spots various berry-bearing plants are found, but no trace of wood. Drift-wood not being met with, bones of whales are used in its place by the natives. Rein-deer, arctic foxes, and hares inhabit the land, but the polar-bear and wolf are not often seen. The birds are ptarmigan, ravens, and snipe; only one variety of seal is abundant; whales, which were formerly numerous, are becoming fewer. The climate is very severe, but in calm summer weather the heat is at times very great, and then mosquitoes abound.

" We proceeded in two boats northwards along the eastern

* Hudson, the Navigator, by G. M. Asher, LL.D., p. 100.

shore of the inlet for about 140 miles, and saw traces of dwellings, especially near the entrance of the most northerly fjords, named *Nurujarschuit*, meaning the corner of a promontory. This place is covered with the remains of houses, whose frame-work is composed of bones. We wintered in the harbour of *Tornait*, and early in May were visited by 150 Eskimos, who are all retained in the service of the ships, to assist in the whale-fishery."—*Warman*.

Sir John Ross examined the part of the coast which Davis fell in with at the first, and reports that Mount Raleigh is pyramidal and very high. It stands in latitude 61° 14′ N., longitude 61½° W. Cape Walsingham is exactly where Davis placed it, in latitude 66° 00′ N., and longitude 60° 50′ W. It is the eastermost land, and the distance from it across to Greenland is about one hundred and sixty miles.*

In the second voyage Davis coasted the American shore from 67 to 57 degrees of latitude. To the north he found only barren islands with abundance of natives, who were gentle and friendly, but marvellous thieves. They cut his boats from the ship's stern, injured his cable and carried away an anchor. To oblige them to restore the latter, he made two captives, whom he brought to England, the anchor not being returned. He seems to have been very diligent in exploring the sounds leading to the westward, with the hope of finding a passage, but in vain. On the Labrador coast he went inland, and found fair woods, firs, spruces, alder, yew, birch, and willows. There he saw a *black* bear, and great store of birds, among the rest ptarmigan and pin-tailed grouse, of which many were killed with the bow and arrow. In the harbour, called Davis' Inlet on the Admiralty chart, there were plenty of cod-fish,

* Ross; Voyage to Baffin's Bay, 1818, p. 215.

so that in one hour the seamen caught a hundred. This harbour lies in latitude 56° N., runs ten leagues into the land, and is two leagues wide.

On the 4th of September Davis anchored in a very good road among many isles, being prevented by head-winds from entering an inlet which he saw, and which gave him great hopes of a passage. This inlet, which is in $54\frac{1}{2}°$ of north latitude, he does not name, but it is called "Ivucktock" on the recent Admiralty chart. The Mermaid and the North Star had forsaken him soon after crossing from Greenland, and Davis explored the coast down to Ivucktock in his bark the Moonshine, of thirty tons and nineteen hands. Here, however, two of his men were slain by a sudden and unprovoked attack of the Eskimos, and the bark having ridden out a severe storm, in which it narrowly escaped shipwreck, sailed homewards on the 11th of September. The Sunshine also reached England, but the North Star was never heard of again.

In the earlier part of his third voyage, Davis, keeping near the Greenland coast on his voyage northwards, as the whalers of the present day are accustomed to do, reached latitude $72\frac{1}{4}°$ N., and on the 30th of June 1587 had the sun five degrees above the horizon at midnight, having fairly entered Baffin's Bay. The Greenland country to the east of him he named the "London Coast," and a passage among Women's Islands, called by him "Sanderson's Hope," has been identified with the *Kosarsuik* of the Greenland Eskimos.

Of the entrance into Baffin's Bay, now known universally as *Davis' Straits*, Davis himself says in his "Hydrographical Description of the World."—"I departed from the (London) coast (of West Greenland), thinking to discouer the north parts of America, and after I had sailed towards the west forty leagues, I fel vpon a great banke of yce (the middle ice of the whalers) : the wind blew north and blew much, and I was constrained to coast the same towards the south, not seeing any shore west from me, neither was there any yce towards the north, but a great sea, free, large, very salt and blew, and of an vnsearchable depth. . . . By this last discovery it seemed most manifest that the passage was free and without impediment towards the north; but by reason of the Spanish fleet, and vnfortunate time of Mr. Secretarie (Walsingham's) death, the voyage was omitted, and niver sithens attempted."*

Laurent Ferrer Maldonado is reported to have sailed up Davis' Strait at this period (1588), till he reached the 75th degree of latitude, and then to have steered south-west till he attained the strait of Anian which separates America from Asia. His narrative, which did not appear till twenty years after the date of the suppositious voyage, received little credit at the time, and is totally at variance with what is now known of the configuration of the northern coasts of America.

In 1602 the Muscovy Company sent out two vessels under the command of Captain George Weymouth in search of the north-west passage. He crossed Davis' Straits, and on the 28th of June reached the western shore, in latitude 63°·53' N., or Warwick Island between Cumberland and Frobisher's Straits. Sailing northwards round Cape Walsingham, he had nearly reached the 69th parallel, when the crew, instigated by John Cartwright, a minister of the gospel, mutinying, he was compelled to turn to the south, and on the 25th of July he came to Hatton's Headland, a promontory of Resolution Island. Turning round this he sailed a con-

* Voyage to North-west, edited by T. Rundall for the Hakluyt Society, p. 50.

siderable way up Hudson's Strait, and then returned to England, where he arrived so early as the 5th of August. Weymouth is therefore one of the navigators who entered the strait leading to Hudson's Bay before Hudson's time.

The voyage of Master John Knight in the Hopewell for the "Discouery of the Nor'west passage," in 1606 came to an early and disastrous termination on the coast of Labrador, by the death of Knight himself, his mate, and three of his men, who were surprised and slain by the Eskimos. The remainder of the crew, after patching up the vessel, which had been shattered in a storm, reached England after enduring many hardships.

Yet this disaster and the previous failures did not extinguish in England the hope of a north-west passage. Henry Hudson, in 1607, made a bold attempt to cross the polar sea.* After passing the latitude of Iceland, he made the east coast of Greenland, or, as he writes it, *Gröneland*, in latitude $67\frac{1}{2}°$, on the 13th of June, in a thick fog, but steering northwards six or eight leagues, he saw very high land, for the most part covered with snow. The headland he called "Young's Cape," and a very high mount, like a round castle standing near it, he named "the Mount of God's Mercie." He then steered or lay-to as stormy weather and thick fogs permitted, with the view of ascertaining whether the land he had seen was "an iland or part of Gröneland. But then (on the 18th p.m.) the fogge encreased very much, with much wind at south, which made us alter our course and shorten our sayle, and we steered away north-east. Being then, as we supposed, in

---

* Henry Hudson, the Navigator, by G. M. Asher, LL.D., London 1860. Dr. Asher is of opinion that the narrative of Hudson's first voyage was written by John Playse or Pleyce, with additions by Hudson himself.

the meridian of the same land, having no observation since the 11th day, and lying a hull from the 15th to the 17th, we perceived a *current setting to the south-west.*" On the 20th, Hudson steered north-north-east hoping to fall in with the body of Newland (Spitzbergen), and on the 22d he saw "mayne high land nothing at all covered with snow, and the north part of that mayne high land was very high mountaynes, but we could see no snow on them. We accounted by our observation, the part of the mayne land lay neerest hand in 73 degrees. The many fogs and calmes, with contrary winds and much ice neere the shoare, held us from farther discovery of it. It may be objected against us, as a fault, for haling so westerly a course. The chief cause that moved us thereunto, was our desire to see that part of Gröneland, which (for ought that we know), was to any Christian unknowne; and we thought it might as well have beene open sea as land, and by that meanes *our passage would have been larger to the Pole;* and the hope of a westerly wind, which would be to us a landerly wind if wee found land. And considering we found land, contrarie to that which our cards make mention, we accounted our labour so much the more worth. And for aught that we could see, it is like to be good land and worth seeing." On the 21st Hudson saw land on the larboard or left hand in 73 degrees of latitude, the sun being on the meridian on the south part of the compass. He named this land "Hold-with-Hope," it was the most northerly point of Greenland that he saw,* and he fixed its latitude by an observation.

* Judging from the abridgment of Hudson's narrative published by Hakluyt, land seen subsequently by Hudson, reaching, as he supposed, beyond 82° N. lat., was conjectured to be part of Greenland, and as such it is indicated by

Steering various courses for nearly a week, but always making much northing, on the 27th of June Hudson saw Newland, or, as it is termed in a marginal note, Greenland. He reserves for Greenland proper the Danish name of Gröneland. Coasting the Spitzbergen shore in a smooth sea, at noon he was, by dead reckoning, in 78 degrees, and near *Vogelhock*, the north point of Prince Charles Island, according to Lord Mulgrave, synonymous with "Fair Foreland," which, by Dr. Scoresby's observations, lies in 78° 53' N. For a fortnight afterwards, Hudson seemed to have tacked about, as the vicinity of the ice and state of the wind and weather permitted, near Prince Charles Island, between latitude $77\frac{1}{2}°$ and $80\frac{1}{2}°$. On the 14th of July he was off a very high and rugged land, on the north side of which there is a small island, which he named after his boatswain, Collins' Cape. On the 15th Collins' Cape bore south-east, and land on the starboard, trending north-east and by east, to the distance of eighteen or twenty leagues, stretched by account into 81 degrees. This land was very high and mountainous, like rugged rocks with snow between them. In Pellham's map of Greenland (Spitzbergen) there is a Castlin's point, on the north side of West Spitzbergen, which is probably the cape named Collins by Hudson, the situation corresponding. In that map its latitude is 79° 50'. On the 16th, having neared the northern land, Hudson saw more land joining the same, and trending north, stretching into 82 degrees, and by the bowing or shewing of the sky much farther. As no part of Spitzbergen lies so far north, Admiral Beechey believed that Hudson committed

the chart which accompanied this compilation in the form it had in the Encyclopædia Britannica. An attentive consideration of the more extended narrative, published by Dr. Asher, shews that this was an error.

an error in his observation of the sun, off Cape Collins, and that the northern land he saw was the Seven Islands. Hudson " meant to have compassed this land by the north ; but now finding, by proof, it was impossible, by means of the abundance of ice compassing us about, and joyning to the land, and seeing that God did blesse us with a fair wind to sayle by the south, etc., we returned bearing up the helme." "And this I can assure at this present, that between 78 degrees and ¼, and 82 degrees by this way, there is no passage."

After this our navigator passed down the west side of Newland (Spitzbergen), and found by "the icy skie and our nearness to Gröneland, that there is no passage that way, which, if there had, I meant to have made my returne by the north of Gröneland to Davis his Streights, and so to England." He came again, however, in sight of Newland, and on the 31st of July bore up for England, passing near Cherie Island, and arriving in Tilbury Hope on the 15th of September.

Part of the coast of Greenland proper seen by Hudson, that is, from Gale Hanke's Bay, in latitude 75°, so named in 1654, down to Cape Barclay in 69¹°, was carefully surveyed by Dr. William Scoresby, when he was master of a whaler. Hudson's Hold-with-Hope he identifies with Broer Ruys Land, and he names an island in its vicinity Bontekoe. For the names of the other capes, islands, and sounds, and for a good description of the country, the reader is referred to Scoresby's West Greenland.* The south district of the same coast is laid down, and described in detail, by Captain Graah of the Danish Navy, whose writings have been quoted in a former page.

Another voyage of Hudson, performed in the course of

* Journal of a Voyage to the Northern Whale Fishery. Edinburgh, 1823.

the following summer, in the service of the Dutch, had for its result, the discovery of the magnificent river, which still bears his name, and at whose mouth the most important commercial city of the New World has arisen. Before this, he had tried for the north-east passage, but did not get beyond Novaya Zemlya.

By these voyages Hudson's reputation as a skilful and enterprising seaman had been so firmly established, that when Sir John Wolstenholme, and Sir Dudley Digges resolved, in 1610, to employ the Discovery, of fifty-five tons, in searching for the north-west passage, Henric Hudson was nominated to the command. On this, the last of his voyages, his fame chiefly rests, because of its disasters, for his previous adventures were not less hazardous, nor less deficient in displays of nautical skill. On this voyage he sighted Cape Desolation, on the western side of South Greenland, and passed what he supposed to be the western outlet of Frobisher's Straits; for in the charts which he used, Frobisher's discoveries were supposed to lie between North and South Greenland. On the 15th of June 1610, he says in his Journal, "we were in sight of the land (in latitude 59°, 27')* which was called by Captain John Davis, Desolation, and found the errour of the former layings down of that land." Continuing to sail to the north-westward, across Davis' Straits, he encountered much ice, with many riplings, or overfals, and ascertained that, in latitude 60° 42', there is a *strong stream setting from east-south-east* to west-north-west.

On the 24th, or eleven days after passing Cape Farewell, Hudson saw land to the north, but suddenly lost sight of it

* This is probably the latitude of the ship at noon, as Cape Farewell lies farther north, though in sight.

again, being then, as he states in a marginal note to his journal, at the east entrance into the straits, into which he continued running to the westward, on the parallel of 62° 17′. This was, therefore, what he called Lumley's Inlet, but which was more properly Frobisher's Strait.

Dr. Asher* remarks that Hudson had two years before that time, when making his second voyage, entertained the design of sailing a hundred leagues, either into *Lumley's Inlet*, or into the *Furious Overfal*, thereby to seek a passage to the north-west. He now embraced the opportunity he had of putting his intentions into practice. Leaving Frobishers' Strait, probably on account of obstructions from ice, or adverse winds, he crossed the entrance of Hudson's Strait, and "plyed upon the souther side" in Ungava Bay, amid much ice, until the 8th of July, when he was on the 60th parallel, and saw "a champagne land" covered with snow, reaching round from the north-west by west into south-west by west. This, which he called "Desire Provoketh," is the Island of Akpatok. Plying westward to the 11th, he reached the Isles of God's Mercies, where he found the flood tide coming from the north, and rising four fathoms. From thence he coasted the south side of Hudson's Strait, naming a part of the shore in $61\frac{1}{2}°$, *Hold-with-Hope,* and the neighbouring extremity of Labrador, *Magna Britannia*. On the 2d of August he had fairly reached the western end of the strait, and named a "faire headland on the norther shoare, six leagues distant 'Salisburies Foreland,' being a cape of Salisbury Island. From thence he ran west-south-west into a great whirling sea, and sailing seven leagues further, was in the mouth of a strait about two leagues broad, and distant from the easter

* Hudson the Navigator, cxcvi.

part of *Fretum Davis* two hundred and fifty leagues, or thereabouts." The southern head of this entrance he named "Cape Worsenholme," the north-western shore he called "Cape Digs," being an island, one of a small group. This is distinct from the Cape Digges of modern charts. After this the land is mentioned as falling away to the southward. Our navigator observed for the latitude in 61° 20′ with a sea to the westward, and then his journal closes on the 3d of August. The rest of his melancholy history is told by Abacuk Prickett, who states that, after sailing for three months in a labyrinth of islands, they were frozen in on the 10th of November in the south-east corner of James' Bay.* Dissensions had early in the voyage sprung up among the crew, some of whom were men of evil passions, and in the June following, a mutiny was brought to a head under the leadership of Robert Juet and Henry Greene, the latter a prodigal and profligate man, who had been rescued from utter ruin by the kindness of Hudson. On the 21st of the month Hudson was seized by the conspirators, bound, and driven with his young son into the shallop. The carpenter, John King, whose name ought to be held in honourable remembrance, made a determined resistance, which being overcome, he leapt into the shallop, being resolved to share the fate of his master. Six sick and infirm men were also forced into the boat which was then cut adrift. None of the party thus inhumanely abandoned were ever heard of again; but retribution speedily overtook the leading mutineers, who were slain in an assault of the Eskimos at Cape Digges. After suffering greatly from famine, the survivors reached England, Robert Bylot (or Billet), who afterwards became celebrated

* Dr. Asher, l.c. p. 110.

as a pilot, having taken charge of the vessel on the death of Juet.

Sir Thomas Button, accompanied by Bylot and Prickett, prosecuted the discovery in 1612-13, taking the route of Hudson through the straits. A group of islands at the southern portal, within Cape Chidley, bears the appellation of *Button's Isles*. A southern point of Southampton Island, which lies on the north side of the entrance into Hudson's Bay, was named *Cary's Swan's Nest;* and on reaching the western side of the bay, Button called the land "Hopes checked," because it arrested his progress on a promising course. Turning southwards he entered *Nelson River* in latitude 57° 10′ N., and there wintered. The estuary of the river was named *Button's Bay*, and the adjacent country *New Wales*. His crew living on salt provisions with limited rations, experienced the usual miseries of scurvy, but procuring large quantities of birds and fish in the spring, were greatly recruited, so as to be able to resume the voyage northwards in the summer of 1613. In this respect the voyage is a memorable one. None of the crews who had wintered in the high latitudes before this time, were in a condition to pursue the object of the voyage in the ensuing season. In advancing northwards, Button's pilot observed a strong tide off the mouth of the Missinippi or Churchill river, and considering that to be a favourable omen of a passage, called the locality *Hubbart his Hope*, but the river was not entered, nor indeed discovered. The voyage out ended in latitude 65° N. somewhere near Whale Point, and the land lying southwards of that projection was termed *Ut Ultra*. On the homeward voyage, commenced on the 30th of July, Cape Southampton was doubled, and an island lying to the eastward of Southampton Island was named Mansel, erroneously written Mansfield on our charts.*

An extract from Prince Henry's instructions for Button's voyage, dated 5th of April 1612, is worth inserting here, as shewing the correct knowledge then possessed of Hudsons Straits and the route to them, so as to have advantage of the currents. 7. "We think your surest way wil be to stand upp to ISELAND, and soe over to GROINLAND, in the heighte of 61°, soe to fall downe with the current to the most southerlie cape of that land, lyeing in about 59°, called CAPE FAREWELL, which pointe, as the ice will give you leave, you must double, and from thence, or rather from some 20 or 30 leagues to the northward of it, you shall fall over DAVIS HIS STRAIGHTS to the western maine; in the height of 62 degrees or thereabouts, you shall finde HUDSON'S STREIGHTS, which you may knowe by the furious course of the sea and ice into it, and by certain islands in the northern side thereof, as your carde shows."

8. "Being in; we holde it best for you to keep the northerne side, as most free from the pester of ice, at least till you be past CAPE HENRY, from thence follow the leading ice, between KING JAMES and QUEEN ANNE'S FORELAND, the distance of which two capes observe if you can, and what harbour or rode is near them, but yet make all the haste you maie to SALISBURY HIS ISLAND, between which and the northerne continent you are like to meet a great and hollowe

---

* Rundall's Edition of Voyages to the North-west, published for the Hakluyt Society, has been chiefly followed in this and the other voyages in his volume. He had his information respecting Button's voyage chiefly from "North-west Foxe," printed A.D. 1635. Copies of instructions from Prince Henry exist in the British Museum; and a broadside containing "Motives for the Discovery of the North Pole," is now in the Smithsonian Institution, inserted in a fine copy of "Davis's World's Hydrographical Description," 1593.

billowe from an opening, and flowing sea from thence. Therefore remembering that your end is west, etc."*

In the same year of 1612, further acquaintance was made with the coast of Greenland by James Hall and William Baffin, who went thither to look for a gold mine, reported to have been worked by the Danes, probably under Admiral Munck. This was sought for at Cunningham's river or fiord, in the district of Holsteinburg, on the arctic circle, but no metal was found, though traces were perceived of former diggings.

In 1615 Baffin, associated with Bylot, passed through Hudson's Strait. Mr. Rundall has given an unmutilated version of Baffin's Journal of this voyage, taken from the autograph original, preserved in the British Museum.

After traversing Hudson's Strait, Baffin passed Mill Island, lying in latitude 64°, and traced the north-east coast of Southampton Island, from *Sea-horse Point* to *Cape Comfort*, which last, was the limit of the voyage, and is situated, according to his observations, in latitude 65° N., longitude 85° 22′ W. Having doubled this cape, the tide was found to set differently from what he expected, and the hope of a passage in that direction failing, he turned back. In the Admiralty charts, the part of Southampton Island traced by Baffin is not filled in, but he mentions his daily position at noon, and his map (published by Mr. Rundall) gives a tracing of the coast-line. His journal proves that he was an experienced nautical astronomer, and an able seaman: it concludes with the following opinion :—

"And now it may be that som expect I should give my

* Hanrott's fac-simile, etc., and Athenæum, 1834. Bibliographical Miscellany, November 1853.

opynion concerning the passadge. To those my answere must be, that doubtles theare is a passadge. But within this strayte, whome is called Hudson's Straytes, I am doubtfull, supposinge the contrarye. But whether there be or no, I will not affirme. But this I will affirme, that we have not beene in any tyde then that from Resolutyon Iland, and the greatest indraft of that commeth from Davis' Straytes; and my judgment is, if any passadge within Resolution Iland, it is but som creeke or in lett, but the mayne will be up *fretum Davis.*"

In accordance with the opinion here expressed, Baffin, in his voyage of 1616, in which he was again accompanied by Bylot, sailed up Davis' Straits, with the intention, according to his instructions, of proceeding into the 80th parallel of latitude, if he could, before turning to the westward. On the 30th of May, Hope Sanderson, the northern limit of Davis's explorations, was left behind, and the vessel proceeded onwards, passed Women's Islands, in latitude $72\frac{3}{4}°$, to Horn Sound, in latitude $74°$, where it was detained until a barrier of ice, which impeded further progress, gave way. Here our navigators had free intercourse with a band of Eskimos, from whom they obtained many narwhal teeth or horns, as they thought them to be, whence the appellation they gave to the inlet. Two centuries later Sir John Ross sailed past *Horn Sound*, and had interviews, in a bay sixty miles farther north, with Eskimos, whom he named Arctic Highlanders. Had Sir John had an efficient interpreter, he might have learnt whether any tradition of the existence of ships, manned by white men, remained among the natives of that coast. He states that these northern Skrællings believed that they were the only inhabitants of the universe, but this supposed ignorance is inconsistent with their speaking a dialect of Eskimo, common in latitude 73° where Sacheuse, Sir John's interpreter, had acquired it. Had they been found cognisant of the meaning of the word *Kabloonacht*, signifying "white people," the question would have been solved, but we are not told that Sacheuse was directed to inquire.

When liberated from Horn Sound, Baffin continued his northern course by Digges Cape, in latitude 76° 35′, and, twelve leagues onward, sailed past *Wolstenholme Sound*, having an island in its entrance. He was next embayed, during a storm, at the mouth of Whale Sound, in latitude $77\frac{1}{2}°$ N., and after passing *Hakluyt's Island*, which is remarkable for a rocky pinnacle rising to the height of six hundred feet, he saw another great sound extending to the north of the 78th parallel, and observed, with surprise, that the compass varied five degrees to the westward. This sound he named after Sir Thomas Smith; it is the northern continuation of Baffin's Bay, and on this voyage its offing, abounding in whales, was the most northerly position attained.

Turning southwards down the western coast of the bay, *Carey's Islands* were next seen; and on the 10th of July, the boat being sent on shore at the entrance of a fair sound, on which the name of Alderman Jones was bestowed, brought back a report of plenty of sea-morses, but no inhabitants. On the 12th of July, *Sir James Lancaster's Sound* was discovered, but a ledge of ice lying athwart it, prevented Baffin from crossing the true threshold of the north-west passage.

In his letter to Sir John Wolstenholme, this able and adventurous navigator explains the causes of these various sounds not having been explored. Off Wolstenholme the ship drove with two anchors ahead, and was obliged to haul off shore under a low sail. At *Whale Sound* an anchor and cable were lost, and the wind continued to blow so strongly when he was off *Sir Thomas Smith's Sound*, that the ship could not remain at anchor. Two centuries and a half afterwards, Captain Inglefield trying to push up the same sound, was driven out by violent gusts of wind. After passing Lancaster Sound, Baffin could not approach the shore because of an intervening ledge of ice, and on the 27th, his crew being sick and weak, he struck over to the Greenland coast, and anchored in *Cockins' Sound*, in latitude $65\frac{1}{2}°$ N., where he saw great sculls of salmon swimming to and fro.

Baffin's report made to Sir John Wolstenholme on his return to England was, that having coasted all, or nearly all, the circumference, he found it to be no other than "a great bay, as the voyage doth truely shew." But he gave such an account of the numbers of whales that he saw, as to encourage the establishment of the Davis' Strait whale-fishery, which the efforts of two centuries and a half have not yet exhausted.

The Danes had not seen unmoved the efforts of England, and Jens Munck, who had previously made some voyages to Greenland, was despatched in quest of a north-west passage. He entered Hudson's Bay and wintered in Churchill river,* whose estuary he named *Munckene's Vinterhavn*. The neighbouring coasts were called New Denmark, and there he spent a miserable winter, owing to ignorance of the methods for economising the resources of the country. That estuary abounds in fish; American hares, which are easily trapped, are plentiful in the willow thickets on its shores, there is no lack of grouse and ptarmigan, and rein-deer might have been

---

* When the Hudson's Bay Company established their fort on this river, one of Munck's cannon, marked C.iv, was found in a cove on the south side of the estuary, from thence named Munk's Cove.

killed by hunters. But from living on salt provisions, scurvy assailed the crew, their beer and wine were frozen, and death was busy among them as the winter wore on. Towards the spring Munck himself lay in a hut four days without food, and when he at length crawled forth, he found only two survivors out of a crew of fifty-two. These, digging under the snow, found some herbs and grass which they ate, and strength enough returned to enable them to fish and shoot. The spring migration of birds yielding them plenty of food, their vigour was restored, so that they were enabled to fit out the smaller of the two vessels that had left Denmark, and, after a hazardous voyage across the north sea, to reach their native country. A romantic story is current of this stout seaman having died of grief on being harshly received by Christian the Fourth, but Forster states that he did not die till eight years after his return to Denmark, and that he continued to be employed by government till that event.

The voyage of Captain Luke Foxe, or as he quaintly called himself "North-West Foxe," promoted also by the Muscovy Company, negatived a westerly outlet from Hudson's Bay, below the highest latitude to which Foxe attained. He sailed from Deptford in May 1631 in the Charles, a pinnace of seventy tons burthen, well stored with eighteen months' provisions of the first quality, and of kinds which he enumerates at length. Crossing from Greenland to the western lands, he reached the north side of Lumley's Inlet, where he obtained a good observation for latitude in 62° 25′ N. on the 20th of June. This shews that Foxe, who was perfectly conversant with what Davis had performed, identifies the Inlet so named with the Strait discovered by Frobisher, and named after himself. Lord Lumley, as Foxe tells us, was "an especial furtherer to Davis in his voyages." Standing across Lumley's Inlet for two leagues, he had an observation in 62° 12' N., and at ten that night saw Cape Warwick; on the 22d he entered Hudson's Strait in smooth water, between Cape Chidley and Cape Warwick, near by the Island of Resolution, by whom named he knew not. Having passed through the Straits he landed at Cary's Swan's Nest, and then doubling the southern cape of Southampton Island, proceeded to survey the channel that separates that island from the main shore.

On the 27th of July, in latitude 64° 10' N., he saw, an island which he took to be the east side of Sir Thomas Button's *Ut Ultra*. This was his northern limit on that meridian, for his instructions directed him to search the western land for a passage from latitude 63° southwards, until he came to Hudson's Bay. Having landed on the island and found that it was an Eskimo burial place, he named it SIR THOMAS ROE'S WELCOME, an appellation which has since been transferred to the entire channel. Rounding the Welcome by the north, and passing southwards on the west side of it, he came to another island resembling it, being a high mass of white quartz, which he called after one of his patrons BROOKE COBHAM, and on a neighbouring group of islands he bestowed the appellation of BRIGGES HIS MATHE-MATICKES. Continuing his course to the south along the American coast, he came to an island in latitude 61° 10' N., which he identifies with the *Hopes Check'd* of Sir Thomas Button, and four days afterwards coming to *Hubart his Hope*, he pronounces it to be "a vaine hope." Standing along a green, pleasantly wooded coast, in latitude 59°, he came to the mouth of a great river (the *Missinippi*) with a cliff at its south entrance like unto Balsea cliff near Harwich. This was *Munckene's Vinterhavn*, where the Hudson's Bay Company's post of Churchill was afterwards established. On the 10th of August Foxe entered Nelson River, and having found that the cross with an inscription which Sir Thomas Button erected, had fallen down, he restored it to its position, with the addition of a notice of his own arrival there, and of his having called the land, in the right of his Sovereign King Charles, NEW WALES.

Keeping his southerly course for a fortnight longer without detecting the slightest indication of the desired passage, he met the Maria, commanded by Captain James of Bristol, and an interchange of courtesy ensued. Foxe, however, a shrewd, intelligent Yorkshireman, and skilful as a navigator in all points, expresses his conviction that Captain James was a good nautical astronomer though but an indifferent seaman.

Having now convinced himself that there was no western passage between the parallels of 65° 30' and 55° 10' N., Foxe expresses an opinion that Sir John will expend no more money in the search, and he therefore names the land on the latter parallel of latitude WOLSTENHOLME'S ULTIMA VALE.

Leaving this coast, Foxe turned his ship's head northwards, and proceeded to explore the opening on the east side of Southampton Island. Having ascertained the positions of several of the salient points of this island, he crossed the North Bay of Baffin and Bylot and attained latitude 66° 47' or FOXE'S FARTHEST. The various projections of the land that crosses the western extremities of Frobisher's and Cumberland's Straits were named by him in succession KING CHARLES HIS PROMONTORY, CAPE MARIA, LORD WESTON'S PORTLAND, and POINT PEREGRINE, this last being the northern extremity of the land that he saw. It has been ascertained of late years that there is an Eskimo route from Point Peregrine through a large sheet of water named Lake Kennedy, to a westerly arm of Frobisher's and Cumberland Sound called by the natives *Tenudiakbeck.**

Captain James gives a most doleful account of his mishaps on the voyage, and the sufferings he endured during a winter in a harbour which he discovered in latitude 52° N. This was the only discovery he made, and was of some importance, as it became and continues to be the winter harbour of such Hudson's Bay ships as are cleared too late in the season to return through the straits.

In a former page it has been mentioned that Hall and Baffin went to look for a gold mine in Greenland, supposed to have been worked by the Danes. In 1636 the Danish Chancellor Fries having been informed that the English had found gold, sent out two vessels to ascertain the truth of the report, but they returned with samples of iron pyrites.

In most, if not in all the north-west voyages alluded to in the preceding pages, the pilots looked for a flood-tide setting from the west, and when that was not found, thought the inlet to be unpromising, and desisted from the search at that place.

For almost a century after Foxe and James' voyages, or from the time of Charles the First till the accession of George the First, civil wars and revolutions at home, and the wars of Marlborough abroad, engaged the attention of the nation to the exclusion of the prosecution of maritime discovery, the only effort in that cause being an abortive attempt by Captain James Wood, in 1676, undertaken through the influence of the Royal Society, to make the north-east passage by way of Novaya Zemlya.

* See page 91.

CHAPTER VI.

AMERICAN CONTINENT, ETC.—A. D. 1668-1790.

Hudson's Bay Company—De la Potherie—Jean Bourdon—Knight—Barlow and Vaughan wrecked and died of famine on Marble Island—Fruitless search for them—Remains discovered after the lapse of half a century—Middleton—Repulse Bay and frozen strait—Moor and Smith—Ranken's Inlet, or Douglass' Bay, or Corbet's Inlet—Chesterfield Inlet—Hearne—Coppermine river and Arctic sea—Phipps—Spitzbergen—Cook—North-west America—Bering's Straits—Icy Cape—Sir Alexander Mackenzie—Mackenzie River—Whale Island—Arctic sea.

Foxe's voyage, mentioned in the preceding chapter, was the last of the north-west expeditions sent out at the expense of the members of the Muscovy Company, but in process of time a new company grew up and entered on the field of discovery as an essential part of its constitution, just as the "Discoverie of New Trades" had been of the older corporation. In 1670, on the 2d of May, King Charles the Second granted a charter to Prince Rupert and several other noble personages, giving them and their successors the exclusive right to the territories drained by rivers falling into Hudson's Bay, and the trade thereof, on certain conditions, one of which was the promotion of geographical discovery. This company is said to have been formed on the immediate representations to Prince Rupert of a Captain Gillam, who with two adventurers that had been employed in the Canadian fur-trade, named Groiseleiz and Ratisson, had sailed from Gravesend in 1668 to Rupert's river in Hudson's Bay. Soon after the company had begun to establish their posts on the rivers that fall into the bay, the Canadians, hearing that the English were enriching themselves, formed *Une Compagnie du Nord* (about the year 1676) with the view of contesting the possession of the country, and Groiseilliers and Radisson, having obtained a pardon from the French monarch, returned to Canada to give their countrymen the aid of their experience.

M. de Bacqueville de la Potherie, in his *l'Histoire de l'Amerique Septentrionale*, says that Jean Bourdon, who sailed from Canada in 1656, was the first of his countrymen who visited Hudson's Bay. In 1686, when the English company had five forts on the bay, and though there was peace between the nations at the time, an expedition coming by land from Canada under M. le Chevalier de Troye took three of the forts, and in 1690 M. d'Iberville failed in an attempt to take York Fort, on Hayes river, near the mouth of the Nelson, but succeeded in 1694. For some years afterwards several of the forts changed masters more than once, and the Hudson's Bay Company lost their ships in a severe action amid the ice. After the peace of Utrecht, York Fort was delivered up to the company in 1714, before which time the only fort the company had been able to keep possession of for some years, was Albany Fort, on the river of that name.

Even during the hot contest with France, the yearly instructions of the governor and committee of the Company in London to their agents in Rupert's Land, urged the sending of intelligent men to discover the inland country, and a lad named Henry Kelsey, having made several overland journeys, in the company of the natives, was patronised and promoted. The first northern voyage by sea, however, which the Company provided for, was undertaken in 1719. The expedition consisted of a frigate, commanded by Captain George Barlow, and a sloop, by Captain David Vaughan, the chief command over both these seamen being intrusted to Mr. James Knight, ex-governor of several of their factories, and described as being most zealous in the cause of discovery, but whose age had reached the mature period of eighty years. This expedition, which sailed from Gravesend in June (the month in which the Hudson's Bay ships always leave the Thames), consisted of the Albany and the Discovery, which were well stored with provisions, a house in frame, and a good stock of trading goods. The instructions were for them to proceed to the northward (by Sir Thomas Roe's Welcome as far as latitude 64 degrees) in search of the Anian strait. As neither of the ships returned to England at the close of the summer of 1720, and no intelligence was received of them, great fears were entertained for their safety; orders were therefore sent out by the next ship for the Governor of Churchill to despatch the Whalebone, John Scroggs, master, to search for them. These instructions reached Churchill too late in the year to be acted upon that season, but Scroggs sailed in 1722. He seems to have been neither judicious nor enterprising, and was greatly embarrassed by the shoals and rocks that skirt that coast. Though he picked up some fragments of ships' fittings on *White* or *Marble Island*, which is the same with the *Brooke Cobham* of Foxe, he made no effective search, and returned, believing that the articles he had found were merely indications of some trifling accident.* People clung

* Arthur Dobbs, fancying that the Hudson's Bay Company kept the events of Scroggs' voyage concealed, because of the discoveries he had made tending to prove the existence of a north-west passage at Whalebone point, took much credit for having published some particulars of it.—*Ellis, Voy. to Huds. Bay*, p. 80.

to the notion that Knight had made his way into the Pacific, and many years elapsed before his fate and that of his companions came to be fully known, though Captain Smith, in 1747, found some traces of shipwreck on the same island.

Every season a sloop was sent to the Welcome to trade with the Eskimos, and in 1767 a whale fishery was carried on in the vicinity of Marble Island. It happened that the boats of the Success, Joseph Stephens, when on the look-out for fish, rowed close to the island, and discovered a harbour near its east end, which had until then escaped notice. At the head of this haven, guns, anchors, cables, bricks, a smith's anvil, and many other articles, were found. The wrecks of the ships lay sunk in five fathoms water, and the remains of the house from which the Eskimos had extracted the nails were still in existence. The figure-head of the Albany, the guns, and some other things were sent home.

In the summer of 1769, fifty years after the catastrophe, Hearne visited Marble Island, which he describes as a barren rock, destitute of every kind of herbage except moss and grass, lying nearly sixteen miles from the mainland, having a like character—the woods there, he says, being several hundreds of miles from the sea-side. While prosecuting the whale fishery in that quarter, he met several Eskimos, greatly advanced in years, and, with the aid of one of their countrymen employed in the Company's service as an interpreter, he extracted from them the following account:—

"When the vessels arrived it was very late in the fall (close of summer), and the largest received much damage in getting into the harbour. Immediately afterwards the white men began to build their house, their numbers being at that time about fifty. Next summer the Eskimos paid them another visit, and found their numbers greatly reduced, and the survivors unhealthy. Their carpenters were then at work on a boat. At the beginning of the second winter only twenty were living. That winter the Eskimos built their houses on the opposite side of the harbour, and frequently supplied the English with whale's blubber, seal's flesh, train oil, and such other provisions as they could spare. The Eskimos left in the spring, and on returning later in the summer of 1721, found only five Englishmen alive, who were in such distress for provisions, that they ate eagerly of the seals' flesh and whales' blubber quite raw as they purchased it. This diet so disordered them that three died within a few days, and the other two, though extremely weak, made a shift to bury them. The two survived the others many days, and frequently went to the top of a rock, and looked earnestly to the south and east, and afterwards sat down together and wept bitterly. At length one of these melancholy men died, and the other, in attempting to dig a grave for his companion, fell down and died also." The longest liver, probably the armourer, was always employed in working iron into implements, for trade with the Eskimos. When Hearne was there the skulls of the two men were lying above ground near the house.*

The disastrous termination of Knight's voyage gave the Hudson's Bay Company a dislike of sea-expeditions of discovery for a length of time, though they gradually extended their trade in peltries into the interior, and gained a knowledge of the country and its inhabitants. At the instance, however, of Arthur Dobbs, Esq., a gentleman who interested

* Hearne, xxx.

himself greatly in the discovery of the north-west passage, the Company sent out two vessels in 1737, which effected nothing, not going farther north than $62\frac{1}{2}°$, or to Whale Cove, short of Marble Island.

In 1741, Captain Middleton, an able seaman and a good nautical astronomer, who had been long employed in the service of the Hudson's Bay Company, and who had received the Copley medal from the Royal Society, for several papers on the variation of the compass, occultations of Jupiter's satellites, and on the cold of Hudson's Bay,* was selected by the Admiralty to conduct an expedition of discovery up the Welcome. Middleton was appointed commander of the Furnace Bomb-ketch, having under him William Moor, master of the Discovery Pink, and was instructed to make the best of his way to Carey's Swan's Nest, and from thence to steer north-westerly, so as to fall in with the north-west land at *Sir Thomas Roe's Welcome* or *Ne ultra*, near latitude 65° north, and having reached Whalebone Point, to try for a passage westward or eastward, directing his course to that side from whence the tide of flood came. Having thus found the passage, he was to proceed onwards, keeping the American shore on his larboard, till he arrived at California, and so on. The Dolphin, man-of-war, was appointed to convoy the discovery vessels as far as the Orkney Islands, for security against the enemy's privateers. Middleton wintered the first year in Churchill River, and in 1742 proceeded northwards, discovered the wide and deep inlet which he named *Wager River*, and entered *Repulse Bay*, the south headland of which he called *Cape Hope*. His further progress northwards being impeded by ice, Middleton landed, and having walked fifteen

* Phil. Trans. Papers, 393, 465.

miles across the high point at the north-eastern end of the bay, he beheld a *frozen strait* turning round the north end of Southampton Island towards Cape Comfort and the North Bay of Baffin, or Foxe's Farthest. Eighty years afterwards, Sir Edward Parry, entering by the frozen strait, then open, proved the perfect correctness of Middleton's survey, and Sir George Back, in 1836, found the strait encumbered with ice, as Middleton had seen it. The flood tide came round Southampton through the Frozen Strait, which is four or five leagues across at the narrowest part.

On his way northward up the Welcome, Middleton examined the Wager Inlet, having its entrance in latitude 65° 23′ N., longitude 88° 37′ W., its southern cape, named *Dobbs*, being in 65° 12′. Eighteen days, or from the 13th of July to the 1st of August, were spent in this inlet, in exploring it by boats and in trying the tides. These were very strong, running five or six miles an hour, and setting in, during flood from the Welcome, with a rise of from ten to fifteen feet. The inlet was found to narrow towards the mouth of a river at its western extremity.

On Middleton's return to England, Arthur Dobbs, who had been a chief instigator of the expedition, and a correspondent of Middleton's on the subject for six years before it left England, was grievously disappointed with the result, and preferred charges to the Admiralty against the captain of his own choice, accusing him of want of honesty in the report of his proceedings, and of concealing everything that told in favour of a passage, so that he might serve the interests of the Hudson's Bay Company, which he alleged would be injured by the discovery of the north-west passage. Middleton's honest and seaman-like reply, and the evidences which he adduced of the truth of his statements, satisfied the Admiralty,*

But Dobbs, who seems to have been a man of much energy, though wanting either in fairness or in judgment, had influence enough, through his publications, to procure the passing of an Act of Parliament, offering a reward of £20,000 for the discovery of a north-west passage; and was mainly instrumental in raising £10,000, by subscriptions of £100 each share, towards defraying the outfit of an expedition that might earn the national reward. On their part, the Admiralty promised protection from impress for three years to all seamen who volunteered for the ships to be fitted out; and a scale of premiums, in the nature of prize-money, was settled for the officers and men, in case of success. The command of the Dobbs galley of 180 tons, was conferred on Captain William Moor, who, having been Middleton's second, had been won over to espouse Dobbs' side in the controversy which that gentleman had stirred up. On board the Dobbs, Henry Ellis, gentleman, agent for the proprietors of the expedition, was embarked; and wrote a narrative of the proceedings. The California, of 140 tons, was placed under the command of Captain Francis Smith, and a journal of his doings was kept by the clerk. In the instructions issued to the Captains, Wager Inlet is denominated a strait, through which they are told to push westward, and when by that route they get into an open sea, they are encouraged to depend on an open passage, and to proceed boldly, keeping America on the

---

* Vindication of the Conduct of Captain Christopher Middleton, F.R.S., etc., London, 1743. According to Mr. Goldson, Middleton, being neglected by the Admiralty, retired to a village near Gainsborough, where he died in pecuniary distress, having previously sold his Copley medal for his support.

left hand. They had also the option of trying Pistol Bay, or Rankin's inlet, near Marble Island, and on finding no obstruction, were to winter on the Pacific in 50° north latitude, and to rendezvous in any harbour nearest to 40°, on the back of California. In these instructions, we can trace a belief in the existence of the fabulous strait of Anian, and the charts afterwards published to illustrate the two narratives, and which it is not uncharitable to suppose embody Mr. Dobbs' study of the older north-west voyages, erroneously transfer Frobisher's strait to the south end of Greenland.

When the discovery ships left the Orkneys on the 12th of June 1746, in company with the Hudson Bay vessels, by a singular fortune a Captain Middleton, then commanding His Majesty's ship Shark, was appointed by his senior officer Commodore Smith, to convoy them to the westward, which he did for six days. The ships wintered at Port Nelson, and next year proceeded to fulfil the objects of their mission. Dobbs seemed to have infused his captious spirit into the officers of the two ships, as we have a double set of names imposed on the several headlands and inlets; and though frequent councils were held, no hearty co-operation ensued, and finally, two polemical narratives were published. In one thing the two captains agreed, namely, that they were not instructed to examine Repulse Bay, and the Frozen Strait; and in another, that Wager Inlet, after an accurate examination, is entirely shut off from having any communication with any place but the Welcome.

The name of Rankin's Inlet was changed to that of James Douglas's Bay, in honour of a merchant of the city of London, one of the adventurers in the undertaking. Ellis calls this same opening Corbet's Inlet. These are the only returns the adventurers had for their £10,000. On the 25th of August, a council was held, and "a definitive resolution was taken to bear away without further delay for England." . . "The discovery being finished," as the other narrative has it.

Chesterfield Inlet had been entered by both ships, and examined as far as an overfall or cascade, by Captain Smith, who changed its name to Bowden's Inlet; but his account of it was not thought in England to be satisfactory, and it was supposed that an unexplored passage existed from its western extremity. To set this question at rest, the Hudson's Bay Company sent Captain Christopher in a sloop to examine it anew, in 1761. On his return he reported that he had navigated the inlet for more than 150 miles in a westerly direction, until he found the water fresh, but had not seen its end. On this, Mr. Norton was sent, in 1762, to trace it to its extremity, which he did, and found it to end at the distance of 170 miles from its entrance, in a fresh-water lake twenty-four leagues in length, and seven or eight wide. A river flows into the western extremity of this lake.* In 1791, Captain Duncan was sent by the same Company to examine Corbet's or Rankin's Inlet, which proved to be a bay, and Chesterfield Inlet, which he found to agree with Norton's description. He traced the river, which enters its extremity, for thirty miles, when finding that it flowed from the northward, he turned back.† Dr. Rae, chief factor of the Hudson's Bay Company, in 1854, entered the inlet and ascended the river Quoich, which falls into its north side. This stream he navigated in a boat for two degrees and

* See Introduction to Cook's Third Voyage, where justice is rendered to Middleton and to the Hudson's Bay Company.

† Goldson on the Passage between the Atlantic and Pacific Oceans. Portsmouth, 1793. Pp. 45 and 53.

a half of latitude, until he had crossed the parallel of Wager Inlet at the distance of twenty miles from its head waters, and about eighty geographical miles from the nearest bend of Back's Great Fish River. The country was difficult, mountainous, and barren.

As early as the year 1715, the northern red Indians, or *Tinné*, who brought peltries to the Hudson's Bay Company's factory, called Prince of Wales Fort, on the Missinippi or Churchill River, described a river in the west flowing northwards to the sea, on whose banks there was abundance of native copper, pieces of which they produced. The disastrous expedition of Knight, Barlow, and Vaughan, of 1719, had reference to the discovery of this river, and Mr. Dobbs had kept it in view in the instructions drawn up for the discovery vessels sent out through his exertions. In 1769, Mr. Norton, governor of the fort or factory of Churchill, proposed an overland expedition to find out the Coppermine River, as it came to be called, but which the Tinné named *Neetha-sansan-tessy* (the far-off metal river). Samuel Hearne was the traveller chosen, two seamen being appointed to accompany him, together with two leading chiefs and eight picked men of the "home-guard,"—that is, of the Cree Indians, or Nathew inyuwuck, living in the vicinity of the fort, and trading constantly with it. Some "very portable astronomical instruments were sent out for his use;" and he was instructed to trace the Coppermine River to its mouth, and to note what advantages it offered for a settlement of the Company. On the 6th of November 1769, Hearne and his party set out, hauling their baggage, which was very light, on sledges, and directing their course to the north-west. The limit of the woods inclines to the north-west, and does not reach the coast to the north of Churchill river. They crossed Seal River, and the Indian chief Chawchinahaw, who assumed the direction of the party, assured Hearne that they would reach the woods in four or five days. In the meantime the cold was great and the party suffered from want of fuel. It appears to have been Chawchinahaw's design to disgust the Europeans with the enterprise, and with that view to keep on the barren grounds, but not far from the woods which they occasionally saw looming in the distance. Finding this plan not to answer his end, he first induced several of the Indians to desert, and then telling Hearne that it was not prudent to go further, he and the remainder went off laughing heartily at their own devices and the difficulties in which he had involved the English, after leading them 200 miles from the fort. Chawchinahaw, however, had the humanity before quitting Hearne to shew him the best course he could take homeward. Hearne and his two men reached Prince of Wales Fort in safety, after an absence of thirty-six days, and no little hard living.

In February 1770, Hearne set out a second time, taking with him no Europeans, because the two men of the former expedition had been harshly treated by the Indians. On this occasion his native guides led him into a labyrinth of lakes, extending, if his distances and latitudes are correct, nearly as far north as the parallel of Chesterfield Inlet, travelling as was convenient to their pursuit of deer, and without the slightest intention of going to the Coppermine River. Though the land is described by Hearne as entirely barren, and destitute of trees and shrubs, except the *Wishacapucca* (*Arbutus uva-ursi*), deer were abundant, and at least 600 Indians were assembled in one encampment by the middle of July. On the 11th of August, having set up his quadrant for the purpose of taking a meridian altitude, it was blown down and broken, which accident determined him to return a second time to the fort, and there was indeed little prospect of the Indians he was then associated with, leading him to the river he was seeking. The instrument thus broken was a Hadley's Quadrant, with a bubble attached to it, instead of an artificial horizon, made by Daniel Scatlif of Wapping. With this, Hearne, if he knew how to observe correctly, might have ascertained his latitudes within moderate limits, and his chart of this second journey is more likely to be right than that of his third and principal one, in which the instrument he had was little used, and of little use. The day after the accident to the quadrant, a party of strange Indians, plundered Hearne and his native "home-guard" companions, of every useful article. In the way back to the fort, however, assistance was obtained from parties of natives going thither with furs, and above all, from Matonabbee, a famous leader of the Tinné, whose aid Hearne had been instructed to seek. On the 25th of November he reached the fort a second time, after an absence of nine months.

The third journey, conducted on a plan sketched out by Matonabbee, and under his guidance, was successful. This time Hearne eschewed the companionship of the home-guard Crees, as he had done on the second occasion, that of European servants of the Company, and threw himself wholly on Matonabbee, as the most influential leader of the *Tinné* nation. The third start was made on the 7th of December 1770. His first winter was spent within the verge of the woods, Matonabbee moving from place to place in quest of deer and fish to supply the party, which, including the women and children, was generally large. The movement was, however, on the whole to the westward, and about the middle of April the party had reached Little-Fish hill, on the banks of a small lake, in latitude $61\frac{1}{2}°$ north, and longitude 112° W. Here preparations were made for crossing the barren grounds by providing a stock of dried provisions, and on the 18th of the month, the course was changed to north. On the 3d of May a halt was made at *Clowey* or (*Thlueh*) Lake,* for the purpose of building canoes, which were not finished till the 20th. Upwards of 200 Indians came to the same place for the same purpose, while the party remained there, and Hearne remarks, that being under the protection of a principal man, none of them offered to molest him.

In reading Hearne's narrative, it is necessary to advert to an overstatement of distances, a very usual circumstance with pedestrians, and the cumbrous Elton's quadrant which had been lying thirty years at the fort, and was the only astronomical instrument he had on this last journey, was not likely to aid him in correcting his reckoning. On collating his chart by aid of two or three ascertained geographical positions, his differences of latitude and longitude are invariably found to be in excess.†

During the stay of Hearne at Clowey, a war-party was organised to attack the Eskimos that frequent the mouth of the Coppermine River, for which sixty men volunteered; and on arriving at *Cogead* Lake, as he calls it, but which now bears, with the Copper Indians, the name of *Contwoy-to*

---

* The *Thlueh* or Trout Lake, discharges itself by the *Tchu-tessy*, into the south side of the eastern extremity of Great Slave Lake, and by information obtained by Sir George Back, must lie between the 62d and 63d parallels of latitude, but nearer the former. In Hearne's map, it is on the 63d parallel.

† For a critique on Hearne's geographical positions, see " Voyage down the Great Fish River by Captain (since Rear Admiral Sir) George Back," p. 144.

or Rum Lake, the women, children, dogs, and heavy baggage were left there until the return of the party, the situation being good for fishing not far from the woods, and, moreover, a common summer resort of the Copper Indians.

Travelling without encumbrance, the war-party, with Hearne in company, reached a river of some size, called *Congecawthawachaga*, on the 21st of June, and there they met a large body of the Copper Indians or Red Knives, one of whom, then a boy named Cascathry, was well-known in 1820-21 to Sir John Franklin. This boy joined the war-party, and in his old age remembered the circumstances well. Hearne says that he ascertained with his Elton's quadrant the position of the ferry over the river to be 68° 46′ north, and 118° 15′ west of London. According to Sir John Franklin's observations it lies in 66° 14′ N., long. 112° W.

Under the guidance of the Copper Indians who knew the country, the party crossed the Stoney mountains, which seemed to Hearne to be at first sight totally impassable, so craggy did they appear. This ridge of mountains is granitic, and terminates in Cape Barrow on the Arctic sea, about ninety miles east of the mouth of the Coppermine River.

Fourteen days' march from Congecawthawachaga, including some detention by bad weather, brought Matonabbee and his party to the Coppermine River, and Hearne's plan of it, though not accurate, is sufficient proof that he had personally inspected it for twenty-five or thirty miles.

The Indians having ascertained by scouts that some families of unsuspecting Eskimos, were encamped on the west side of the river near a cascade, stole upon them when they were asleep, and butchered upwards of twenty men, women, and children. In 1821 some human skulls lying on the spot bore testimony to this cruel slaughter. Hearne's description of the country, and of the abundant fishery at the waterfall, are correct, but it is impossible to give him credit for endeavouring to speak the truth, when he says that at the mouth of the river, on the 17th and 18th of July, he had sunshine the whole night. "The sun," he says, "was at midnight certainly some height above the horizon, how much, as I did not *then* remark, I will not *now* take upon me to say, but it proves that the latitude was considerably more than Mr. Dalrymple will admit of."* In his written report to the Company, Hearne had said that the sun was "a handspike high at midnight." On the same days in the year Sir John Franklin saw the sun set, and the mouth of the river was ascertained to be in latitude 67° 48′ N., longitude, 115° 47′ W. Hearne, by dead reckoning carried on from Congecawthawachaga, places it more than four degrees too far north, and nearly five too far west.

The Admiralty expedition in 1773 to Spitzbergen, under Captain Phipps, afterwards Lord Mulgrave, made no discovery; and the hopes of the advocates for further search were then fixed on our great navigator Cook, who was induced by the Earl of Sandwich to leave his honourable retirement in Greenwich Hospital to undertake the third and last of his voyages, mainly for the purpose of ascertaining whether or not a passage existed between the Northern Pacific and Atlantic Oceans. The Expedition sailed in 1776, and returned to England in 1780, though, alas! without its distinguished commander. Bering's† discovery of the Strait which retains

---

* Hearne, preface, p. vii.

† Bering was a Dane, and his family retain the orthography of his name, which we have adopted.—(*Baer Nachricht,* etc.)

his name, was kept in view in framing Cook's instructions, as well as Hearne's journey, which negatived a passage in a low latitude ; and Cook was enjoined not to lose time in exploring rivers or inlets till he got into the latitude of 65°. Though Cook considered the reports of the pretended *Strait of Da Fonte*, supposed to lie between the 50th and 55th parallels to be improbable stories that carry their own refutation with them, he would have examined the coast there, had he not been prevented by a gale of wind. His judgment, however, was fully confirmed by Vancouver's accurate survey, made twenty years later.

Cook's careful examination of the American coast, from the 58th parallel of latitude northwards, proved that there was no passage below *Icy Cape*, which was the limit of his voyage within Bering's Strait. The Russian surveyor Gwosden had seen the American side of Bering's Strait in 1730 ; and Bering Tchirikow and De Lisle had rounded the peninsula of *Alaska*, and touched, in 1771, the main land near *Mount St. Elias*, as well as in latitude 55° 30′ ; but Cook was the first who made a continuous and effective survey of those coasts. The failure of Phipps in the Spitzbergen seas, of Cook by way of Bering's Straits, and of the vessels sent on two successive seasons to Davis' Straits to co-operate with him, satisfied the Admiralty of the day, and for forty years the North-west Passage was unheard of in the government bureaus.

In 1789, Sir Alexander Mackenzie, a member of the North-West Fur Company, trading from Canada, descended the great river which bears his name, and traced it to its termination in the Arctic sea. Though this traveller says that he was not supplied with the necessary books and instructions, and with much modesty adds that he was deficient in the sciences of astronomy and navigation, his survey was in the main highly creditable, and the position of *Whale Island*, his extreme point, is very nearly accurate. He had actually reached the sea-coast, but the Mackenzie pours out such volumes of fresh waters from its various mouths that the sea does not become salt till near Garry Island, which lies about thirty miles out from the coast of the river-delta. The rising of the tide was however observed. The latitude of Whale Island was found to be $69\frac{1}{4}°$ by a meridional observation of the sun, and its longitude, by dead reckoning, 135° west. Before leaving the island many Belugas or white whales were seen—whence the name given to the island by Mackenzie. His descriptions of the channels he followed in the delta are so complete, that they were readily recognised when Sir John Franklin afterwards surveyed the river.

CHAPTER VII.

RUSSIAN VOYAGES ALONG THE SIBERIAN COAST—A.D. 1598-1843.

Dyakow—Jenissei River—Piissidi River—The Lena—Jelissei Busa—The Lena and Olekma—Tunguses—The Jana—The Tshéndoma—The Jukahirs—Ivanoio—The Indigirka—The Alascia—Staduchin—The Kolyma—The Tchuktchi—Cape Chelagskoi or Erri-nos—Tchaun Bay—Rein-deer Tchuktchi or Tuski—Deschnew doubled the North-east Cape, and passed Bering's Strait to Anadyr—Alexiew—Svatoi-nos—Bering—East Cape—St. Lawrence Island—American Shore of Bering's Strait—Kurile Islands—Kamtschatka—America—St. Elias—Aleutian Islands—Tchirikow—Cape Edgecumbe—Kotzebue—Kotzebue's Sound—Permäkow—Liakhow Islands—Svatoi-nos—Indigirka—Eterikan—Liakhow—Anjou—Kotelnoi Island—New Siberia—Fadejevkoi Island—Sieveroi vostochnoi-nos—Laptew—River Olenek—Bay of Nordvich—Bay of Chotanga—Cape St. Faddei—River Taimura—River Chotanga or Khotanga—Taimura Lake—Tunguses.

WE now turn to the progress of Arctic discovery on the coasts of the Russian empire; and it may simplify our statements, if we premise, that the "North-east Cape," or *Sieveroi Vostochnoi Nos* of Wrangell, or the Cape *Cheliuskin* of Middendorf and Peterman, is the northernmost point of Asia. It lies on the 100th meridian, and reaches the 78th parallel of latitude, extending higher than Cape Taimura, and about 7° to the north of any part of the American continent. It has never been doubled, either by a boat or a sailing vessel.

No ship coming from Europe has passed eastward beyond the *Sea of Kara* or *Karskoie*, and the whole of the northern coast of Asia has been discovered exclusively by Russian subjects, mostly Cossacks, employed to subject the Samojeds, Ostiaks, Tunguses, Jakuts (or *Linzacha*, in their own tongue), and Tchuktches, to the imperial authority, and make them pay the customary *yassak* (tribute). The coast of the icy sea was partially known to Russian navigators in the middle of the sixteenth century. They were accustomed to sail in small flat vessels, *lodji*, from the White Sea and from the Petchora, across the Sea of Karskoie, to the entrances of the Obi and Jenissei, sometimes performing the whole voyage by sea, sometimes shortening the distance by drawing their boats across the isthmus which separates the Gulf of Obi from the Karskoie Sea. In the latter case, they ascended the river Mutnaia to two lakes, where they made a portage of only 200 fathoms into Lake Silenoi and the river of that name, by which they reached the Obi, the whole voyage out from Archangel occupying about three weeks.*

In 1598, Fedor Dyakow was sent from Tobolsk to demand *yassak* from the Samojedes of the Jenissei; and in 1600, a town named Mangaseia was built on the river Jasa, in the country of the Samojedes. This town being removed to the river *Turuchanka*, the Cossacks made constant excursions from it for the purpose of rendering the Samojedes, Ostiaks, and Tunguses tributaries. In 1610, the predatory Cossacks reached the mouth of the Jenissei, and in the same year a body of *promischlenniki* (fur-hunters), traced the coast between the Jenissei and the *Pässida*,† a river which joins the sea more to the eastward.

* Lutke's Voy., i. p. 76.

† *Pässida* was a general name of the country about the Lower Jenissei in the Samojed language, and means a barren plain, called in Russian, *tundra*.— Wrangell, *Siberian Arctic Sea*, p. 390.

In 1630 the Cossacks of the Jenissei discovered the Lena, and in 1636, Jelissei Busa one of their number, was commissioned to examine the Lena and other rivers that fall into the Polar Ocean, and to impose *yassak* on all the natives of those quarters. He reached the western mouth of the Lena, and after navigating the sea for twenty-four hours, came to the *Olekma*, which he ascended, and wintering among the Tunguses, imposed the *yassak* upon his hosts. In 1638, the same Busa, navigating the ocean eastward for five days from the Lena, discovered the *Jana*, on whose banks he passed another winter. In the following year, resuming his voyage eastward by sea, he reached the river *Tshéndoma*, and wintering for two years with the Jukahirs, dwelling in half-subterranean huts on its banks, made them also tributary to Russia.

In the same years, Ivanoio discovered the *Indigirka*, and carried the coast-survey onwards to the *Alaseia*, 163° east of Greenwich.

In 1644 the Cossack Michael Staduchin formed a winter establishment on the delta of the *Kolyma*, which has since expanded into the town of *Nijnei Kolymsk;* and this adventurer was the first who got intelligence of the warlike *Tchuktchi*, who inhabit the north-eastern corner of Siberia, and who resisted the imposition of *yassak* with bravery and successful pertinacity. In consequence of Staduchin's report, a party of *Promischlenniki* descended the Kolyma, and navigating the sea to the eastward for forty-eight hours, found a party of Tchuktchi encamped in a bay. With them they entered into traffic, by exposing merchandise on the strand, and retiring to some distance. The Tchuktchi then approached, took what pleased them, and put down in exchange seahorse teeth. This mode of traffic is carried on at this day in

Arctic America, by the Red Indians and Eskimos, between whom there are blood feuds. Staduchin afterwards navigated the sea eastward to *Cape Chelagskoi* or *Erri-nos*, which rises on the east side of *Tchaun* Bay, a little beyond the 70th parallel of latitude, and may be considered as the north-eastern cape of Siberia. Between it and Bering's Strait, is the proper country of the Rein-deer, *Tchuktchi*, called by Lieutenant Hooper of the Plover, who wintered among them in 1848-9, *Tuski*, which he gives as the native pronunciation of their designation.

In 1648, Semen Deschnew, Gerasim Ankudinow, and Fedot Alexeiew, in three vessels of the kind, called *kotchy*, sailed from the Kolyma, with the intention of reaching the *Anadyr* by sea. They passed the *Svatoi-nos* (or holy promontory), as he calls the *Chelagskoi-nos* of Wrangell, on which Ankudinow's vessel was wrecked, and on the 20th of September had a battle with the *Tchuktchi*, in which Alexeiew was wounded, and shortly afterwards his vessel separated in a violent storm, and did not join company again. Deshnew's kotchy was driven about by contrary winds, and at length cast ashore in the Gulf of Anadyr, considerably to the south of the river of that name. By this remarkable voyage, Deschnew passed through the strait subsequently traversed by Bering, whose name it continues to bear. He accomplished this feat, however, at the expense of great suffering. Thrown ashore with twenty-five companions, after a painful march of ten weeks he reached the mouth of the Anadyr, but being unprovided with food or the necessary apparatus for fishing, part of his men perished of hunger; and the twenty survivors reached a native tribe named *Anauli*, with whom they dwelt for a considerable time, but finally on their refusing *yassak*, Deshnew put them to death. That year, Deshnew laid the foundation of *Anadyrskoi Ostrog*. In the meantime, efforts were made to reach the Anadyr overland, and in 1650, a party under Semun Motóra discovered Deshnew and his men on the Anadyr, to the mutual joy of both parties. Deshnew afterwards established a walrus-fishery from the Anadyr, and in one of his voyages landing at some Koriak huts, he learnt from a Jakut woman that Alexeiew's vessel had been driven on shore, that one of his men died of scurvy, others were killed in fighting with the Tchuktchi, and the remainder, it was afterwards ascertained, reached the Kamtschatka River, and lived some time with the Kamtschatdales, but were at length put to death by that people in a quarrel. Both Deshnew and Alexeiew, therefore, passed through Bering's Strait from the north in the middle of the seventeenth century. The *Svaitoi-nos* of Deshnew, is to be distinguished from one of the same name which stretches to the 73d parallel of latitude, between the Indigirka and the Lena, and is called *Svaitoi* (accursed or sacred), because of the danger of doubling it. Liakhow Island lies due north of the latter.

Baron Wrangell in 1820-1824 surveyed the coast accurately, from the mouth of the Kolyma eastward to Cape Chelagskoi, and from the account given in his volume of previous Russian discoveries on the Arctic coasts, much of the preceding sketch has been taken.*

In 1728, Captain Vītus Bering and Lieutenants Tchirikow

---

* Voyages from Asia to America by Thomas Jeffreys, 1761. And a Chronological History of North-eastern Voyages of Discovery, by Captain James Burney, 1819, have also been consulted.

Verchnei Ostrog, on the River of Kamtschatka, was built in 1699 by Atlasoff, a Cossack officer who came from Jakutsk.

and Spanberg, having, by orders of the Empress Catherine, built a vessel named the Gabriel, at Nischnei Kamtschatkoi Ostrog, on the east side of the peninsula of Kamtschatka, and being joined by the Fortuna from Ochotsk, sailed northwards on the 14th of July. On the 8th of August, in latitude $64\frac{1}{2}°$ N., Bering had some intercourse with the Tchuktchi, and saw the Island of St. Lawrence, which lies in the strait. On the 15th of the same month he reached latitude $67° 18'$, or East Cape, beyond which he did not go, being satisfied with the westerly trending of the coast beyond the cape.

In 1730, a Cossack officer named Tryphon Krupischew, and the Geodetiste (surveyor) Gwosdew, sailed in the Fortuna to the Tchuktchi coast, and being driven in a gale of wind back from Bering's extreme point, they steered east until they saw an Island (St. Lawrence), and beyond it a large land which they coasted to the southward for two days, when another storm compelled them to put back to Kamtschatka. This voyage completed the discovery of both sides of Bering's Strait, and excited much interest when its results were made known in Europe. The Russian Government especially was roused to renewed activity in exploring all parts of Siberia, and of the seas washing its coasts.

In 1740, Captain Vitus Bering again sailed from Ochotsk in the St. Paul, on a voyage of discovery, having as second in command his former companion, Lieutenant Alexei Tchirikow, in charge of the St. Peter. They passed through a channel between the first *Kurili Island,* and the south point of the peninsula of Kamtschatka, named *Lopatka,* to *Awatchka* Bay, where they wintered. In 1741, they set out again, and on the 18th of July, Bering having separated from Tchirikow, made the American coast in latitude $58° 28'$ N., or to the northward of the archipelago of King George, and in sight of the lofty volcanic mountain of *Saint Elias*. One cape was named after Saint Hermogenes, another after Saint Elias, and the gulf between them has since been called Prince William's Sound. From this place, Bering steered for Kamtschatka, but was much hampered by the difficult navigation down the peninsula of Alaska, and among the Aleutian Islands, one of which he named Schumagin. The season was advanced before he got clear of this intricate navigation, scurvy appeared among his men, and on the 5th of November the Saint Paul was driven on the rocks of Bering's Isle, which lies off the coast of Kamtschatka. There Bering died. The survivors of the crew after wintering on the island, and living chiefly on white foxes, built a boat out of the wreck of their ship, and reached Awatchka Bay next summer.

Tchirikow, after separating from Bering, saw the American coast in latitude 55° 36′ N., on the 15th of July, by Russian reckoning, or the 26th, new style. Captain Cook identifies this part of the coast with his Cape Edgecumbe, and a mountain in sight he named Mount Edgecumbe. At this place Tchirikow lost his two boats, all that he had—their crews having been cut off by the natives. He ran along the coast, however, for 100 German miles, and succeeded in carrying his ship back to Awatchka Bay on the 9th of October; but of his crew of seventy men, twenty-one died, including M. de Lisle de la Croyere, from whose journals the details of the voyage became known.

To complete all that is necessary to be said of the later Russian discoveries in America, we may notice that in the years 1815-17, Lieutenant Kotzebue of the Russian Imperial Navy, made a voyage, at the cost of Count Romanzoff, from the Baltic to Bering's Straits, in which he discovered Kotzebue's Sound, which is crossed by the Arctic circle and lies within Bering's Strait, but added nothing else to the previous discoveries of Cook.

In giving this summary view of the successive discoveries of the northern coasts of the widely spread empire of the Czar, we have, to avoid breaking in upon the narrative of the progress eastward of the explorers, omitted some remarkable efforts made northward by various traders or government officials, among whose names those of Demetrius Laptew and of the merchant Liakhow stand out prominently.

In 1710, Jakow Permäkow made the first report of the existence of an island (*Liakhow*) lying off the *Svätoi-nos*, a promontory between the Lena and Indigirka; and in consequence thereof various attempts were made to explore the island and the neighbouring sea, but at first unsuccessfully, its situation having been only indistinctly indicated. In 1760 the Jakut Eterikan of Ustiansk again saw the island, and several stories got into circulation respecting the existence of a large country to the north of Siberia. A merchant named Liakhow, happening to visit Svätoi-nos on business in March 1770, saw a large herd of deer coming over the ice from the north, and was induced thereby to start with sledges early in April to trace the tracks the deer had made. After travelling seventy versts (about forty-seven miles) he came to an island, and twenty versts (thirteen miles) further he reached a second island, at which, owing to the roughness of the ice his excursion terminated. He saw enough, however, of the richness of the two islands in mammoth teeth, to shew him that another visit would be a valuable speculation; and on making his report to the government he obtained an exclusive privilege to dig for mammoth bones and hunt arctic foxes on the islands he had discovered, which the Empress Catherine, moreover, directed should be called after his name. In the summer of 1773, Liakhow visited his discovery with a more complete outfit, and ascertained the existence of a third island, much larger than the others, mountainous, and having its coast covered with drift-wood. He then went back to the first island, built a good hut, and having wintered there, returned to Ustiansk in the spring with a valuable cargo of furs and mammoth-tusks. One of Liakhow's companions described to M. Sauer the extraordinary collection of fossil remains of animals he saw on these islands exceeding in variety and extent the mammalian remains in any other quarter of the world, and in 1775 the Russian government sent a land-surveyor to make a regular survey of the archipelago. It was not, however, until 1823 that Lieutenant Anjou having travelled round the group, ascertained their correct positions. The largest of them reaches about ten miles beyond the seventy-sixth parallel of latitude, and is named *Kotelnoi* (or Kettle Island), on account of a kettle having been left on it by some unknown visitors. It exceeds 100 miles in length from north to south, and is about sixty miles across at the widest. *New Siberia*, the next island in extent, lies from twelve to twenty degrees of longitude more to the eastward, and between them intervenes the third in size, *Fadejevkoi Island*. *Liakhow* Island and some adjacent islets are more to the south and nearer the Svätoi-nos. From the western point of Kotelnoi Island to the eastern cape of New Siberia, the distance is 205 miles. We shall have occasion to mention these islands again, in a subsequent chapter to be devoted to the Geology of the Arctic regions.

We have still, in order to complete our sketch of the Russian surveys, to mention some of the attempts made to double *Cape Taimur* and the *Sieveroi Vostochnoi-nos*. After failures in 1735, 1736 and 1737 by others, the project was entrusted to Lieutenant Laptew, who sailed from Jakutsk, on the Lena, in 1739, passed the mouth of the *Olenek*, the Bay of *Nordvich*, the Bay of *Chotanga*, and on the 20th of August anchored off *Cape Saint Faddei*, in latitude 76° 47′ N. by reckoning. Fogs prevented a party which he sent on shore from acquiring a knowledge of the interior, or finding the river Taimura, which they were desired to seek ; but mountains seen in the north, with snowy summits, were concluded to be those discovered in 1736 by a previous navigator. He then returned southward, and with difficulty, owing to embarrassments from drift-ice, succeeded in entering the river *Chotanga*, on which he wintered near the *Bludnaia*, one of its affluents, and in the neighbourhood of a tribe of Tunguses named *Sidätshi*, because they are not nomades but have fixed residences. A party sent out across the promontory in March 1740, found the *Taimur River*, and traced it to the sea, and about sixty miles of the coast of the *Taimura Bay*. In attempting to return to the Lena, the vessel was wrecked in the ice ; the crew reached the shore with difficulty, and many of them perished from fatigue and famine before the rivers were sufficiently frozen to enable the feeble survivors to return to their former winter station at Chotanga (or Khotanga). Notwithstanding the hardships he and his party had endured, Laptew divided his men into three sledge-parties to prosecute the survey of the promontory. Laptew himself set off on the 24th of April 1741 to the north-west across the *Tundren* (or Barren Grounds) towards Taimura Lake, distant above 190 miles, and reached it on the 30th. He found the Taimura River as it flowed from the lake to be about a mile and three-quarters wide, and having followed it to the sea, ascertained its mouth to be situated in latitude 75° 36′ N. He then proceeded to survey the west coast of the promontory southwards to meet his mate, whom he had directed to travel northwards. On the 2d of July the two parties met and continued their survey together to a Tonguse settlement on the *Pässina*, in latitude 73° 39′ N. The third party did nothing in the way of discovery, but succeeded in reaching the Jenissëi by a direct overland march. Lieutenant Chariton Laptew deserves to rank very high among arctic discoverers for the resolution with which he overcame difficulties, and his perseverance amid the severest distresses.*

An open sea was found at Cape Taimura by Middendorf, in 1843, but no one has as yet succeeded in doubling the *Sieveroi Vostochnoi-nos*. Some authors have supposed this to be the cape called *Tabin*,† by Pliny; but all attempts to identify any of the arctic promontories with capes known to the Romans, are futile.

* He must not be confounded with another very persevering Russian Lieutenant, Demetrius Laptew, who sailed from the Lena eastward, doubled the *Svatoi-nos*, wintered in the Indigirka, surveyed the Bear Islands, passed a second winter in the Kolyma; and in a fourth season extended his survey of the coast to the Baranof rock, including in all, 37° of longitude.

† "*Iterumque deserta cum belluis, usque ad jugum incubans mari quod vocant* Tabin."—*Plinii, Hist. Nat.*, lib. vi., cap. xx.

CHAPTER VIII.

NINETEENTH CENTURY—ENGLAND.—A.D. 1817-1845.

Dr. Scoresby—Sir John Barrow—Buchan and Franklin—Sir John Ross—Sir Edward William Parry, and Lieutenant Liddon—Sir John Franklin—His second Expedition—Dr. Richardson—Captain Beechey—Sir John Ross—Sir James Clark Ross—Rear Admiral Sir George Back—Peter Warren Dease and Thomas Simpson—Dr. John Rae.

THE efforts made by England in the present century to explore the Arctic American sea, have exceeded all former example, and being chiefly instigated by the humane desire of rescuing the crews of two ships shut up in the ice, redound more to the national credit than if discovery alone had prompted the expenditure of men and money. From the time of Cook till 1817, England had been quiescent in the search for the Northwest Passage, until Captain Scoresby, the able and scientific master of a whaler, published his account of the Greenland seas, and drew the attention of Europe to that quarter; and Sir John Barrow, secretary of the Admiralty, by his writings and personal influence roused the British Government to undertake a new series of enterprises on a scale commensurate with improvements in ship-building, and in the art of navigation. It was urged that a great disruption of ice having taken place in the north in the past year, this would be a favourable time to renew the north and north-west projects, and that several very important scientific researches in nautical astronomy and magnetism might be carried on by the parties employed in the search. The Government entering into these views, four stout vessels were selected, and strengthened to resist the shocks and pressure of the ice by diagonal timbers and double planking, in a manner which had never before been attempted.

Two of these vessels were destined to proceed northwards by way of Spitzbergen and to endeavour to cross the polar sea—reports of the high latitudes that had been reached without difficulty by various whalers giving hope of success in that direction. Of these, the Dorothea was given to Commander David Buchan, and the Trent to Lieutenant John Franklin. They sailed in 1818, and reached the seas between Spitzbergen and Greenland; but being embayed in a heavy storm with an ice-pack under their lee, they were compelled, as the only chance of safety, to the dangerous expedient in such weather, of "taking the ice"—that is, of thrusting the ships into any opening among the moving masses that could be perceived. In this very hazardous operation the Dorothea having received so much injury, that she was in danger of sinking, was therefore turned homewards as soon as the storm subsided, and the Trent of necessity accompanied her.*

The other two ships, the Isabella, commanded by Captain John Ross, and the Alexander, by Lieutenant William Edward Parry, were appointed to perform their voyage of discovery through Davis' Straits. This voyage was so far successful, that it established the accuracy of Baffin's survey of the bay which bears his name, the various sounds and islands that he described being found in the positions that he had assigned to them. But Captain Ross was satisfied with a still more super-

* An account of the voyage was published by Captain Beechey in 1843.

ficial examination than circumstances had compelled Baffin in his small and ill-found vessels to make. Lancaster Sound, which Baffin was prevented from entering by a barrier of ice, was the only inlet attempted by Captain Ross, who sailed a short way within its headlands, Capes Charlotte and Fanshawe. But when he had reached longitude 81½°, in latitude 74° 3' N. on so promising a course, he was arrested by a vision of a range of mountains closing the bottom of the sound, seen by few or none in the ship except himself. Without taking the precaution to ascertain, by a nearer approach, whether the *Croker Mountains* were firm land or merely one of the atmospheric deceptions so common in those seas, he forthwith returned to England.

Doubts of the reality of the Croker Mountains being entertained by most of Captain Ross' associates, and the report made by that officer not being thought to be conclusive by the Admiralty, the Hecla and Griper were commissioned in 1819 for the purpose of exploring the sound, whose entrance only had been seen by Baffin and Ross. The former ship was placed under the command of Lieutenant William Edward Parry, and the latter under that of Lieutenant Mathew Liddon. These ships sailed triumphantly over the site of the fancied range of Croker Mountains, and holding a due westerly course, leaving Regent Inlet to the southward, and Wellington Sound and Byam Martin Channel, with the intermediate islands, on the north, reached the south side of Melville Island. There, in a haven which was named Winter Harbour, the two ships remained for the ten months in which the navigation of those seas is closed by frost; and after making another but fruitless attempt to penetrate the icy barrier which shuts up the strait between Melville and

Banks' Island, returned to England in October 1820. The success of this voyage so far exceeding expectation, the perfection of the commanding officer's general arrangements, and the consequent preservation of the health of the crews during the long arctic winter, placed Lieutenant Parry at once in the van of Arctic discoverers, and he was speedily raised through the grade of commander to that of captain, amid the gratulations of his countrymen. Reckoning in round numbers the distance between Baffin's Bay and Bering's Strait at 110 degrees of longitude (viz. from 60° W. to 170° W.), Captain Sir William Edward Parry explored his way up to Cape Dundas through fifty-four of these degrees, or nearly a half of the whole distance; and he saw on the verge of his western horizon, Banks' Land lying two degrees and a half farther off, his view reaching beyond the middle distance between the Atlantic and Pacific outlets. He also laid down on the north of his tract the chain of islands bearing the names of North Devon, Cornwallis, Bathurst, and Melville Island; and in the south, North Somerset, Cape Walker, and Banks' Land or Island, as it has since proved to be. The sea south of Melville Island is now called on the charts *Melville Sound*, and the strait leading westward from it Banks' Strait.

In 1821-23, Sir William E. Parry, commanding the Fury, and having as second captain G. F. Lyon in the Hecla, was employed on discovery through Hudson's Strait and past Foxe's Farthest. In this voyage he examined Repulse Bay, and proved the accuracy and good faith of Middleton, so recklessly attacked by Dobbs, as mentioned in a former chapter; and after one winter passed at Winter Island and another at Igloolik, he traced the Fury and Hecla Strait to its junction with Regent Inlet. This strait was obstructed by ice for two summers, so that he could not pass through it with his ships. His return to England with his crews in health, after two winters in the high latitudes, was another triumph of judgment and discipline. An attempt made by the same distinguished navigator to find a passage through Regent's Inlet, was terminated by the shipwreck of one of his ships—the Fury, commanded by Captain H. P. Hoppner. With great forethought, Sir Edward had all the provisions landed from the wreck and safely housed on Fury Point, off North Somerset.

Finally, being frustrated in his endeavours to find a passage westward, Sir Edward Parry attempted the Polar voyage in boats, starting in 1827 from the north end of Spitzbergen. He actually, by an unexampled boat-voyage, reached 82° 40′ 30″ of north latitude, which is beyond the highest authenticated position of any previous navigator; and he would have gone much further, but the current which set continuously to the south, carried back the boats during the hours necessarily allotted to the repose of the crews, and the daily advance, notwithstanding great exertion, was consequently small. At length fatigue and diminution of fuel and food compelled him to fall back on his ship, the Hecla, which awaited his return under Captain Forster in Treurenberg Bay. Ross's Islet, the most northerly rock seen on this most singular and adventurous voyage, lies in latitude 80° 49′ N., and until recently was the most northerly land known. This was the fourth and last of Sir Edward Parry's northern voyages.

Whilst Parry was so employed by sea, Lieutenant Franklin, afterwards promoted through the grades of commander, and captain, and knighted, was engaged in tracing the northern coast of the continent by land. Trained to be an accurate nautical surveyor, under his kinsman Captain Flinders, in the

L

Australian seas, and known as an active first-lieutenant and able seaman in his long course of ordinary service in the royal navy; during which he had distinguished himself in the action between Sir Nathaniel Dance and Admiral Linois, in the great sea-fight of Trafalgar, and in boarding and carrying an American gun-boat at New Orleans. Franklin's reputation, backed by the high opinion of his abilities entertained by Sir Joseph Banks, the President of the Royal Society, secured his nomination to employment on the expeditions of discovery, and his first appointment was to the Trent, as mentioned above.

From 1819 to 1822, Franklin was employed in leading an overland expedition from Hudson's Bay to the mouth of the Coppermine River, and the adjoining coast of the Arctic Sea. When this expedition was planned, the Admiralty, by whom it was organized, knew more of the condition of the country through which it had to pass, in the days of Hearne, than of its actual state in 1819, and relied solely on the aid that could be given to it by the Hudson's Bay Company, until it was on the eve of embarking. A letter of recommendation was then obtained from one of the agents of the North-west Fur Company, trading from Montreal to their chief factors and chief traders employed in the north, but too late for any effectual provision being made that year to assist the Government party. The fact was, that the two rival companies were carrying on a deadly warfare with each other. The Hudson's Bay Company claimed an exclusive right to occupy and trade in Rupert's Land, by virtue of a charter from Charles the Second; and their opponents based their rights of trade in the interior, on prior occupation, while they, at the same time, denied the validity of the charter.

In the contest the Indians were demoralized by the free distribution of spirituous liquors, and when parties of the two companies met, the stronger attacked the weaker one. None travelled unarmed. In the conflicts that ensued, life was held of little value; and in a single action, Governor Semple of the Red River Colony and twenty-one settlers were slain by a party of half-caste natives, brought from the Saskatchewan by the North-west Company. In another part of the country, fifteen or sixteen men were starved to death by the opposite party carrying away the Indian hunters on whom they depended for support. There were other deaths on a smaller scale, yet the war was carried on under the semblance of legal authority—the Hudson's Bay Company, in terms of their charter, empowering their local governors to commission magistrates and constables; while the chief factors and chief traders of the North-west Company, laying informations at Montreal of the violence done in the north, were made justices of the peace, and had authority to appoint constables, and put in force the warrants of apprehension that were issued. Such was the state of matters, when Franklin and his party landed at York Factory with three officers and one English seaman. On entering the Factory or Fort as it is called, for it is stockaded, he found four of the leading partners of the North-west Company captives within its walls, having been taken a few weeks previously at the distance of four or five hundred miles inland. As some of these gentlemen had wintered on the Mackenzie further north than the Hudson's Bay posts had at that time extended, the information they gave of the country through which the expedition had to travel was very important. After wintering at Cumberland House on the Saskatchewan, 650 miles from York Factory, and engaging a party of Canadian voyageurs that had been in the employ of the North-west Company, the expedition travelled northwards in canoes to Fort Chepewyan, the chief northern post of the Company just named. Up to this place, the provisions obtained from the two companies lasted, but no further supply could be given; and the expedition party, consisting then of twenty-five individuals, started with one full day's supply of food. It was joined on the north side of Great Slave Lake by a band of Copper Indians, under their chief Akaitcho, and finally reached the second winter quarters at Fort Franklin, in latitude 64½ N., on the 19th of August 1821. This place is 553 miles from Fort Chepewyan, and from thence to Cumberland House the distance exceeds 800 miles, so that the party travelled from the first year's wintering place to the second, 1350 miles. Every effort was made to procure ammunition and other supplies from Fort Chepewyan; Lieutenant Back having travelled thither in the middle of winter for the purpose, but his cold journey over the snow of 1100 miles there and back, was productive of a contribution very inadequate to the wants of the party. During the winter, the inmates of Fort Enterprize were supplied with venison by Indian hunters; but spring found their store exhausted, and the journey over the barren grounds to the mouth of the Coppermine River, was performed on the casual and often very scanty products of the chase. Of that distance, amounting to 334 miles from Fort Enterprize, 120 miles had been performed by dragging the canoes and baggage over snow and ice. The mouth of the river was found, by meridional observations of the sun, to be in 67° 48′ north latitude, and by chronometers in 115° 37′ west longitude.* On the 21st of July, the two

* Hearne, as mentioned in Chapter VI., gives the position of the mouth of

canoes were launched on the Arctic Sea, and by the 16th of August, the coasts of Bathurst Inlet and of the rest of Coronation Gulf were surveyed eastward to Point Turnagain, whose latitude is 68° 19′ N., and longitude 109° 25′ W. At this, the extreme eastern point of the expedition, the canoes were detained for some days by a heavy snow storm, and on the weather moderating, the advanced period of the season made a return southwards indispensable. The way was therefore retraced to Bathurst Inlet, and up Hood's River for a little way —and then the canoes being reduced so that one of them could be carried on a man's head, and the men's baggage being restricted to a blanket a piece, their guns and ammunition, a course was shaped for Point Lake, and the march commenced over the barren grounds. This disastrous retreat was made, except for the first two or three days, over snow; game was scarce, the strength of the party rapidly failed under the conjoint influences of cold, famine, and fatigue, and more than half the party perished, among the rest a young officer of extraordinary promise, Lieutenant Hood, under most distressing circumstances, which are related at length in Sir John Franklin's narrative of the journey. The survivors being succoured by the Indians of Akaitcho's band on the 7th of November, with their aid reached the Hudson's Bay Company's post on the north side of Great Slave Lake on the 11th of December, and England in the following October 1822.

In the years 1825, 1826, and 1827, Captain Franklin having received the honour of knighthood from his sovereign, again left England to resume the survey of the Arctic coasts

the river, corrected for publication as 71° 54′ N., 120° 30′ W.; by reckoning from Congecathewachaga, more than four degrees of latitude too far north.

of the American Continent, but under much happier circumstances than before. The Hudson's Bay Company had now amalgamated with the North-west Company, and the two having no rival, were carrying on a peaceful commerce throughout the length and breadth of the fur-countries. The Indians well-treated and happy, acquiesced in the absence of the "fire-water," which was no longer carried to the north, and were beginning to listen to the missionaries, as well as becoming gradually more amenable to the influence of the traders, which has always been beneficial when not perverted by commercial rivalry. On this expedition, Sir John Franklin wintered in 1825 on Great Bear Lake, and during the following summer descended the Mackenzie, and surveyed the coastline to the westward as far as Return Reef, more than 1000 miles distant from his winter quarters on Great Bear Lake. In connection with this survey, Captain Beechey in the Blossom had entered Bering's Straits, and by his boats explored the coast considerably beyond the Icy Cape of Cook, as far as Point Barrow, lying on the highest parallel of latitude to which the American Continent reaches, and constituting therefore the north-west cape of America. Its position is 71° 38′ N., and 156° 15′ W., the distance between it and Return Reef being 160 miles. In his advance to the last-named locality, Sir John Franklin had rounded the northern extremity of the great chain, named the Rocky Mountains, and consisting, as he perceived from sea, of several parallel ranges.

At the same time that Sir John was occupied to the westward of the Mackenzie, a division of his party in two boats, commanded by Dr. Richardson and Lieutenant Kendall, were performing a voyage eastward from that river to the mouth of the Coppermine River. They doubled Cape Bathurst in latitude 70° 31′ N., Cape Parry in 70° 6′ N., and passing through the Dolphin and Union Strait between Wollaston Land and Cape Krusenstern, reached the Coppermine River; thus connecting Sir John Franklin's former discoveries to the eastward in Coronation Gulf with those made by him on this occasion to the westward of the Mackenzie; and with the exception of the unexplored 160 miles adjacent to Point Barrow, carrying, in conjunction with Captain Beechey, the northern outline of the American Continent from Bering's Straits eastward, through sixty degrees of longitude. Sir Edward Parry had previously penetrated on a higher parallel, from the entrance of Lancaster Sound, on the eightieth meridian, to Cape Dundas, on the one hundred and fourteenth, or through thirty-four degrees of longitude, the two surveys overlapping each other by six degrees, and requiring merely the discovery of a connecting channel running north and south to complete the long-sought-for North-west Passage. The eastern division of Franklin's party also circumnavigated and laid down Great Bear Lake, with the exception of one bay, afterwards surveyed by Mr. Thomas Simpson.

In the years 1829-33, Captain John Ross, being laudably desirous of obliterating the reproach of former failure by some worthy achievement, and having through the munificence of Sir Felix Booth, Baronet, been provided with funds for fitting out a vessel, named the Victory, of 150 tons, sailed in her with the intention of seeking a passage through Regent's Inlet. The Victory was set fast in the ice, and finally abandoned in Victoria Harbour, near the seventieth parallel of latitude, and on the opposite side of Regent's Inlet to the Strait of the Fury and Hecla that was discovered by Sir Edward

Parry in 1822-1823. This expedition of Captain Ross was remarkable for the number of winters spent within the Arctic circle, three of them in the Victory, and the fourth after abandoning her, on Fury Beach, where the provisions stored up by Sir Edward Parry were found serviceable. The party at length escaped in good health in 1833, in their boats, and fortunately reached a whaler in Lancaster Sound. During his stay in Regent's Inlet, Sir John Ross surveyed the country immediately adjoining his winter harbours, and gave to the lower part of the Inlet the appellation of the Gulf of Boothia.

But the chief discoveries were made by Lieutenant James Clark Ross (now Rear-Admiral Sir James), who, by several well executed extensive sledge journeys, traced a portion of the coast-line of King William Island, and of the west side of the Peninsula of Boothia up to the Magnetic Pole, and Cape Nikolai; he also surveyed Lords Mayor's Bay and its vicinity, in the Gulf of Boothia. In travelling over the ice with the boats to Fury Beach, the retreating party entered Brentford Bay, and crossed it more than once, yet they failed to perceive Bellot's Strait, the subsequent discovery of Mr. Kennedy. This is one instance out of many, that might be adduced of the difficulty of laying down a rocky coast-line correctly, without rigidly following the shore. On returning from his long absence, Captain John Ross received the honour of knighthood.

After two winters had passed without tidings of the Victory, public attention was called by several writers in the Transactions of the Geographical Society, to the necessity of sending out a party in search of the crew. Government was applied to, and at first entertained the propositions of the writer of this compilation, who, having been made acquainted in confidence by Mr. (afterwards Sir Felix) Booth, with the route Captain

Ross intended to pursue, drew up a scheme of sending relief down the Great Fish River, whose general course and approximate outlet were known to him by communications from the Indians and Eskimos. Afterwards, however, circumstances connected with the temporary resignation of the ministry caused the loss of a season, and threw the expense of the outfit of the searching party on a public subscription, and Captain (now Rear-Admiral Sir George) Back, was appointed to the command with the concurrence of Government, which eventually gave pecuniary aid also. Captain Back wintered at the east end of Great Slave Lake, and in 1834 descended the Great Fish River to its mouth. He also surveyed the coasts of its estuary as far as Cape Britannia on the one side, and Point Richardson on the other, leaving but a small space unexamined between his northern extreme and the tract of Lieutenant James Ross's southern sledge journey. Having been made acquainted by an express from England with Sir John Ross's safe return, Sir George Back had only geographical research to plead for extending his voyage further, and this the advanced and stormy season, and the ricketty condition of his boat, rendered inexpedient. But this able officer had by this voyage touched on a part of the Arctic Sea thirteen degrees of longitude to the eastward of Sir John Franklin's Point Turnagain, and as much nearer to the Gulf of Boothia, so that at this stage of the progress of discovery, only the intermediate part of the coast required to be explored, together with the 160 miles to the westward of the Mackenzie, between Return Reef and Point Barrow, nearly to complete the North-west Passage.

This the Hudson's Bay Company prepared to do, by an expedition under the direction of Peter Warren Dease, Esq., one of their chief factors, and Mr. Thomas Simpson, an accurate astronomical observer, on whom the survey devolved. They first completed the western part of the coast, chiefly by the personal exertions of Mr. Simpson, who, after the boats were arrested by ice, prolonged the voyage to Point Barrow in an Eskimo Baidar. This was in the year 1837.

In 1838-9, the survey of the eastern part of the coast between Franklin's Point Turnagain and the estuary of Sir George Back's Great Fish River was accomplished by the same zealous discoverers, and the south side of Wollaston Land, or, as it was renamed by Mr. Simpson, Victoria Land, was laid down, together with the south side of King William's Island, from Cape Herschel* to Point Booth. The passage between this island and Adelaide Peninsula of the main land, has been named Simpson's Strait, and another narrow channel between Kent Peninsula and Cape Colborne, has been called Dease Strait. The boat voyages by which all this was effected were the longest that had hitherto been made on the Arctic Sea, and embraced the extremes of the sixty-two degrees of longitude that intervene between Point Barrow and Simpson's eastern extreme, the Castor and Pollux River. The advanced season, however, prevented Messrs. Dease and Simpson from ascertaining whether a passage existed between the estuary of the Great Fish River and the Gulf of Boothia, or with the King William's Sea of Sir James Clark Ross.

In 1845-7, Dr. John Rae, also in the employment of the Hudson's Bay Company, was selected to complete the eastern part of the survey which Dease and Simpson had left unfinished owing to the advance of winter. Dr. Rae, by a most

---

* A cairn was erected by Dease and Simpson on Cape Herschel on the 25th of August 1829, and revisited by Captain M'Clintock in 1859.

hazardous enterprize, considering the means at his command, after wintering in Repulse Bay, and by his skill in the chace supplying his party with provisions for ten months, surveyed the bottom of the Gulf of Boothia, otherwise Regent's Inlet, up to the Fury and Hecla Strait of Parry, on the east, and on the west to Sir James Ross's Lords Mayor's Bay, thereby ascertaining that an isthmus of the width of four degrees of longitude, intervened between the bottom of Regent's Inlet and the eastern bay of the sea explored by Dease and Simpson.

CHAPTER IX.

Last Voyage of Sir John Franklin in the Erebus, accompanied by the Terror—Discovery of the ice-encumbered North-west Passage—Abandonment of the Ships—Death of all the Crew.

THE Admiralty expeditions, having been intermitted for eight years,* were resumed in 1845, and the Erebus and Terror were commissioned to make a new attempt at the North-west Passage, under the command of Captain Sir John Franklin, who had recently returned from Tasmania, of which colony he had been Lieutenant-Governor for five years. Captain Crozier, of the Terror, the second in command, had been second to Sir James Clark Ross in the arduous voyage towards the southern pole ; and Commander Fitzjames, Lieutenant Fairholme, and other officers of high reputation, were joined to the expedition. The two ships were made as strong as the skill of the shipwrights, improved by previous experience, could effect, and were filled with stores and provisions ; the latter, however, scarcely sufficing for the consumption of three entire years, at full allowance of the ordinary rations. A warming and ventilating apparatus of the most approved kind was

---

* Captain (now Rear-Admiral) Sir George Back, made a voyage in the Terror to the north end of Southampton Island, for the purpose of prosecuting discovery from Repulse Bay, but his ship being severely crushed by the ice, he was compelled to return to England without reaching the scene of his intended operations.

fitted, and abundance of fuel provided, both for that purpose and for working the engines belonging to the auxiliary screws.

The expedition sailed from England on the 19th of May 1845, and reached Whale-fish Islands, near Disco, early in July. There a transport which had accompanied them was cleared of its stores and sent back to England, bringing the last letters that have been received from the discovery ships. The letters of the officers were written in the most cheerful spirit, speaking of their happiness in the society of one another, and with respect and love, of the intelligence, activity, and kindness of the commanding officer. Sir John, in his last despatch to the Admiralty, which bore the date of the 12th of July, says, " The ships are now complete with supplies of every kind for three years ; they are therefore very deep, but happily we have no reason to expect much sea as we proceed further." Lieutenant Fairholme wrote, in a private letter, " Sir John is in much better health than when we left England, and really looks ten years younger. He takes an active part in everything that goes on, and his long experience in such services makes him a most valuable adviser. We are very much crowded ; in fact, not an inch of stowage has been lost, and the decks are still covered with casks. Our supply of coals has encroached seriously on the ship's stowage ; but as we consume both fuel and provisions as we go, the evil will be continually lessening."

The Expedition was seen again on the 26th of the same month, waiting for a favourable opportunity of crossing " the middle ice," on the way to Lancaster Sound, then distant 220 geographical miles. On that afternoon, a boat manned by seven officers, boarded the Prince of Wales, whaler, and invited the master, Captain Dannett, to dine with Sir John on the following day; but a favourable breeze springing up, the Expedition pursued its course, and the opportunity of sending letters was lost.

A year and a half subsequent to the last mentioned date, Sir John Ross addressed letters to the Admiralty, and to the Royal and Geographical Societies, in which he expressed his conviction that the discovery ships were frozen up at the *Western end of Melville Island*, from whence their return would be for ever prevented by the ice accumulating behind them. The Lords Commissioners of the Admiralty (judging from the duration of Sir Edward Parry's, and Sir John Ross's voyages) thought that the second winter of Sir John Franklin's absence was too early a period to give rise to well-founded apprehensions for his safety, but lost no time in calling for the opinions of naval officers who had been employed on Arctic expeditions. The officers who were consulted, reported that they did not apprehend that the expedition had foundered in Baffin's Bay, as some naval men of high rank, but not of arctic experience, had suggested; that it had not as yet passed Bering's Strait; and that until two winters without tidings had elapsed, serious fears for its safety need not be felt; but that immediate preparations for its relief ought to be made, to be carried out in the event of the summer closing without intelligence arriving. As to the direction of the relief parties, the most rational opinion was adopted, namely, that the discovery ships were to be sought for on the route that Sir John Franklin was instructed to pursue. These instructions were, that he was to proceed to about *latitude* $74\frac{1}{4}°$ *N., and longitude* 98° *W.,* in the vicinity of Cape Walker, and from thence to penetrate to *the southward and westward in a course as direct to Bering's Strait* as the position of the ice and existence of land at present unknown, may admit. He was, moreover, expressly forbidden to attempt the passage westward from Melville Island, the experience of Sir Edward Parry having been considered conclusive as to the inexpedience of trying again on that parallel; and to none of the officers with whom Sir John Franklin had conversed on the subject of his voyage before sailing, had he intimated an intention of taking that direction, so that Sir John Ross's opinion of the ships being fast there, must have been founded on misapprehension.

No time was lost by the Admiralty after receiving the reports from the officers who were consulted. With a view to ulterior measures, 17,424 pounds of pemican* were prepared and secured in water-tight canisters, four light boats were constructed and sent, together with competent crews, to Hudson's Bay in the Company's ships, which sailed in June 1847. Two ships were also ordered to be immediately strengthened to proceed in 1848 on Franklin's route, and others were sent at the close of 1847 to meet him with supplies at Bering's Strait, should he succeed in getting thither.

These were the beginnings of a series of *Searching Expeditions* fitted out by the Admiralty on a scale of magnitude and expense exceeding all former example, and persevered in year after year until tidings were obtained. The earnest appeals of Lady Franklin roused the sympathies not only of our own country and its colonies, but of the whole civilized world; and devoting her fortune to the search, she sent out

---

* Amounting to 9000 rations on full allowance, or to 75 days' provision for the crews of 120 men, which was the complement of the ships on entering Lancaster Strait. This was thought to be more than sufficient to take the crews to a fishing station on the Mackenzie, if the boats had found them.

ship after ship, furnished at her own cost, and manned by humane and zealous volunteers. America also sent forth able and active officers on the mission of humanity, the munificence of one of her merchants providing the funds. France, likewise, added a Bellot and a De Bray.

The object of this summary is to trace the progress of arctic discovery, passing over, cursorily or silently, the expeditions which added nothing new to geography. The searching expeditions explored a large extent of coast, which might have remained for ever concealed, and made their discoveries known before the fate of Franklin and his companions was revealed, but it agrees better with our plan to follow in chronological order the route of the Erebus and Terror, as far as it has become known, before mentioning in detail the proceedings of the searching parties.

The principal source of our information respecting the doings and fate of the perished expedition is a brief record found on King William's Island in 1859, of which a transcript is presented on the opposite page. From it we learn that the Erebus and Terror must have had a quick passage from the spot where they parted from Captain Dannett to Lancaster Sound, for Sir John Franklin had time, before the close of the navigation, to sail up Wellington Channel to the seventy-seventh parallel, to descend by the west side of Cornwallis Island,* and to return to Beechey Island at the mouth

---

* The survey of Mr. Goodsir for a time led to the belief of Cornwallis Island being joined to Bathurst Island; but the more accurate explorations of Captain M'Clintock and Lieutenant M'Dougall, shewed that a passage existed between them, down which the Erebus and Terror must have come—see afterwards—and Journal of the Voyage of the Lady Franklin and Sophia, in search, etc., under the command of Mr. William Penny, by Dr. P. C. Sutherland, 1852, ii. p. 145.

M. S.hips Erebus and Terror
Wintered in the Ice
184) Lat. 70° 5' N. Long. 9
ed in 1846—7 at Beechey
° 43' 28" N. Long 91° 39' 15" W
llington Channel to Lat 77
side of Cornwallis Island
Franklin commanding th

finds this paper is requested to forward it to the S
, London, *with a note of the time and place at w*
aore convenient, to deliver it for that purpose to
nearest Port.

E trouvera ce papier est prié d'y marquer le tem
, et de le faire parvenir au plutot au Secretaire de
Londres.

A que hallare este Papel, se le suplica de enviarlo a
zgo, en Londrés, con una nota del tiempo y de

lie dit Papier mogt vinden, wordt hiermede verz
edigste, te willen zenden aan den Heer Mini
ederlanden in 's Gravenhage, of wel aan den Se
iraliteit, te London, en daar by te voegen
tyd en de plaats alwaar dit Papier is gevonden

af dette Papiir ombedes, naar Leilighed give
iralitets Secretairen i London, eller nærmeste Er
orge, eller Sverrig. Tiden og Stædit hvor dett
beligt paategnet.

n Zettel findet, wird hier durch ersucht dens

of Wellington Channel, where he wintered in 1845-6. It is presumed that he made this divergence northward from his direct course to Bering's Strait in virtue of a clause in his instructions authorizing him, in the event of his finding the southern route obstructed by ice, to attempt Wellington Channel should it be open. Before he sailed northwards he had doubtless discovered the harbour in Erebus and Terror Bay, behind Beechey Island, to which he returned on the approach of winter. We are also entitled to assume that he had convinced himself of the hopelessness of penetrating to the westward on the high parallel of 77°, for he might have found a harbour to the north of the islands had he designed to try for a passage there next summer.

Except the two circumstances of the passing of the discovery ships up Wellington Channel to latitude 77°, and their return by the west side of Cornwallis Island, we have no hint of that summer's proceedings. The winter of 1845-6 was spent in Erebus and Terror Bay, but the date of the entry of the ships into that harbour is unknown. On Beechey Island many traces of a winter residence were found, the sites of a large store-house and workshop, of observatories, and of the blacksmith's forge, a great many coal-bags, scraps of cording and clothing, coal-dust, cinders, and many preserved meat tins, regularly piled, and filled with gravel. Several mounds, two feet high, were raised of these canisters, varying in breadth from three to four yards; and Dr. Sutherland states[*] that six or seven hundred were counted, besides many more that were dug up and emptied in search of documents. These cases were labelled "Goldner's Patent." So large a quantity of preserved meat as they were calculated to hold could not have been

[*] Lib. cit. i, p. 306.

needed during the first winter from England; but it is known that a vast quantity of preserved meat, supplied by Goldner to the Royal Navy, being found putrid, was condemned by survey at Portsmouth, and thrown into the sea. It is, therefore, most probable that the defective condition of this portion of the provisions having been discovered, a survey was ordered, according to the custom of the navy, and that the bad cases, readily known by the convex form which their ends take when putrefaction is going on within, were opened and piled in the order in which they were found, so as to be the more easily counted by the surveying officers. The loss of so large a proportion of the supply of provisions, was doubtless a main cause of the disastrous fate of the expedition two winters afterwards.

Traces of excursion or hunting parties were discovered at Cape Spencer and other places within easy distances of the winter harbour. A cairn was likewise discovered on the southwest cape of Beechey Island, but, though it was twice taken down, and its site carefully dug over, and the whole island repeatedly searched, no papers relating to the ships were found, except a fragment of a note, and some leaves of a book. It cannot for a moment be thought that the Erebus and Terror left their winter harbour before a careful record of the year's proceedings had been prepared and deposited by the commanding officer; but as no recent traces of Eskimos existed, the record, in its tin case, was most probably placed where it would be most readily found, exposed on the top of the cairn. The voyagers did not know that the polar bear is in the habit of carrying off and knawing such unusual objects, a fact subsequently learnt by the searching parties.*

* In the course of Barentzoon's memorable winter on Novaya Zemlya, the

Three men of the expedition died in the first winter, two of them belonging to the Erebus and one to the Terror. On head-boards erected at their graves, their names, ages, and day of death were recorded, but not the cause of death.

Recurring to the record found in King William's Island, we find as follows, "Lieutenant Graham Gore and Mr. Charles F. des Vœux, mate, left the ships on Monday the 24th May 1847, with six men (to deposit papers on King William's Island);"—"the Erebus and Terror wintered in the ice, in latitude 70° 5', longitude 98° 23' W., having been beset since the 12th of September 1846. After previously wintering at Beechey Island, ascending Wellington Channel to latitude 77°, and returning by the west side of Cornwallis Island, (Sir) John Franklin commanding the expedition,—All well!!" *

The second winter, though spent in the pack, and conse-

---

polar bears actually dragged the empty cook-safe out of the ship and carried it to the shore, and these animals are now well known to take pleasure in tearing to pieces any canvas, cloth, or other prominent and uncommon object that they find: nor do they hesitate to swallow a tin-box.

* From the context of the record, we learn that Lieutenant Gore landed with the intention of depositing a record at the cairn erected by Sir James Clark Ross in the year 1831, but that not having found the cairn, he placed it in a spot from whence it was removed in the following year to the geographical position of the cairn, by orders of Captain Crozier, and the additional note was then written by Captain Fitzjames. Lieutenant Gore deposited a second record at a place further along the coast, where it was found by Lieutenant Hobson in 1857, not having been moved by Captain Crozier. This second record was a transcript of the first, with the exception of course of Captain Fitzjames's addition. Both contain a clerical error in the wintering date being 1846-7 instead of 1845-6. The discrepancies in the other dates are easily reconcilable. It is probable that Sir John Franklin signed the written notice on board; that Lieut. Gore left the ship on the 24th of May, deposited the record on the 28th, having spent the intermediate time in searching for Ross's cairn, and did not return till June. The "Sir" prefixed to Franklin's signature is not of his writing, but is perhaps Lieut. Gore's, as are evidently the dates of 24th and 28th May.

quently in as dreary circumstances as can well be imagined, had not crushed the hopes nor damped the zeal of the gallant band. They doubtless hoped that the opening of the navigation in July would free them from the prison in which they had been shut up for eight or nine months, and that twelve hours' easy sail would then carry them into the navigable channel known to exist on the continental coast, and thus crown their voyage with complete success. No unusual mortality had as yet taken place, as the "All's well" shews, and they had great reason to congratulate themselves on what they had accomplished in their first summer, and on the discovery of the channel leading to the south, in the mouth of which they were arrested, but as they had reason to hope, only for a season.

The wintering position of the Erebus and Terror in 1846-7-8, was off the north end of King William's Island, five or six leagues from the shore, and a question may be raised respecting the route of the ships in going thither from Beechey Island, whether they descended the strait on the east side of Prince of Wales' Island, now known by the names of Peel Sound, and Franklin's Strait, or by the wider passage on the west side leading from Melville Sound, called in the most recent charts M'Clintock Channel. Down the western one the ships might naturally have been led by Sir John Franklin's endeavour to comply with his instructions to steer south-west as far as the trending of the land would allow; but Sir Leopold M'Clintock and Captain Allen Young, whose local experience give weight to their opinions, believe that Peel Sound was the passage traversed by the discovery ships, although its numerous islands render it less promising when viewed from a distance.

Referring again to the Point Victory record for information respecting other incidents in this eventful history, we find that a most mournful addition was made to it on the 25th of April 1848, after the lapse of another winter in the ice, or the third after leaving England, denoting that the Erebus and Terror were deserted five leagues north-north-west of Point Victory (having drifted only twelve or fourteen miles southwards since they were first beset). The total loss by deaths in the expedition up to that date, is stated to have been nine officers and fifteen men.* Sir John Franklin died on the 11th of June 1847; a period, it must be remarked, at which the hope of extrication from the ice in the course of the following month must have been as strong as ever, and the consequent anticipation of ultimate success most cheering. This most excellent officer and humble and sincere christian, was mercifully spared the distress of mind which he must have endured, had he survived to see the summer waste away with the ships still fast in the packed ice, and to mourn the fate of officers and men perishing around him. Two months before his death he had completed the sixty-first year of an active, eventful, and honourable life.

A preceding page contains Sir John Franklin's statement that on the 12th of July 1845 the ships had on board a supply of provisions calculated to last three years: but making no allowance for defect in the preserved meats. The number of deaths indicates that officers and men had gone on short allowance, an expedient which, however needful, cannot be resorted to in Arctic climates without inducing scurvy; and

---

* One hundred and thirty-four individuals left England in the Erebus and Terror, of whom five were sent home from Greenland, leaving one hundred and twenty-nine on board, from which deducting the twenty-four who died, we have the remainder of one hundred and five mentioned in the record.

the low and barren shore of King William's Island would yield little game, nor indeed did Captain M'Clintock discover traces of hunting parties having been sent on shore there. Even on the ordinary full allowance of the navy, scurvy has almost invariably assailed the crews of ships after a second winter within the Arctic circle, the expeditions that have escaped that scourge having had either a large supply of preserved meats and pemican to resort to, and plenty of dried vegetables and vegetable acids, or been successful in adding deer, musk-oxen, and bears, to their stock of provisions.

One hundred and five souls landed from the ships on the 25th of April 1848, Captain Crozier, who had succeeded to the command of the expedition, Captain Fitzjames, and Lieutenant Irving being among the survivors, while Acting Commander Gore was numbered with the dead. The intention of the commanding-officer was to lead his party over the ice to Back's Great Fish River, in the hope of killing deer there or procuring fish for their support, as the Eskimos, who met the weary and wasted travellers later on the march, were told. Boats placed on runners and sledges had been prepared. The distance to the mouth of the Great Fish River from the spot where the ships were abandoned is about two hundred and fifty miles, and even in May much cold and stormy weather was to be encountered in travelling over the ice. Such a sledge-journey would scarcely have been attempted by enfeebled crews, had not the failure of their provisions rendered it a matter of absolute necessity, since by waiting for two months they knew well that lanes of water would open near the shore, by which much of the distance to the navigable channel of Dease and Simpson might be traversed in boats, carrying all their supplies with comparatively small fatigue.

That their provisions were actually exhausted before they got beyond King William's Island, was made known to Dr. Rae by a party of Eskimos, who sold them some seal's flesh; and a band of the same people told Captain M'Clintock that the Englishmen dropt from the drag ropes on the march, and died where they fell—their track being marked by a line of dead bodies discovered in the following season.

It is characteristic of scurvy, that its victims are not aware of their weakness, and the near approach of death, until on some sudden exposure or unusual exertion, they expire without warning. Captain M'Clintock, who traced the route of the doomed men backwards from the estuary of the Great Fish River, found only three skeletons of those who had perished by the way; but he remarks that the line of march being over ice, near the shore, many of the bodies would be washed away on the channel opening. He discovered one skeleton near Cape Herschel, which is the southwest corner of King William's Island, and two in a boat, that had been left about half-way between that place and Point Victory. One of these sad memorials was lying in the hinder part of the boat, carefully covered with a load of clothing, evidently the remains of one who had been unable to proceed further. The other skeleton occupied the bow of the boat, and two loaded muskets were standing upright beside it.*

The Eskimos said, that about forty Englishmen reached the vicinity of the Great Fish River, where they all died. That none survived, was the unanimous report of the several bands of Eskimos that were met in succession by Dr. Rae, Mr. Anderson, and Captain M'Clintock, but it was added, that

---

* Lieutenant Hobson, travelling from the north, **came to this boat some days sooner than Captain M'Clintock.**

the spring birds had arrived before the last of the men perished. Brent, Wavies, and Eskimo geese, reach the Arctic Sea about the end of May, and the Great Northern Divers, which make their appearance as soon as rapid rivers partially open, are about a fortnight earlier. The party must still have numbered a good many men to have been able to drag a boat as far as Montreal Island,* where the Eskimos found it and broke it up; but strength had failed them before they could accomplish the remaining forty miles to the mouth of the river, where, with nets, they might have caught salmon-trout.

From the entrance of Back's Great Fish River to the nearest fur-post, Fort Resolution, on Great Slave Lake, the marching distance, even to the Red Indians who know how to avoid the numerous and extensive intervening lakes, exceeds a thousand miles, most of the way being through "barren grounds." We cannot doubt that this fact was well known to Captain Crozier, and duly considered by him and the other officers. It is therefore probable, that provisions having totally failed in the ships, the object in going so early in the spring to Back's River, was to kill a sufficiency of fish, birds, and deer, to enable the party to pursue the voyage to the Coppermine or Mackenzie River, in their boats, during the summer.

Dr. Rae says, that the Eskimos of Great Fish River are named by themselves *Utku-hikalik*, and they were called

* Mr. Anderson, a chief factor of the Hudson's Bay Company, went to the place on Montreal Island where the Eskimos said the boat had been abandoned, and there found the shavings and fragments left by that people in converting the wood to their own uses. The half-caste natives who accompanied Mr. Anderson, being skilful in drawing conclusions from signs and objects, could not be mistaken when they asserted that the boat had been broken up there.

more at length by Augustus, the excellent interpreter who accompanied Sir John Franklin on his first and second overland expeditions, *Utku-sik kaling mœut* (stone-kettle people).* They are an inoffensive tribe, and no reason whatever exists for supposing that they would offer violence to the remnant of white men, however feeble and helpless. On the other hand, active philanthropy is not an attribute of the Eskimos, and little or no effort would be made by them to prolong the lives of strangers perishing on their lands.

By the wearying and fatal journey down the western and southern sides of King William's Island, the party who reached Montreal Island to die there, or in its neighbourhood, connected Lancaster Strait with the navigable channel that extends along the continent to Bering's Strait, thereby proving the existence of the long-sought-for North-west Passage. That Victoria Strait, which they traversed on the ice, is rarely, if ever, navigable for ships, is probable ; Banks's Strait and the entrance of Prince of Wales Strait into Melville Sound are always, as far as is known, in the same condition, which Professor Haughton attributes to the meeting of the Atlantic and Pacific tides producing perennial accumulations of ice in these several localities.

The conduct of the brave, resolute, and persevering but unfortunate men, who perished in accomplishing this discovery, which for so many centuries had been pursued by their country, seems to have been most excellent throughout. Sir Leopold M'Clintock testifies that every trace of their proceedings which he could perceive, gave evidence of all their movements being in perfect order, and under the guidance of their officers.

* Narrative of a Journey to the Polar Sea, p. 264.

CHAPTER X.

SEARCHING EXPEDITIONS—A.D. 1847-1859.

List of searching parties—Sir John Richardson's boat voyage—Dr. Rae's survey of the south and east sides of Wollaston Island, proving Victoria Land to be a part of it—Sir James Clark Ross' survey of the west coast of North Somerset, or Peel's Strait—Mr. Saunders' survey of Wolstenholme Sound—Captain Henry Kellet's discovery of the Herald Islands—Lieutenant W. S. J. Pullen's boat voyage from Bering's Strait to the Mackenzie—Captain Collinson and Commander M'Clure sail from England—Captain Horatio Austin's squadron sail—Captain Penny surveys Wellington Channel—Lieutenant De Haven, U. S. Navy, enters Wellington Channel; his ships are caught in the pack, and drift with it during the winter—Captain Austin's sledge-parties survey the north and south-east shores of Melville Sound, with M'Dougall and Byam Martin Channels—Commander Inglefield enters Smith's, and Whale Sounds—Mr. Kennedy and Lieutenant Bellot discover Bellot's Strait, and cross Prince of Wales' Island—Sir Edward Belcher's squadron finish the survey of Wellington Channel, and of the Parry Islands—Lieutenant Pym sent by Captain Kellet, opens a communication with Captain M'Clure—Captain M'Clure discovers Prince of Wales Strait, and sails round the west side of Banks's Land; abandoning his ship, he travels over the ice with his crew, and thus makes a passage from Bering's Strait to Baffin's Bay—Captain Collinson makes the voyage from Bering's Strait to Cambridge Bay, and returns by the same way. Captains Belcher and Kellet return to England, bringing Captain M'Clure and the crew of the Investigator—Dr. Rae obtains proofs of the fate of the crews of the Erebus and Terror—Mr. Anderson visits the estuary of the Great Fish River for further intelligence—Captain M'Clintock, in the Fox, discovers a record paper, and authentic evidence of the movements of the Erebus and Terror, their abandonment, and death of the whole of the crews—Summary of coast line searched—Dr. Kane's investigation of Smith's Sound.

Though the successive tracings of coast by the searching expeditions were known in England before the previous discoveries made by Sir John Franklin, of whom they were in quest, were ascertained, yet the chronological order of Arctic research, has been adhered to in giving the first place to his proceedings. We cannot attempt even a brief account of all the searching parties, and therefore, purpose merely to mention the extent of coast traced for the first time in successive years; but that the reader may have a general idea of the noble efforts made in the cause of humanity, we commence this chapter with a list of the various searching expeditions.

- 1847-1850, Sir John Richardson, C.B., and Dr. Rae, overland, and along the coast in boats, from the Mackenzie to the Coppermine.
- 1848-1852, Captain Thomas Moore, of H.M.S. Plover, to Bering's Straits.
- 1848-1850, Captain Henry Kellet, of H.M.S. Herald, to Bering's Straits.
- 1848-1850, Robert Shedden, Esq., in the private yacht, Nancy Dawson, to Bering's Straits.
- 1848-1849, Captain Sir James Clark Ross, of H.M.S. Enterprise, to Lancaster Strait.
- 1848-1849, Captain E. J. Bird, of H.M.S. Investigator, to accompany the Enterprise.
- 1849-1850, James Saunders, Esq., Master of H.M.S. North Star, to Wolstenholme Sound and Pond's Bay.
- 1849, Dr. Robert Anstruther Goodsir, in the Advice, whaler, to Baffin's Bay.
- 1849, Lieutenant (now Captain) W. J. S. Pullen, of H.M.S. Herald, boat voyage from Bering's Straits to the Mackenzie.
- 1850-1851, Lieutenant De Haven of the United States Navy, in the Advance, fitted at the expense of Henry Grinnell, Esq., of New York, to Lancaster Strait and Wellington Channel.
- 1850-1851, S. P. Griffin, Esq., United States Navy, in the Rescue, at the expense of Mr. Grinnell, to Lancaster Strait and Wellington Channel.
- 1850-1851, Captain Horatio Austin, of H.M.S. Resolute, to Lancaster Strait and Cornwallis Island.

1850-1851, Captain Ommaney, H.M.S. Assistance, to accompany Captain Austin.

1850-1851, William Penny, Esq., Master of the Lady Franklin, under Admiralty orders to Lancaster Strait and Wellington Channel.

1850-1851, Alexander Stewart, Esq., Master of the Sophia, under Admiralty orders to Lancaster Strait and Wellington Channel.

1850-1851, Rear-Admiral Sir John Ross, in the Felix yacht, fitted at the expense of the Hudson's Bay Company, to Lancaster Sound.

1850, Captain C. C. Forsyth, R.N., commanding the Prince Albert, belonging to Lady Franklin, to Regent's Inlet and Beechey Island.

1850-1854, Commander (now Captain Sir) Robert M'Clure, of H.M.S. Investigator, to Bering's Straits, Banks's Island, and Lancaster Straits. The crew abandoned the ship, and by walking over the ice to Beechey Island, made the northern Northwest Passage.

1850-1855, Captain Richard Collinson, C.B., of H.M.S. Enterprise, to Bering's Straits, Banks's Island, and along the Southern or Continental Channel to Cambridge Bay, in Wollaston (or Victoria) Island, near King William's Island; whence he retraced his course to England.

1851, Dr. John Rae, employed by the Admiralty, descended the Coppermine, and traced Wollaston Land, from its eastern extremity to its junction with Victoria Land, and up to the parallel of the north end of King William Island, in Victoria Straits.

1851-1852, William Kennedy, Esq., Master of the Prince Albert, belonging to Lady Franklin, to Prince Regent Inlet, Bellot's Strait, and Prince of Wales Island.

1852, Captain Charles Frederick, of H.M.S. Amphitrite, to Bering's Straits.

1852, Captain Edward A. Inglefield in the Isabel, Lady Franklin's vessel, to Lancaster Sound.

1852-1855, Captain Rochfort Maguire, H.M.S. Plover, Bering's Straits.

1852, Dr. R. M'Cormick, a boat excursion in Wellington Channel.

1852-1854, Captain (Rear Admiral) Sir Edward Belcher, C.B., of H.M.S. Assistance, to Wellington Channel.

1852-1854, Captain Henry Kellett, C.B., of H.M.S. Resolute, Lancaster Strait, Melville and Banks's Islands.

1852-1854, Lieutenant (now Captain) Sherard Osborn, of H.M.S. Pioneer, to Wellington Channel.

1852-1854, Captain Francis Leopold M'Clintock, of H.M.S. Intrepid, to Lancaster Strait and Prince Arthur Island.

1852-1854, Captain William Samuel John Pullen, of H.M.S. North Star, Beechey Island.

1853, William H. Fawckner, Esq., Master R.N. Breadalbane Transport, Beechey Island; crushed in the ice and foundered.

1853, Lieutenant Elliott, of the store ship Diligence.

1853, Captain Edward A. Inglefield, of H.M.S. Phœnix, to Beechey Island.

1853, Dr. John Rae, of the Hudson's Bay Company, acting under Admiralty orders, by sledge, to Wollaston Land, and boat voyage to Victoria Strait, between that Island and King William's Land.

1854, Captain Edward A. Inglefield, of H.M.S. Phœnix, to Beechey Island.

1854, Commander Jenkins, of the Talbot, to Beechey Island.

1853-1854, Dr. John Rae, chief factor of the Hudson's Bay Company, boat expedition at the expense of the Company to Repulse Bay, and east side of King William's Island, bringing the first intelligence of the loss of the Erebus and Terror, and all their crews.

1853-1855, Dr. Elisha Kent Kane, of the United States Navy, to Smith's Sound, Humboldt Glacier, and Grinnell Land.

1855, Chief factor John Anderson of the Hudson's Bay Company, canoe voyage down the Great Fish River, to Montreal Island, and Point Ogle, procuring further relics of the Erebus and Terror.

1857-1859, Captain Francis Leopold M'Clintock, R.N., in the Fox, Lady Franklin's yacht, to Peel Sound, Regent's Inlet, Bellot Strait, King William's Island, and Montreal Island, bringing precise intelligence of the fate of the Erebus and Terror, and a short record of their proceedings.

This long list of ships and overland parties sent out in search of the lost expedition shews, without possibility of misconstruction, that the nation was bent on recovering her seamen, regardless of expense; and the reader will perceive that no less than five times did Lady Franklin dispatch a well-fitted vessel to explore quarters that she thought would otherwise be neglected.* To the influence, also, of her ardent and moving appeals may be fairly attributed Mr. Grinnell's humane engagement in the search, and the volunteering of such men as Kane, Bellot, De Haven, De Bray, Shedden, Griffin, Forsyth, Kennedy, Inglefield, M'Clintock, Young, Hobson, and others, to give gratuitous aid to the cause.

Though the labours of several of the parties were directed to quarters whither the discovery ships were not likely to have gone, they were productive of additions to our geographical knowledge which would otherwise have remained perhaps for ever concealed, and the magnitude of the general movement may serve to cheer shipwrecked men in future times, as shewing the efforts that will be made to carry them relief as long as hope remains.

Now that the course actually pursued by the Erebus and Terror is known, we see that the first scheme of relief organised by the Admiralty was devised with correct judgment. It was founded on the belief that Sir John Franklin would follow his instructions as closely as circumstances permitted, and having reached the meridian of Cape Walker ($97\frac{1}{4}°$) on the parallel of $74\frac{1}{4}°$, would then seek a channel to the southward and westward, leading into the open passage known to bound the northern shores of the continent. In attempting to execute this project, his ships might be arrested by ice in some channel south of Melville Sound, or having passed into the continental channel, might have been wrecked there; or, thirdly, after two winters, might succeed in reaching Bering's Straits or their vicinity, but be in distress for provisions and other aid. The overland searching party, under Sir John

* Including one arrested in the Pacific on its way to Bering's Strait by the desertion of the crew.

Richardson, was to descend the Mackenzie, and examine the coast between that river and the Coppermine River, and also the south coast of Wollaston's Land, to meet the second contingency. The surveys of Dease and Simpson, conjoined with the more distant view of Wollaston's Land by Sir John Richardson and Lieutenant Kendall, had rendered it almost certain that channels leading southwards from Lancaster Strait and Melville Sound could make their exits only in three localities, viz., at the west end of Wollaston's Land, between that land and the Victoria Land of Simpson, if these were separated, or to the eastward of the latter. Sir John Richardson's party might therefore fully calculate on getting tidings from the natives, on the coast between the Mackenzie and Coppermine, if the discovery ships had passed that way, or of discovering the channels in the two westernmost localities in which they might be shut up. That this searching party might not be without the means of affording relief should it fall in with the ships, or with crews retreating from them, it carried out from England 17,400 pounds of pemican, in hermetically sealed tin canisters, being upwards of sixty days' provision for the entire crews of the Erebus and Terror, at the full allowance of two pounds a man. This is mentioned here because a naval officer,* now no more, in a pamphlet on the Arctic searching expeditions, characterizes this one as especially useless, and unable to lend effectual aid, had it found the ships; but though the whole of the pemican was not carried to the sea, three boat loads were—quite sufficient to have fed the ships' companies until they could be conducted to productive fishing stations in Great Bear or Slave Lakes, where they could winter in safety. As the length of the

* Rear-Admiral Sir John Ross.

interior navigation from Montreal to the Arctic Sea is, in round numbers, 4400 miles, which could not be performed in loaded boats, in one season, without exhausting the provisions, the men, boats, pemican, and other stores were sent out in June 1847 by the annual Hudson's Bay ships, to advance as far as the season permitted, and the two officers followed in light canoes in the next spring to overtake them, which they did before they reached the Mackenzie.* Owing to unavoid-

* This searching party reached Great Slave Lake on the 17th of July 1848, and had it been actually known that Captain Crozier had left the ships for the mouth of the Great Fish River, and Sir John Richardson had gone in that direction to meet him, the transport of the boats and stores over the height of land at the east end of Great Slave Lake, and their navigation down the Great Fish River, would have occupied a month, or more probably six weeks, and could not, therefore, have reached Montreal Island till some months after the last of the discovery party had perished.

Matters would not have been mended by despatching a party in 1847, *via* Canada, to travel in a light canoe, starting from Lake Superior as soon as the navigation opened. On its arrival at Great Slave Lake, without a year's previous notice to the Hudson's Bay Company, its stores would be exhausted, and the voyageurs, instead of being in a condition to descend the river, would be under the necessity of establishing fisheries for their subsistence. Much skill in hunting and fishing would be required to provide for the subsistence of the party during the winter, and supposing that an adequate supply of pemican could have been got from the Hudson's Bay Company before the spring of 1848, and that boats to carry the party to be relieved were built in the winter, and deposited at the sources of the Great Fish River, that river is not navigable throughout till late in July. On the 20th of that month Sir George Back was arrested in his voyage down the river by firm ice covering Lake Pelly, and on the 22d he witnessed the disruption of the ice on Lake Macdougall. Thus, even under the most favourable circumstances, and supposing that boat-building materials, provisions, and Indian hunters could have been assembled at the source of the Great Fish River, the means of transporting an enfeebled party up the river, and through Slave Lake to the nearest Hudson's Bay post, being a distance of 1000 miles, could not have reached the sea-coast till after the date assigned to the death of the whole of the discovery party. This statement is made in answer to the expressed regrets of some writers that a searching party had not been sent in that direction. To have given such a party a fair chance of success, arrangements should have been made in 1846 with the Hudson's Bay

able delays, which could not be previously calculated upon, the mouth of the river was not attained till about a week later than was expected, and the short time for boat navigation in the Arctic Sea was abridged to that extent, or by one-sixth. The coast was, however, closely examined between the above-named rivers, and communication opened with the various bands of Eskimos, without observing traces or obtaining tidings of the expedition. The unusually early setting in of winter caused the boats to be abandoned with their stock of pemican, ammunition, etc., at a considerable distance seaward of the mouth of the Coppermine, and the remainder of the journey to the winter quarters to be made on foot. This necessarily restricted the outfit of the following year (1849) to a single boat, in which Dr. Rae descended the Coppermine River, but was frustrated in his attempts to cross over to Wollaston Land because Coronation Gulf was for the whole summer filled with impracticable ice.

In 1851, however, Dr. Rae, acting under direct instructions from the Admiralty, succeeded in tracing the south side of Wollaston Land, from Prince Albert Sound, at its eastern extremity, situated to the westward of Dolphin and Union Straits, to its junction with Victoria Land of Simpson, which name, therefore, becomes merged in the prior one of Wollaston. Dr. Rae, then passing through Dease Strait, examined the eastern extremity of the island formed by the conjoined lands, and ascended the western coast of Victoria Strait to Pelly Point, situated to the north of the parallel of Cape

Company to have men and boats, with a sufficiency of provisions, on the Great Fish River, so that advanced depôts of boats and stores could be made on the river in 1847, thus enabling a detachment to travel to sea on the ice early in 1848. All this would appear preposterous without express knowledge of the course intended to be pursued by the crews of the ships when beset.

Felix, the northern extremity of King William's Island, and in a higher latitude than the places where the discovery ships were arrested and abandoned. The exact distance from the capes on the west side of Victoria Strait to either of the stations of the discovery ships did not exceed thirty miles, yet Dr. Rae, on that voyage, found only a single fragment of wreck that could be supposed to have belonged to them. This boat voyage gave us a knowledge of considerable tracts of island coast-line, and limited the possible outlets from Melville Sound to the east end of Wollaston Island, on the 101st meridian, and the west end, beyond the 118th or 119th.

To Sir James Clark Ross, in the Enterprise, accompanied by Captain Bird, in the Investigator, was committed the duty of following up the track of the discovery ships, should that be ascertained, and of searching for traces of it in Wellington Channel, along the northern shore of North Somerset, from Leopold Island or Cape Clarence, onwards to Cape Walker, and having placed the ship in a convenient harbour, of exploring, by boat or sledge parties, the west side of North Somerset and Boothia; also, if the state of the ice permitted, of sending a steam-launch to Melville Island, from whence parties might be carried to Banks's and Wollaston Lands. No better plan, as we now know, could have been devised for tracing the lost ships, and the Enterprise was well placed in the harbour at Leopold Island, but too late in the season (September 11) to use the steam-barge with effect, the ice having shut up the harbour the day after the ship entered it. The officer sent across Lancaster Sound on the ice in spring, reached Cape Hurd, but was prevented, by the hummocky condition of the ice, from going to Cape Riley or Beechey Island, where he would have found traces. An examination of the east coast of North

Somerset, down to Fury Beach, taken in conjunction with Dr. Rae's surveys of the bottom of the Gulf of Boothia in 1847, rendered it morally certain that the discovery ships had not passed down Regent's Inlet. Sir James Ross himself travelled with a sledge along the north side of North Somerset, and down its western coast to latitude 72° 38', thus adding to our maps the passage since known by the name of Peel Sound. From thence he was obliged to retrace his steps to Port Leopold, owing to his provisions being exhausted. On the breaking up of the ice, Sir James left his harbour, with the intention of going on to Cape Walker, and making further search, but his ships were suddenly enclosed in the pack, and, with it, drifted out of Lancaster Sound, nor were they released till the 25th of September, before which date all navigation within the strait has been found to close. He therefore bore up for England. The failure of the earliest searching parties did not, however, occasion loss of life in the discovery ships, whose crews had perished to a man before the earliest of the searching parties could have reached the scene of the disaster, even had they been able to have gone straight to the spot.

In the same summer in which Sir James Ross was drifted into Baffin's Bay, the North Star, J. Saunders, Esq., master, sent after him with supplies (being unable to cross that bay on account of the ice), wintered in Wolstenholme Sound, on the west coast of Greenland, and ascertained that it was merely an inlet.

In the summer of 1849, also, Captain Kellett discovered a group of high islands within Bering's Strait, on the Asiatic coast, in latitude 71° 20' N., and longitude 175° 16' W. The Herald's Islands, as they were called, are on the bearing of lands occasionally seen, in certain states of the atmosphere, from Cape Jakan in Siberia, and thought by the hunters of that coast to be inhabited.*

Before Captain Kellett made this discovery, he had dispatched Lieutenant Pullen and Mr. Hooper, with two whale boats, to search the coast between Point Barrow and the Mackenzie. They were convoyed beyond the point by the Nancy Dawson, Mr. Shedden's yacht. The voyage to the Mackenzie was successfully made, whereby, in conjunction with Sir John Richardson's and Dr. Rae's boat-voyages above mentioned, the whole continental coast-line between Bering's and Victoria Strait, was examined without any trace of the Erebus and Terror being found. After wintering on the Mackenzie, Lieutenant Pullen tried, in 1850, to reach Bank's Land, but got no further than Cape Bathurst.

On the return to England of Sir James Ross and Sir John Richardson in 1849, the hope of the safety of the expedition which, though mingled with fears, had been previously cherished, gave place to the certainty of some serious misfortune having occurred, and the Admiralty determined to renew the search on a more extensive scale, regardless of expense. The Enterprise and Investigator having been refitted with celerity, were sent round Cape Horn under command of Captains Collinson and M'Clure, to make the passage eastward from Bering's Strait; and in time for the opening of the navigation in Baffin's Bay, two stout ships, with two steam tenders, were dispatched under command of Captain Horatio Austin of the Resolute; Captain Ommaney of the Assistance, was second in command, and the steam tenders were officered by Lieu-

* The land seen by Serjeant Andreef or Andreyer, in 1762, is farther to the west, having been discerned from the northernmost *Mevidji*, or Bear Island.

tenant Sherard Osborn, and Lieutenant F. L. M'Clintock. To this squadron, but with separate orders, the Admiralty added the Lady Franklin and Sophia, commanded by Captains Penny and Stewart, experienced masters of whalers. The association of officers of the royal navy with masters of the merchant marine, having independent commands, in pursuit of the same object, in the same place, was ill-advised, and sure to lead to misunderstandings and bickerings, which did not fail to follow. Rear-Admiral Sir John Ross, in the Felix, joined himself to the Admiralty expedition, and the American vessels under Lieutenants De Haven and Griffin, also followed the same route, so that the nine vessels were congregated at one time in Lancaster Strait. Had definite fields of search been selected, by mutual consent, Peel Sound would probably have fallen to the lot of some one of the parties, notwithstanding that the absence of cairns or any other trace of the Erebus and Terror, observed by Sir James Ross, discouraged a search in that direction.

The commanders of the two American schooners, being the first to perceive the impolicy of so many ships pressing to the westward on one parallel, turned back, and were by winds and currents carried some distance up Wellington Channel, where, with the privilege always allowed to discoverers, they named the headlands, being of course in total ignorance of Franklin having passed up that channel before them. Afterwards, in making their way out of Lancaster Sound, the Advance and Rescue were shut up in the ice-pack, and with it they drifted down Baffin's Bay and Davis' Straits the whole winter, yet without damping the zeal of the officers and crews, for, on being released in the spring, they again went northward to aid in the search.

Captain Ommaney in proceeding westward after Captain Austin, saw unmistakeable traces of the Erebus and Terror at Cape Riley and on Beechey Island; and Captain Penny, following on the same route, made a more extensive search on Beechey Island, by which the first knowledge was obtained of the discovery ships having wintered there in 1845-6.

Captain Penny's two ships became fast in the ice at the southern extremity of Cornwallis Island, which bounds the western side of Wellington Channel. Being so convenient to his position, the survey of that channel was performed by his sledge-parties in the spring, and carried on northwards to Cape Beecher, and on the south, in a westerly direction, till Houston Stewart Island bore north-east. The inexperience of some of his officers employed in surveying, led to errors, and an over-estimation of the distance travelled, by which, in the maps constructed on data furnished by them, Cornwallis and Bathurst Islands were supposed to be connected by a narrow isthmus, at the place where the Erebus and Terror are believed to have descended into Macdougall Bay, and Lancaster Sound, in accordance with the obvious meaning of the record found on King William's Island. Captain Penny having exhausted the strength of his crews, and the means at his disposal in this survey, applied to Captain Austin for assistance in men and boats, to enable him to extend the examination of Wellington Channel further to the north; but that officer, as was natural and proper, preferred the employment of the men under his command, in carrying out his own scheme of search.

By a skilfully combined system of well-organized sledge-expeditions, Captain Austin examined the whole north shore of Melville Sound, Macdougall Bay, and the openings on each side of Byam Martin Island, Lieutenant M'Clintock commanding one of these parties, having made a journey of extraordinary length in the exploration of Melville Island. The south coast of Melville Sound was also surveyed to a great extent by Captain Austin's parties, Prince of Wales Island having been traced down its west shore to the 72d parallel, by Lieutenant (now Captain) Sherard Osborn; and along its east coast to the 73d parallel, including a thorough survey of Cape Walker, proved to be the extremity of a small island, which then was named Russell. The examination of Prince of Wales' Island was most important, as it lay exactly in the course that Sir John Franklin was directed to pursue, and these surveys shewed that, as far as they went, there was a passage on either side of it; but the fact of no cairns or other signs of the discovery ships having been perceived, overbalanced the significance of the existence of these passages, in the planning of future schemes of search. To Captain Austin the merit is due of having brought the sledge equipments to a degree of efficiency that they had not previously attained on the Arctic Sea; and he was well seconded by his officers and crews, who made journeys remarkable both for the number of days they were absent from the ships, and for the great extent of coast traversed. On his homeward voyage, Captain Austin examined the entrance of Alderman Jones' Sound.

In the autumn of 1852, Commander Inglefield, in charge of Lady Franklin's screw-steamer the Isabel, looked into Whale Sound and Smith's Sound, so that all the sounds named by Baffin, had by this time been approached or entered.

On the return to England at the close of the summer of 1851, of the ships that had been employed in Lancaster

Sound, Captain Penny complained loudly of his operations having been cramped by the want of the aid he had demanded from Captain Austin; and the discovery on Beechey Island of sure traces of the Erebus and Terror having given eclat to his operations, the public press took up his cause warmly. The line of search he had pursued was pronounced to be the true one, the speculative opinion of the existence of a Polynia or open sea, towards the pole, was brought into play; and it was broadly stated that the discovery ships were sailing in a latitude higher than had yet been reached, and were to be sought for on the north coast of Siberia, or any where but near the continent of America, the fact that Franklin entertained none of these notions, being wholly ignored. Lady Franklin alone, with a sound judgment, sent her vessel, the Albert, in the right direction.

The Albert, commanded by Mr. Kennedy, who had an able assistant in Lieutenant Bellot, of the French Navy, having wintered in 1851-2 in Regent's Inlet, these two gentlemen made a winter journey of sixty-three days duration, in which they discovered Bellot Strait, a channel about a mile in width, separating North Somerset from Boothia Felix. Unfortunately on passing through this strait, Mr. Kennedy, misled by the appearance of Peel Sound, so blocked up by islands as to leave apparently no passage for a vessel, instead of prosecuting the search southwards in the direction of King William's Island, as he had been instructed to do, went to the westward, and after crossing and re-crossing Prince of Wales' Island diagonally, touched at Cape Walker, and regained his ship by rounding the north end of Somerset Island. It is but fair to notice, that Mr. Kennedy did not know of the northern half of Prince of Wales' Island having been already coasted by

Captain Austin's parties, and that, therefore, the importance of visiting Cape Walker and its neighbourhood, would be in his eyes as great as ever. The duration of this journey, performed in the severe cold of the early months of the year, bears testimony to the ability and endurance of the party; and had their efforts been directed to King William's Island, there is every reason to believe that they would have ascertained the fate of the discovery ships, before either Rae or M'Clintock.

The Government preparations for the search in 1852, were more complete than in any previous year, and Captain Sir Edward Belcher, C.B., sailed in command of a most thoroughly efficient squadron, whose strength was wholly directed by him to the survey of the north side of the Parry Islands, in accordance with the then existing popular feeling. Sir Edward carried his own vessel, the Assistance, and her steam-tender, the Pioneer, up Wellington and Queen's Channels to Northumberland Sound, in latitude 76° 52′ N., on the west side of Grinnell Peninsula. Captain Kellett, in the Resolute, accompanied by his steam-tender the Intrepid, established his winter quarters at Dealy Island, in Bridport Inlet, on the south side of Melville Island; while Captain Pullen, in the North Star, was stationed at Beechey Island, in an intermediate position for communicating with either division, and with vessels coming from England. From these three stations, sledge and boat parties were sent out in autumn, spring, and early summer, by which the whole chain of Parry Islands was laid correctly down on the charts, up to its north-western extremity. Ireland's Eye and Prince Patrick Island were examined thoroughly by the indefatigible M'Clintock; while Commanders Richards, and Sherard Osborn, and other officers leading the very numerous sledge-parties, traced the northern coasts of Melville, Bathurst, and Cornwallis Islands, with the straits that separate them, and also corrected errors in the prior survey of Wellington Channel. Sir Edward Belcher himself surveyed the south side of North Cornwall, the channel bearing his own name that leads into Jones' Sound, the north side of Grinnell Peninsula, and the adjoining promontory of North Devon. On the opening of the navigation in the summer of 1853, Captain Belcher ordered the retreat of both divisions of his squadron towards Beechey Island, but his own ship was shut up in the ice off Point Eden, in Wellington Channel, and Captain Kellett's had the same fate in Barrow Strait, south of Austin Channel, which separates Bathurst and Byam Martin's Islands. A most melancholy accident occurred this autumn in connection with Wellington Channel, Lieutenant Bellot, of the French Imperial Navy, having been drowned in attempting to carry dispatches from the North Star to the Assistance. His loss was regretted by all who knew and had learnt to admire his amiable qualities and gallant behaviour.

We revert now to the proceedings of the Enterprise and Investigator, which sailed from England as mentioned above, in January 1850, for Bering's Strait, Commander (now Sir) Robert Le Mesurier M'Clure, of the latter ship, was, through a combination of favourable circumstances and the exercise of a prompt and sound judgment, able to get round Point Barrow time enough before the close of that summer, to push along the north coast of the continent to the south end of Banks' Island, which he doubled. He then sailed through Prince of Wales' Strait, between that Island and Wollaston Land, until the firm ice of Melville Sound stayed his progress, when he retired into the strait for the winter. In the spring of 1851, a travelling party, under the command of Lieutenant Haswell, surveyed the western coast-line of the peninsular part of Wollaston Island, which Captain M'Clure had named Prince Albert Land, down to a deep inlet called Prince Albert's Sound. From the northern side of this inlet he turned back on the 14th of May, and exactly ten days afterwards, Dr. Rae, in prosecuting the survey of the south side of Wollaston's Land as mentioned in page 178, reached the opposite side of the sound. At the same time, Mr. Wynniatt travelling along the north coast of Prince Albert's Peninsula, and rounding Glenelg Bay, attained Reynolds' Point, in latitude 72° 4' N., and longitude 107° 40' W. The coast between this point and Cape Collinson on Gateshead Island, forming the south side of M'Clintock Channel, is not yet explored, and is in fact the only piece of coast-line within the sphere of the searching parties which has not been traced. It comprises a distance of about 160 geographical miles. On the opening of the navigation, on the 14th of July 1851, Captain M'Clure made another fruitless attempt to cross the ice-covered Melville Sound, and then despairing of succeeding at that place, determined to try a more northern route. This he did, by nearly circumnavigating Banks' Island, exposed to frequent imminent danger of shipwreck, from the pressure of the polar pack coming down the west side of Parry's Archipelago, until he found shelter in the Bay of Mercy, on the north side of Banks' Land. There the Investigator remained shut up during the winters of 1851-2, and 1852-3, making three winters in all of her abode in the ice. In the spring of 1853, preparations had been made for abandoning the ship, for sending the weaker part of the crew to the

Hudson's Bay Company's posts on the Mackenzie, and for attempting, with the more able men, to travel over the ice to Lancaster Sound, when Lieutenant Pym most unexpectedly appeared among them, bringing intelligence of relief being at hand. This officer had been sent by Captain Kellett to communicate with the Investigator, whose presence on the coast of Banks' Island he had learnt from a note, deposited by Captain M'Clure at Winter Harbour on Melville Island.

In the early summer of 1853, the Investigator was abandoned, and the ship's company travelling over the ice, were received into the Resolute by Captain Kellett, where they passed their fourth winter, and being in the spring of 1854 transferred to the North Star, at length reached England in the month of October, after an absence of nearly four years. They were the first navigators who had passed from Bering's Strait to Baffin's Bay. Though the Investigator had providentially not been provisioned with Goldner's patent preserved meat, yet three winters had told so severely on the health of the crews, that, except for the aid supplied by the Resolute, the results of the journeys that were contemplated on the abandonment of the ship, could not have been otherwise than most disastrous.

Captain Sir Robert M'Clure by this perilous voyage, prosecuted with undaunted perseverance, found a strait connecting the continental channel with Melville Sound, and thus discovered the North-west Passage, after it had been discovered in another quarter by Captain Crozier, and the survivors of the Erebus and Terror, who perished in accomplishing their object.* A parliamentary grant shewed the national sense of

* The prior discovery of a north-west passage by the survivors of the Erebus and Terror is with great candour allowed in the published narrative of Sir Robert M'Clure's voyage.

the bravery and skill of Sir Robert M'Clure, his officers, and men. There is, however, little prospect of the navigation in the direction of Banks' Land being ever practicable for ships. Sir Edward Parry was stopped there by fast ice in the summers of 1819 and 1820. Sir Robert M'Clure found it to be equally impassable in 1850, 1851, and 1852. In 1853 also, Mr. Krabbè states the ice to have remained firm;* and Captain Austin, in 1850-51, was unable to advance westward beyond Cape Cockburn. Captain Kellett got to Dealy Island, only a little beyond Cape Cockburn, and short of Winter Harbour. Sir Edward Parry, in his report of the state of the ice in this quarter, says, " It now became evident, from the combined experience of this and the preceding year, that there was *something peculiar* about the south-west extremity of Melville Island, which made the icy sea there extremely unfavourable to navigation, and which seemed likely to bid defiance to all our efforts to proceed much farther to the westward in this parallel of latitude." Captain Osborn, in his narrative of M'Clure's voyage, also remarks, that,—"The heavy pack of Melville (Banks') Strait, lying across the head of the channel, was supposed to be the reason of the ice filling Prince of Wales' Strait ceasing to move on to the north-east, and the *impassable* nature of the pack in the same direction in the following year, confirmed this hypothesis." A writer in the *Dublin Natural History Review* for April 1858, attributes the constant packing of the ice in Banks' Strait, to the meeting there of the Atlantic and Pacific tides; and the Rev. Samuel Haughton, (who is understood to be the writer alluded to) in the appendix to Captain M'Clintock's journal, shews in a map the co-tidal curve, passing from the vicinity of the magnetic

* Dr. Armstrong's Personal Narrative, etc., p. 592, and Blue Book for 1855.

pole by the north-end of King William's Island, across Banks' Strait, and eastward along the north side of the Parry Islands, towards Jones' Sound; thus ascribing the packing of the ice in these several quarters to the confluence of the main currents, as has been briefly stated in a preceding chapter (p. 170).

Captain Collinson arrived in Bering's Straits later in the season than M'Clure, and was unable to double Point Barrow in 1850. In 1851, however, he succeeded in getting round that low, and generally ice-encumbered projection, and pursuing the continental channel with the same facility that his precursor had done, followed him through Prince of Wales' Strait; but though he penetrated a few miles further into Melville Sound, he found no passage, and returning to the south end of the strait, passed the winter of 1851-2 in Walker Bay. Next summer he carried his ship through Dolphin and Union Straits, Coronation Gulf, and Dease Strait, to Cambridge Bay, in the Victoria end of Wollaston Island, where he spent his second winter. His sledge-parties explored the west side of Victoria Strait as far as Gateshead Island, some miles beyond Rae, who had preceded them, and whose cairn they found. From the Eskimos who visited the Enterprise in Cambridge Bay, a piece of an iron bolt was purchased, and also a fragment of a hutch-frame, being evidently parts of the wreck of the Erebus or Terror. A deficiency of coals compelled Captain Collinson to return by the way that he came, instead of spending another year in forcing a passage through Victoria Strait, where the attempt would doubtless have been made had he persisted. He did not, however, get round Barrow Point on his return, without passing a third winter on the northern coast of America.

In the meantime, as has been said above, Sir Edward

Belcher, in endeavouring to descend Wellington Channel on his way home, was caught in the ice off Eden Point, and there passed the winter of 1853-4; Captain Kellett, of the Resolute, being enclosed during the same season in the pack between Byam Martin Island and Prince of Wales' Island. In these positions the Assistance and Resolute were abandoned, with all their stores and provisions, and also their steam-tenders, the Pioneer and Intrepid, by command of Sir Edward Belcher, the senior officer. The Resolute, having been previously made snug and the hatches securely battened down by Captain Kellett, drifted afterwards into Baffin's Bay, and being found there by the master of an American whaler, was carried by him to his own country, and finally presented by the United States Government to the British Admiralty.

The loss of five fine vessels (besides a transport) closed the Admiralty search by sea, but the Hudson's Bay Company again dispatched Dr. Rae to Repulse Bay, one object of his mission being to ascertain, beyond cavil, the continuity of the isthmus which separates Regent's Inlet or the Gulf of Boothia from the estuary of the Fish River and the southern extremity of James Ross's Strait. On his way northwards, Dr. Rae entered Chesterfield Inlet, and, in the hope of finding a route from thence to the estuary of the Great Fish River, ascended the river Quoich, which falls into the north side of the inlet on the 94th meridian. After navigating the Quoich, however, which is full of rapids, up to the 66th parallel of latitude, as has already been mentioned at page 121, he found the country to be too mountainous for the passage of boats, and, therefore, descending the stream, he left the inlet, and pursued his way to his former winter quarters in Repulse Bay. From thence, in the spring, he crossed the neck of Simpson Peninsula and the Boothian Isthmus, whose western coast he traced from the Castor and Pollux River of Dease and Simpson, up to Cape Porter of Sir James Ross, fully establishing the insularity of King William's Island.

Dr. Rae also obtained, on this journey, unquestionable evidence of the melancholy fate of the crews of the Erebus and Terror. In the spring (four winters past, as he was told, but actually) six winters past, whilst some Eskimos were killing seals near the north end of King William's Island, about *forty* white men were seen dragging a boat and sledges over the ice, on the west side of the island. None could speak the Eskimo language so as to be understood, but by signs they gave the natives to understand that their ships had been crushed in the ice, and that they were going where they expected *to find deer to shoot*. All the men hauled the drag ropes except one tall, stout, middle-aged officer. They were looking thin, and seemed to want provisions. At night they slept in tents.

At a later date in the same season, but previous to the disruption of the ice, the corpses of some thirty persons and some graves were discovered on the continent, and five dead bodies on an island near it, about a long day's journey to the north-west of the mouth of the *Utku-hikalik-kok*, or Back's Great Fish River. Some of the bodies were lying in tents, and one, supposed to have been an officer, lay on his double-barrelled gun, with his telescope strapped to his shoulders.

Dr. Rae's report, and the numerous relics of the deceased purchased from the natives, were adjudged by the Admiralty to be certain testimony of the entire loss of the Franklin expedition, and £10,000 were paid to him and his party, being the sum promised to any one who should find and relieve the missing mariners, or bring correct intelligence of their fate.

But in the hope of receiving some fuller details of the sad event, Government requested the Hudson's Bay Company to send a party down the Great Fish River, to explore its estuary, and communicate with the neighbouring Eskimos. Mr. Anderson, one of the Company's chief factors, was accordingly employed on this mission in the summer of 1855. Unfortunately, no interpreter could be procured on so short a notice, there being none within 2000 miles, and the only conversation Mr. Anderson could hold with the Eskimos he saw at the mouth of the river, was by the uncertain medium of signs. From them, however, he obtained many additional articles which they had found on the deceased; and on Montreal Island he discovered the spot where the natives had broken up the boat for its wood and nails. By expressive and unmistakeable pantomime, the Eskimos told him that the white men had died of hunger. A minute and patient search of Montreal Island, of the whole peninsula of Point Ogle, and of an adjacent island to the westward, revealed neither books, scraps of paper, nor arms, nor a single human bone or grave. He supposed that all the dead were concealed by the drift sand which abounds on Point Ogle, but it is more probable that he had not discovered the exact place mentioned by the Eskimos as the spot where the remnant of the crew had breathed their last, or that their tents having been pitched on the strand, their bodies had been swept off by the rising sea on the breaking up of the ice.

Lady Franklin was not satisfied that all had been done that was required for the fame of her gallant husband and his brave companions; and having not yet abandoned all hope of rescuing some forlorn survivor of the catastrophe, urged the Government, in an eloquent letter to the Prime Minister, to send out another searching party by sea to the scene of the catastrophe, the position of which had been so nearly ascertained. A memorial, signed by many eminent scientific men, and also by officers who had been employed on the searching expeditions, was addressed to the same quarter, recommending a renewal of the search. These having failed, Lady Franklin, aided by private subscriptions, but mostly at her own expense, fitted out the Fox yacht in 1857, and placed it under the command of Captain M'Clintock, who volunteered to encounter gratuitously the hardships and hazards of an Arctic search, of which he had had so many years' experience. Another Arctic officer, Lieutenant Hobson, likewise came forward to serve without pay, and Captain Allen W. Young, of the Mercantile Marine, not only gave his own services to the cause, but contributed a very considerable sum to the expenses of the outfit. This generous devotion to the enterprise was shared by Dr. Walker, the surgeon, also a volunteer, and by the petty officers and men who completed the crew of the Fox, making twenty-six souls in all on board.

In attempting in the first summer to cross Baffin's Bay, the Fox was "beset in the middle pack," and drifted with it all the winter, remaining helplessly enclosed for 242 days, during which time the southerly drift was 1385 statute miles, or $5\frac{2}{3}$ miles daily. On the 24th of April 1858, a heavy storm broke up the pack, and the little Fox steamed out from among the rolling masses of ice, escaping almost miraculously without suffering serious damage.

Relieved from this great peril, the noble crew had no thought of retreating, but grieved at the loss of a season and the consequent disappointment to Lady Franklin, used their utmost endeavours to reach the region of their search; and, after touching at Greenland, succeeded in crossing over to Pond's Bay, on the western shore. This inlet was entered and pursued by the Fox, as far as the ice permitted, and from what Captain M'Clintock himself saw, and from a survey made in 1855 by Mr. Gray, master of a whaler, we learn that it is a strait leading to an inland sea named Eclipse Sound, which again communicates with Regent's Inlet. Eclipse Sound has in all three northern entrances, the other two being Admiralty and Navy Board Inlets.*

On the 18th of August, the Fox descended Peel Strait for twenty-five miles, when, being stopped by a bridge of ice, Captain M'Clintock turned about, and rounding North Somerset, went down Regent's Inlet to Bellot Strait. This strait, being twenty miles long, and in some places not above a mile in width, is traversed by very rapid tides, of which the night tides are by much the highest. The flood tide comes from the west, as it does also in the Fury and Hecla Strait, on the other side of Regent's Inlet.† An ice-floe lying across the west end of the strait obstructed the further progress of the Fox, which was therefore housed for the winter in Port Kennedy, a snug harbour within its western entrance.

The proceedings of Captain M'Clintock and his associates cannot be more briefly stated than in his own words:—" Our geographical discoveries amount to nearly 800 miles of coastline; they are interesting not only in consequence of their extent and the important position they occupy, but also from

* See Captain Allen Young's Chart in the "Cornhill Magazine for January 1860, No. 1.

† The tides will be discussed by Professor Haughton, of Trinity College, Dublin.

the great difficulty of access, whether by sea or land, to this newly explored area. With the exception of a comparatively small and unimportant part of the shore of Victoria Land (between Wynniatt's farthest and Cape Collinson), the whole coasts of Arctic America are now delineated.

"My sledge journey to the Magnetic Pole in February completed the discovery of the coast-line of the American continent. The insularity of Prince of Wales Land was ascertained, and the discovery of its coast-line concluded, by a sledge party under the direction of the sailing-master, Captain Allen Young; as also the west coast of North Somerset, between Bellot Strait and Four-river Bay. Lieutenant Hobson and his party completed the discovery of the west coast of King William's Island, picking up the Franklin records; whilst, with my own, I explored its eastern and southern shores, returning northward by its west shore from the Great Fish River.

"Repeated attempts were made in 1858, before the close of the navigable season, to reach the open water visible in the broad channel westward of North Somerset; but a narrow barrier of ice which lay across the western outlet of Bellot Strait, was there hemmed in so firmly by numerous islets as to continue unbroken throughout the autumn gales, and to foil my sanguine hope of carrying the Fox (according to my original plan) southward to the Great Fish River, passing east of King William's Island, and from thence to some wintering position upon Victoria Land. From a very careful survey of the ice during my journeys over it in February, March, April, May, and June, it was evident that in this western sea it had all been broken up; whilst eastward and southward of King William's Island there had been hardly any ice last autumn;

and, therefore, in all probability, we *saw* in that barrier of ice, some three or four miles wide, the only obstruction to our complete success.

"The wide channel between Prince of Wales Land and Victoria Land, upon which I conferred the name of 'Lady Franklin,' admits a vast and continuous stream of very heavy ocean-formed ice from the north-west, which presses upon the western face of King William's Island, and chokes up Victoria Strait.

"I cannot divest myself of the belief that had Sir John Franklin been aware of the existence of a channel eastward of King William's *Land* (so named until 1854), and sheltered from this impenetrable ice-stream, his ships would safely and speedily have passed through it in 1846, and from thence with comparative ease to Bering Strait." *

The very long journeys over the ice mentioned by Captain M'Clintock were not accomplished without much personal suffering by all engaged in them. Lieutenant Hobson, especially, having been previously enfeebled by scurvy, was unable to walk or even to stand before he reached the ship, and the health of Captain Allen Young sustained severe injury. From the Eskimos who were hutted on the west coast of Boothia many interesting relics of the Franklin party were obtained, and also intelligence of the fate of the two ships. One of the ships was seen by the natives to sink in deep water, and they rescued nothing from her. The other was forced on shore by the ice on a point named by them *Utlu-lik*. These events took place in August or September, the white men having some months previously gone away towards Great Fish River. The body of a man was found on

* Proceedings of the Royal Society, x., No. 37, p. 147. Nov. 1859.

board the stranded ship, and from that vessel the natives obtained wood and many of the articles they possessed.

Lieutenant Markham, in a paper which he read at the Dublin Meeting of the British Association in 1857,* sums up the extent of coast line examined by various searching parties, as follows:—Sir James Ross in 1849 explored 990 miles of coast on the eastern side of Peel's Strait, in Lancaster Strait and in Regent's Inlet; Captain Austin traced 6087 miles; Sir Edward Belcher and Captain Kellett, 9432 miles; Sir Robert M'Clure 2350 miles; Captain Collinson in his voyage to Cambridge Bay, and Dr. Rae's previous exploration of the same coasts, included 1030 miles—making in all 21,500 miles of coast line examined, of which 5780 were previously unknown. In this enumeration, the boat expeditions of Sir John Richardson and Captain Pullen are omitted, both being along shores previously well surveyed. The extent of search made by Captains Penny and Stewart, by the American expeditions of De Haven and Kane, and by the commanders of Lady Franklin's several expeditions, are also left out. To the total amount, Captain M'Clintock's survey is to be added, having been made subsequent to the reading of the paper.

To avoid interrupting the narrative of the discovery of the fate of the Erebus and Terror, an account of one of the most remarkable of all the enterprises undertaken in connection with the search for Sir John Franklin, has been postponed to this place, instead of being mentioned in chronological sequence. We allude to Dr. Elisha Kent Kane's wonderful exploration of Smith's Sound. This expedition Dr. Kane says, was based upon the probable extension of the landmasses of Greenland, to the *far north*, a fact not verified at

* Nat. Hist. Rev., Jan. 1858, p. 35.

that time by travel, but sustained by the analogies of physical geography. Believing in the extension of the peninsula of Greenland (in form of a congeries of islands connected by interior glaciers), and feeling that the search for Sir John Franklin would be best promoted by a course that might lead directly to the open sea, of which Dr. Kane had inferred the existence, he chose Smith's Sound as the scene of his operations, thinking that the highest protruding head-land would be most likely to afford some traces of the lost party.*

Dr. Kane left the United States in the Advance, with a crew of seventeen officers and men, to which two inhabitants of Greenland were added. On the 7th of August 1853, he entered Smith's Sound, and after much labour and many narrow escapes from shipwreck, the Advance was secured in Rensselaer Bay, from whence she was destined never to emerge. The geographical position of this place was ascertained to be in latitude 78° 38′, and longitude 70° 40′, determined by astronomical observations, and it is farther north than the wintering place of any other ship, being a degree and forty-six minutes higher than Sir Edward Belcher's harbour, in Wellington Channel. According to Dr. Kane's view of the structure of the coast, Greenland terminates at Cape Agassiz, in latitude 79° 14′, and longitude 65° 14′ W., ascertained by intersecting bearings. North of this the coast line is formed by the stupendous Humboldt glacier, which issues from *a mer de glace*, and presents an unbroken precipitous sea-face of nearly sixty geographical miles. A similar glacier exists farther south, in Melville Bay, presenting an unbroken front, estimated by Captain M'Clintock to be forty or fifty miles in extent. The Eskimos state that herds of rein-deer

* Arctic Explorations, by Elisha Kent Kane, M.D., U.S.N., pp. 16, 17.

retire into the interior across the glacier, whose extent inwards has never been ascertained. Dr. Kane's personal explorations terminated at the great glacier, and so far the geographical positions of the headlands are doubtless correct. It was very nearly at the sacrifice of his life that he went so far. Beyond the Humboldt glacier the coast was explored by William Morton, and the positions being laid down mostly by dead reckoning or cross bearings, cannot lay the same claim to perfect accuracy.

A meridional observation of the sun, however, was obtained on the 21st of June, at Cape Andrew Jackson, in latitude 80° 1' N. Another observation on the 24th gave 80° 41' N. for the latitude, which is the most northerly position ascertained by the meridional altitude of the sun on Mr. Morton's journey. This seems to have been in a bay on the north side of Cape Jefferson. From this spot to Cape Constitution, the most northerly point reached, Mr. Morton travelled on foot carrying a load, and concluded his journey between noon and midnight, but his journal mentions neither the distance travelled nor the number of hours. Making a correction, however, for dead reckoning corresponding to that which was found to be required for Cape Andrew Jackson, Cape Constitution cannot be far short of the 81st parallel of latitude. The western side of the inlet, named by Dr. Kane "Grinnell Land," is laid down almost wholly by cross bearings. Its extreme northern point, Mount Parry, lies in about 82° 14' N., corrected latitude, and is 100 miles to the north of Ross' Inlet, the extreme rock of the Seven Islands in the Spitzbergen group, which was previously the highest land known.

The width of Smith's Sound or its northern prolongation,

Kennedy Channel, is about thirty-three geographical miles across, at the narrowest places. The more southern half was closed by a firm field of ice during the two years that Dr. Kane watched it; but in the month of June, Mr. Morton found open water, traversed by small streams only of brash-ice, extending from Cape Andrew Jackson, northwards, and as far as his vision could take in, when looking from an altitude of 300 feet some way up the cliffs of Cape Independence (which are 2000 feet high). He saw an open sea, frequented by numerous water-fowl and brent geese; on shore he observed considerable vegetation, among which were the *Salix arctica* and *S. uva-ursi*, denoting a climate much like that of Spitzbergen.*

After the lapse of two winters, Dr. Kane was obliged to abandon his ship, not being able to get it out of the ice, and his successful voyage in boats, with his starving party, to Sanderson's Hope, is nearly as memorable as his perseverance amid the dangers and privations of Smith's Sound.

This summer (1860), Dr. Hayes has sailed from America, to complete the survey of Kennedy's Channel.

* Kane's Arctic Explorations, I. p. 299. The elevation of Morton's look-out station is stated to be 500 feet in that seaman's own report.

SECTION II.

PHYSICAL GEOGRAPHY.

CHAPTER XI.

SPITZBERGEN.

Number and Aspect of the Islands—Mountains—Glaciers—Iceberg—Avalanche—Disintegration of Rocks—Vegetation—Animals—Drift-wood—Marine Currents.

IN presenting a summary view of the physical aspect and ethnology of the lands within the arctic circle, it is convenient to begin with Spitzbergen, because of its position, intermediate between the eastern and western hemispheres. It is only to its physical geography that our attention is called, since it has no indigenous human inhabitants; and it is in fact, with the exception perhaps of the still more inhospitable antarctic lands, by far the largest country wherein no traces of mankind met the eyes of its discoverers.

The principal island of the Spitzbergen group has a peculiar shape resembling a pair of trousers with the waist-band deeply indented towards the pole, by Weide and Leifde Bays, and hung up in the north on the parallel of 80°, while the legs fall down 3° of latitude to the southward. The western leg, flanked by Charles Island and many lesser islets, is called *West Spitzbergen* or *King James's Newland*, and often by the older navigators, *Greenland*, on account of their supposing it to be a continuation of the Greenland of the Eskimos. The eastern leg, named *New Friesland* or *East Spitzbergen*, is cut across near the middle of its length by a strait; and the detached southern part designated in old charts as *Witches' Land*, is termed *Staats Iland* by the Dutch, and *Maloy Broun* by the Russians. Another large island of a sub-triangular or pentagonal shape, is named from its position the North-east Land, and is separated from East Spitzbergen by Henlopen Strait or Waigatz. North of it lie the Seven Islands, and Walden Island; Ross Islet, which is the northernmost rock of all, is in latitude 80° 49′ N. The Archipelago of the Thousand Isles is at the entrance of *Weide Jans Water*, which separates the legs of the trousers from each other.

The only account of the geology of Spitzbergen which we have seen, is a brief one by Professor Jameson, drawn up from fragments of rock brought to England by Sir Edward Parry. These specimens consisted of primitive granite, gneiss, and mica slate; gneiss with precious garnets was obtained on the most northern islets. In Henlopen Strait, fetid limestone and a limestone containing madreporites, orthoceratites, and terebratulites, were found. They were detached probably from silurian deposits. Red sandstone, thought by Professor Jameson to be of more recent origin, also exists in Red Bay on the north of West Spitzbergen, and in Henlopen Strait. Tertiary laminated and cubical glance coal, found in small pieces near the beach on the eastern and western shores, a little above the ordinary line of drift-timber, were evidently conveyed thither by marine currents. Some pieces of vesicular lava that were picked up, are also thought to have been floated to Spitzbergen by sea, from Iceland probably, or Jan

Mayen's Island, whose peak, named Beerenberg, is a volcanic cone, rising 6780 feet above the sea level, and is, according to Dr. Scoresby, occasionally active.

Captain Beechey says, that the high ridge of Western Spitzbergen runs north and south, lowering in the latter direction, and that its lateral eastern spurs are also lower, the land generally sinking towards the east. Where the sandstone exists, there are table-topped hills, and Low Island is described by Dr. Irving, who accompanied Captain Phipps (Lord Mulgrave), as being formed of hexagonal stones commodiously placed for walking. This kind of pavement was probably the summits of basaltic columns, or perhaps the faces of horizontal beds of silurian limestone, which is cracked in that manner by the frost. Spitzbergen derives its name from the pointed peaks, seen while coasting its western side. Its mountains there rise steeply from the beach to a very considerable height. Round *Smeerenberg* (Oily Hill) Harbour, many of them exceed 2000 feet in height. The Devil's Thumb on Charles Island, is calculated by Dr. Scoresby to rise 1500 or 2000 feet, and Horn Mount, in the harbour of the same name, he states to be 4400 feet high.

The mountains on the west coast are very steep, many of them inaccessible, and most of them dangerous to climb, from the smooth hard snow by which they are encrusted in summer, and the loose rocks which project through it so poised, that they give way under the slightest pressure of the foot. The views of the coast given by Captain Phipps, shew dark, craggy rocks, projecting every where in summer above the snow, and the Devil's Thumb, a crooked peak, is alike destitute of snow and verdure, but the high rocks are black with lichens.

Almost all the valleys, says Captain Beechey, which have not a southern aspect, are occupied either by glaciers or immense beds of snow, which must be crossed before the summits of the mountain ridges can be gained. Where the valleys open out on the sea, the glaciers shew precipitous cliffs of ice, in some places 400 or 500 feet high, washed by the waves. Dr. Scoresby says, that a little to the northward of Charles Island are *seven icebergs*, each of them occupying a deep valley formed by hills of about 2000 feet elevation, and terminated in the interior by a mountain chain rising above 3000 feet, and running parallel to the coast. The upper surfaces of the glaciers are generally concave, the higher parts always covered with snow, and the lower parts, towards the end of summer, converted into bare ice. They are traversed by many rents, and their ends, as they advance out of the valley into the sea, are continually breaking off, to form icebergs of various and often vast magnitude. There are four glaciers in Magdalena Bay, the smallest having a sea-face of about 200 feet. One called the Waggon-way, is 7000 feet across at its terminal cliff, which is 300 feet high, presenting an awfully grand wall of ice. A concussion of the air is sufficient to detach one of these icy cliffs, and there is the same necessity for preserving silence in passing under them, as the poet inculcates on a traveller over the Swiss Alps—

"Mute lest the air convuls'd by sound,
Rend from above a frozen mass."

An avalanche of this kind, on a magnificent scale, was produced by the purser of the Trent firing a gun from a boat when about half a mile from a glacier in Magdalena Bay. "Immediately after the report of the musket, a noise resembling thunder was heard in the direction of the iceberg, and in a few seconds more an immense piece broke away and fell headlong into the sea. The crew of the launch, supposing themselves to be beyond the reach of its influence, quietly looked upon the scene, when presently a wave rose and rolled towards the shore with such rapidity, that the rowers had no time to take any precautions, and the boat being in consequence washed upon the beach, was completely filled by the succeeding wave. As soon as their astonishment had subsided, the seamen examined the boat, and found her so badly broken, that it became necessary to repair her in order to return to the ship. They had also the curiosity to measure the distance the boat had been carried by the wave, and found it to be ninety-six feet."*

"About Fair Haven," says the same officer from whose writings the preceding paragraph has been quoted, "the mountains which came under our observation appeared to be rapidly disintegrating from the great absorption of wet during the summer, and the dilatation occasioned by frost in the winter. Masses of rock were, in consequence, repeatedly detached from the hills, accompanied by a loud report, and falling from a great height, were shattered to fragments at the base of the mountain, there to undergo a more active disintegration. In consequence of this process, there is at the foot of the hills, and in all places where it will lodge, a tolerably good soil, upon which grow several Alpine plants, grasses, and lichens, that in the more southern aspects flourish in great luxuriance. Nor is this vegetation confined to the bases of the mountains; it is found ascending to a considerable height, so that we have frequently seen the rein-deer browsing at an elevation of 1500 feet." "During three or four

* Voyage towards the North Pole (Dorothea and Trent), by Captain F. W. Beechey, R.N., F.R.S. 1843.

months of the year the radiation of the sun at Spitzbergen is always very intense, and its effect is greatly heightened by the very clear atmosphere that prevails over every extensive mass of snow or ice, so that we find mountains bared at an elevation nearly equal to that of the snow-line of Norway; and as vegetation is not regulated so much by the mean temperature of the situation as by its summer heat, there seems to be nothing anomalous in the fact. Plants which can endure considerable frost and remain at rest during an Arctic winter, vegetate very rapidly in a mild temperature; hence they burst into flower almost as soon as their snowy covering is removed, perfecting their seed, and preparing for a quiescent state again, all within the space of a few weeks." "In some sheltered situations at Spitzbergen the radiation of the sun must be very powerful during two hours on either side of noon, as we have frequently seen the thermometer in the offing, upon the ice, at 58, 62, and 67 degrees, and once at midnight it rose to 73 degrees, although in the shade, at the same time, it was only 36 degrees of Fahrenheit's scale."* During summer, streams of water flow down the inclined surfaces of the glaciers, or make their noisy way through interior tunnels. Dr. Scoresby, travelling over the land round King's Bay in 1818, found large ponds of fresh water derived from melted ice and snow, and near the base of the mountains sunk to the knees in a morass of a moorish aspect, consisting apparently of black alluvial soil mixed with some vegetable remains. In ascending the hill the ground gave way at every step, so that progress could be made only by leaping or running. The first ridge he crossed was of limestone, in loose pieces, and so acute on its summit that he sat

* Beechey, *lib. cit.* pp. 138-9.

astride on it as on horseback. The higher and more inland ridge was surmounted at midnight. Under its brow, at an estimated altitude of 3000 feet, the temperature in the shade was 37° F. On the lower limestone ridge it had been 42°, and on the plain near the sea, when the sun was higher, from 44° to 46° F. The mean temperature of the three summer months (June, July, and August) in Hecla Cove was + 35° F., and on board the Trent, cruising in the latitude of 80° among ice, some years previously, it was only half a degree lower. Dr. Scoresby found the temperature of the sea to increase gradually, and invariably with the depth to which the thermometer was let down. At the surface in the summer it varied from + 31° to + 34°, but at the depth of 700 fathoms (4200 feet) it was + 38°. His observations were confirmed by the experiments made on board the Trent, in one of which water from a depth of 700 fathoms raised the thermometer to 43°, though in the surface water it fell to 33°. The Russians mention their having experienced heavy rain on Maloy Broun on October the 7th, and even in the month of January. (Rain after the second week in October is very rare in Rupert's Land, 20° farther to the south.) A lake in the centre of Moffen Island, latitude 80°, was found by Phipps to be frozen over on the 26th of July.

The isothermal line of 32° F., or of the freezing point for the month of July, curves, in Dove's charts north of Spitzbergen, higher than in any other meridian within the Arctic circle, and descends to cut *Novaya Zemlya*, on the east, and on the west to pass through Melville Island. This line probably coincides nearly with the upper range of the variable snow-line, but not exactly, for the direct radiation of the sun in the high latitudes denudes the rocks of snow, and suffices for the vegetation of lichens in situations where a thermometer, placed in the shade, is constantly below the freezing point.

The flowering plants hitherto discovered in Spitzbergen belong to the following families:—*Ranunculaceæ*, 1; *Papaveraceæ*, 1; *Cruciferæ*, 5; *Caryophylleæ*, 7; *Rosaceæ*, 1; *Saxifragæ*, 7; *Compositæ*, 1; *Campanulaceæ*, 1; *Ericaceæ*, 2; *Scrophularineæ*, 1; *Polygoneæ*, 2; *Salicaceæ*, 1; *Junceæ*, 2; *Cyperaceæ*, 2; *Gramineæ*, 6; in all, forty phenogamous plants. None of these plants are woody, the fine thread-like stem of *Andromeda tetragona*, and the crown of the root of *Salix herbacea*, making the nearest approach to ligneous structure. Of the inferiorly organised vegetables, there have been found in Spitzbergen, of *Lycopodineæ*, 1; *Equisetaceæ*, 1; *Musci*, 19; *Hepaticæ*, 2; *Lichenes*, 23.

In some parts, where there is a superior soil, grass is luxuriant. In Mussel Harbour, on the north-east of Spitzbergen, Martens walked through grass that covered his ankles, and the lichens, especially the *cetrariæ*, are spoken of as abundant.

Of the land quadrupeds that exist in Spitzbergen, the rein-deer is the most important, and it is more abundant all the year than could have been expected. The polar bears are probably the only native enemies it has on these islands, and its fleetness furnishes it with ample means of escape from so clumsy a pursuer. No wolves are mentioned by any of the parties who have wintered on the islands, and who, if any were there, could scarcely have missed seeing them when engaged in the pursuit of rein-deer in the spring. Four Russian sailors, who passed six years on Maloy Broun, lived during that time on the venison they procured in the chase, and on the flesh of ten bears, accumulating, moreover, two thousand pounds of rein-deer fat. Lord Mulgrave's crew killed fifty deer on Vogel Sang, which is a noted hunting-place. On Sir Edward Parry's polar expedition, about seventy deer were killed in Treurenberg Bay by inexperienced deer-stalkers, and without the aid of dogs. Sir James Ross observes, that these animals are very numerous along the northern shores of the islands; and they abound also on the western coasts, where, in Bell Sound, Horn Sound, and other places, many are killed every year by the whalers. One locality has been named Deer-Fell (Hert-berg), because of the herds that frequent it.

The other quadrupeds that have been seen are the polar bear, the arctic fox and lemmings. All these inhabit also the whole circle of arctic Europe, Asia, and America; but the musk ox, whose proper country is the north-east corner of America, does not exist in Spitzbergen, and has not been found alive on the Old Continent. On Low Island, Dr. Irving saw, but did not procure, a creature somewhat larger than a weasel, with short ears, a long tail, and a skin spotted white and black, which cannot easily be identified with any known species by this brief description. The quadrupeds of Spitzbergen pass the winter as well as summer there.

The marine warm-blooded animals, the morses or walruses, the seals and whales of different kinds, are the chief inducements which have drawn Europeans to Spitzbergen. Their earliest resort was to Bear (or Cherie) Island, in search of walruses, which were then so plentiful there, that a thousand could be killed in a few hours. Indiscriminate slaughter, however, drove the marine herds to more secluded northern districts. According to Purchas, the first whale killed in the seas between Spitzbergen and Greenland was by the Biscaynes in 1611. The fishery of the right whale has also declined through the scarcity of the animals. At one time the Hans Towns took the lead in it, and Admiral Beechey saw upwards of one thousand coffins in Smeerenberg harbour, over a few of which boards, with English inscriptions, were erected, but the greater number were Dutch, and had been deposited in the eighteenth century. In Treurenberg Bay, also, Sir Edward Parry found thirty Dutch coffins. None of the marine mammals are peculiar to the Spitzbergen seas.

Of the insessorial birds, the well known snow-bird (*Plectrophanes nivalis*), and the familiar lesser redpole, which winters within the Arctic circle in Norway and America, are the only representatives in Spitzbergen. There are perhaps grass seeds enough in Spitzbergen to nourish the redpole all the winter.

Of rasorial birds, the ptarmigan (*Lagopus albus*) more certainly winters in Spitzbergen, judging from the facility with which it can procure its food under the snow. The officers of the Hecla shot several in Treurenberg Bay, and the species was also met with by the French scientific expedition.

Of the waders, the common ringed-plover (*Charadrius hiaticula*) occurs, a single individual having been killed by Dr. M'Cormick of the Hecla. This may perhaps be the "Ice-bird," which Martens saw in English Harbour, and which he would not shoot at, lest he should spoil its fine plumage, and so, notwithstanding its tameness, he let it fly away. It was, he says, almost equal in size to a small pigeon, and when the sun shone on it looked like gold. Another Spitzbergen wader is the purple sandpiper or snite (*Tringa maritima*). The common sandpiper (*Tringa hypoleuca*) was seen there in flocks by Dr. Scoresby.

The web-footed birds or water-fowl are more various. The Brent goose (*Bernicla brenta*) breeds in large flocks on Walden and Little Table Islands, and a nest of this bird, containing two eggs, was found on the most northern of the rocks, Ross' Islet. The eider duck (*Somateria mollissima*) and the king duck (*S. spectabilis*) also breed on Spitzbergen, as do likewise the great northern diver (*Colymbus glacialis*); the red-throated diver (*C. septentrionalis*); the razor-bill (*Alca torda*), which rears its young on the most northern rocks; the diving parrot, puffin, or coulterneb (*Fratercula arctica*); the looms of British sailors or the guillemot (*Uria troile*, and *U. brunnichii*), called by the Danes *lom* and *loom*, which words appear to be of Finnish origin; the pigeon-diver, dovekie or black guillemot (*U. grylle*); the rotge (so named from its cry), called also the little auk (*Arctica alle*); the fulmar petrel (*Procellaria glacialis*), seen in lanes of water beyond 82 degrees of latitude; the solan goose or John of Ghent (*skua*); the pomarine skua, or strunt-jager (*Stercorarius parasiticus*); Ross' gull (*Rhodostethia Rossii*) seen also beyond 82° N. lat.; the burgermaster (*Larus glaucus*); Sabine's gull (*Xema Sabini*); the kittiwake or mew (*Rissa tridactyla*); the ivory gull or rathsher (*Pagophila eburnea*); the arctic tern or kirmas (*Sterna arctica*). Some of these water fowl, such as the dovekies, remain in the high latitudes all the winter, feeding in the occasional ponds and lanes of water that open among the ice. Others seek milder climates after their young are fledged. No fresh-water fishes are mentioned by any author as having been found in Spitzbergen, although fresh-water ponds of considerable size exist there.

On various parts of the Spitzbergen shores, but more particularly on the northern ones near Henlopen Strait, drift-wood is found. On Moffen Island, Captain Lutwidge of the Carcass saw a piece with its root about three fathoms in length, and as thick as the mizen mast of his ship; and on Low Island Dr. Irving observed several large fir trees lying at the height of sixteen or eighteen feet above the level of the sea; some of these trees were seventy feet long, and had been torn up by the roots, others had been cut down by the axe and notched for twelve feet lengths; this timber was no ways decayed, nor were the strokes of the hatchet in the least effaced. There were likewise some pipe staves, and the beach was formed of old timber, sand, and whalebones—(*Phipps*, p. 58). "All the drift-wood which we saw (except the pipe-staves) was fir, and not worm-eaten. The place of its growth I had no opportunity of ascertaining" (*Ib.* p. 71). Were pieces of the drift-wood brought to England, perhaps the microscope would enable us to ascertain the species, and consequently whether it is of Asiatic, European, or American origin.

During the summer months at least, the prevailing current north of Spitzbergen and along its shores is from the north or north-east. Sir Edward Parry, in his attempt to reach the North Pole in boats, succeeded, with great labour, in attaining 82° 45′ of north latitude, after travelling in direct distance from where he left his ship, 172 miles mostly over ice. Throughout this remarkable journey he had to contend with a general southerly drift, and when the wind was from the northward the loss by drift during the necessary hours of repose sometimes exceeded all the advance that he could make during the hours of labour. At the extremity of the voyage but little ice was in sight. In latitude $81\frac{1}{4}°$ as he was returning, he saw several pieces of drift timber and birch bark, and a still larger number nearer Table Island. On Walden Island driftwood he says, was, "as usual," in great abundance. Of the low limestone shore, to the southward of Low Island, at the northern entrance of Henlopen Strait, Sir Edward remarks— " On this and all the land hereabouts where lagoons occur, *enormous* quantities of drift-wood line the inner beach, which is now quite inaccessible to the sea, and this wood is always more decayed than that which lies on the outer or present sea-beach, by which it appears that the latter has been thrown up to the exclusion of the sea long since the inner wood was landed. A great many small rounded pieces of pumice-stone are also found on this part of the coast, and these generally occur above the inner line of drift-wood, as if they had reached the highest limit to which the sea has ever extended. We found one piece of bituminous wood-coal which burned with a clear bright flame, and emitted a pleasant odour."* At the present date the tidal rise on these parts of the Spitzbergen coast is said to be only eight feet, but a line of drift-timber at more than twice that height above the sea is mentioned by Lord Mulgrave. A secular elevation of the islands is perhaps in progress.

An immense quantity of trunks of birch, pine, and fir are said to be thrown upon the northern shores of Iceland also, especially on the promontory of *Langanès*. On the west side of the island, according to Van Troil, boats of twelve tons' burthen are constructed of this drift-timber and sold to the inhabitants of other districts. This drift-wood probably comes from the Obi or other large rivers that fall into the sea of Kara. Dove observes that "the watershed of the Kara Sea exceeds that of the Mediterranean. A current issues from it

* Attempt to reach the North Pole, by Captain W. E. Parry.

through the *Waigatz* and Strait of *Matochkin Schar* to the westward towards Spitzbergen, is deflected to the southward by the coast of Greenland, and then flows south-westward between Iceland and Greenland to Cape Farewell. This current, carrying with it masses of ice (in which the ship Wilhelmine was enclosed for 108 days in the year 1777, and carried 1300 nautical miles), brings with it everywhere intense cold."*
Rear-Admiral Beechy states that "the south-west drift of the ice between Spitzbergen and Greenland has been ascertained by ships beset in it to move at the rate of about thirteen miles a-day," which is equal to that at which Sir Edward Parry calculated the drift of his boats in latitude 82° on a day when the north wind blew. The utmost exertions of the crews of the Dorothea and Trent were unable to maintain these ships in position on the west coast of Spitzbergen, unless they had favouring winds. It is Admiral Beechey's opinion, however, that this south-westerly current does not reach below the parallel of Bear (or Cherie) Island on the east, nor extend as far as Cape Farewell on the west, and certainly not further; "for there," he says, "a south-easterly current prevails, as proved by the fact of bottles thrown into that sea having been picked up on the shores of Great Britain and Teneriffe, and likewise by the casks of the William Torr, whaler, which was wrecked in Davis Strait, having been found in the Bay of Biscay, off Rockhall, on the west of Scotland, and at intermediate stations between that islet and Newfoundland. "It seems," he says, "that the south-westerly current sets from Davis Strait down the coast of Labrador, and, turning eastward, is met by the drain of the gulf stream, which diverts it

---

* H. W. Dove on the Distribution of Heat, p. 17.
† Beechey, Voyage to North Pole, p. 341, and fcp. 342-3.

to the north-east towards Iceland, the Feröe Islands, and the shores of Britain. Nay, there is an indication of this effect of the gulf stream further to the northward, even beyond the North Cape, and we carried the unusually high temperature of the sea as far as the seventy-fifth degree of latitude."* Sir Edward Parry appears to have passed through this warm stratum of water on his voyage from Sorö in Norway to Spitzbergen. If it co-exists with the current mentioned above, on the authority of Dovè, as running north-westward from the Sea of Kara, the two streams must cross nearly at right angles, one flowing over the other, or perhaps they intermit, one overpowering the other at certain times, or when accelerated by certain winds. Sir James Clark Ross does not hesitate in ascribing a Siberian origin to the drift-wood of the Greenland coast—" It is this current," he says, " that carries the timber of Siberia down between Spitzbergen and the east coast of Greenland to Cape Farewell, whence it takes a north-westerly direction up the western shore of Greenland until it meets the southerly current from Baffin's Bay at Queen Anne's Cape, near the Arctic circle. The drift-timber is frequently cast ashore as high as Holsteinberg, but never to the northward of that place. The breadth of the current at Cape Farewell may be considered to extend one hundred miles from the land, gradually diminishing its extent from the coast until it is entirely lost at Queen Anne's Cape.† During two days while coasting the barren district between the eastern and western Bygds, Captain Graah's vessel was set to the northward at the rate of half a mile an hour. Captain E. Irminger of the Danish

---

* Admiral Beechey supports his statements by reference to his own observations in the Trent, to Dr. Scoresby's authority, and to Commander Beecher's bottle-chart. † Graah's Greenland, Engl.tr. p. 24.

Navy, on the authority of the log-books of the Danish ships trading annually to Geeenland, establishes the course of the drift-ice from the Spitzbergen seas down the east coast of Greenland, round Cape Farewell, and up the west coast in spring. The ice mostly disappears between September and January on the south and south-westerly coasts of Greenland, reappearing towards the close of January. But he gives reasons for affirming that there is no current running in a direct line from East Greenland to the banks of Newfoundland, an assertion quite compatible with Admiral Beechey's observations.*

* Journ. Roy. Geogr. Society, xxvii. p. 36, vol. 26. A.D. 1856.

CHAPTER XII.

CURRENTS OF THE POLAR SEAS.

Spitzbergen Current from North—Gulf Stream—Davis Strait Current—Smith's Sound—Kennedy-Polynia—Elevation of the Coast—Comparison of the Vegetation of Smith's Sound and Spitzbergen—Parry Islands—Bering's Strait—Siberian Marine Currents—Siberian Polynia—Secular Elevation of Coast—Currents on the North Coast of the American Continent—Bellot Strait—Fury and Hecla Strait—Prince of Wales' Strait—Parry Islands—Barrow Strait and Wellington Channel—Jones' Sound—Professor Haughton's Theory of the Polar Tides.

In the preceding chapter, the south-west current, setting along the eastern coast of Greenland out of the Spitzbergen Seas, has been mentioned. The effect of such currents in modifying climate, is discussed at length in Lieutenant Maury's comprehensive work on "The Physical Geography of the Sea." He therein assumes as the most probable causes of the *gulf stream*, the increased saltness of its water coming from the regions of the trade winds, and the inferior saltness of the northern seas, whose consequently lighter waters are displaced by the more saline and heavier southern flood. This current runs northward out of the Gulf of Mexico, like a mighty river, to the banks of Newfoundland, which our author considers to be formed of deposits made at the meeting of the current coming from the north along the coasts of Labrador and Newfoundland, with the warmer but salter stream from the gulf.

Captain Scoresby counted 500 icebergs floating southwards in the Greenland-Labrador current. Many of these loaded with earth, gravel, and boulders, take the ground on the banks, and there deposit their loads. In Lieutenant Maury's chart (plate ix.) the gulf stream is shewn as deflected to the eastward at the Great Bank, and continuing its course to the north-east between Iceland and the northern extremity of Europe, with counter currents of much less breadth setting south-west down the coasts of Norway and Greenland. The data for ascertaining the northern limits of the gulf stream are imperfect. There are some reasons, however, for believing that it continues its course beyond the north cape of Norway to the western coasts of Novaya Zemlya. In the year 1608 Henry Hudson, being a little to the north of the Goose-coast of that island, was drifted in a calm to the northward "by a streame or tide;"[*] and Admiral Lütke traced this northerly current for more than three degrees of latitude further, or to Cape Nassau, lying between Lütke's and Barentz Lands, which was as far as he went in that direction. That the same current is prolonged to the northward and eastward as far as Cape Taimur is also probable, since Middendorf, in 1843, found a polynia or open sea there, and a tidal rise of thirty-six feet in Taimur Bay.[†] Whether the current deflects westward from Cape *Tcheliuiskin* or *Sieveroi Vostochnoi nos*, into the polar basin, is not known, no one having as yet attained that north-western extremity of Asia.[‡] Malte Brun, however, says confidently, on the authority of Olafsen, that the Gulf-stream constantly sets along the north coast of Siberia

---

[*] Second voyage of Master Henry Hudson, Purchas iii., p. 577.
[†] Beke, N.E. voy., Hakl. Soc. map.
[‡] Usually named in English charts the N.E. cape.

from east to west, and carries into all the bays that open to the east, Pernambuco and Campeachy woods, as well as the coniferous trees of Siberia itself. Barentzoon detected no tide in the Sea of Kara, or as it is called from its calmness, *Marmora*, but he found the height of water at its entrance or *Waygatz* to be greatly influenced by the wind. At Spitzbergen in Treurenberg Bay, the highest rise of the spring tides was ascertained by Sir Edward Parry to be only four feet two inches, and to take place at the fourth tide after the full moon. Martens was unable to detect any tide on the western coasts of the Spitzbergen islands. The rise therefore observed in Taimur Bay is very remarkable.*

A quotation from Sir James Ross, in a preceding page, mentions that a branch of the current which flows out of the polar basin to the south-west, down the eastern coast of Greenland, curves round Cape Farewell to run northwards along the land of West Greenland up to the Arctic circle, carrying with it a belt of ice; but the main surface current of Baffin's Bay and Davis' Straits is from the north, to form the Labrador and Newfoundland iceberg-bearing stream above mentioned. This south-going current setting through Davis' Straits, is supposed to have been fully demonstrated

---

* Dr. Wallich, in his "Notes on the Presence of Animal Life at vast Depths in the Sea," drawn up from observations made on Sir Leopold M'Clintock's survey of the sea-bottom of the Northern Atlantic, in the Bulldog, in 1860, says that the presence of the *Globigerina* tribe of FORAMINIFERA in the deep sea deposits is evidently associated, in an intimate manner, with the gulf stream or its offshoots. These organisms in a recent, if not in an actually living condition, were abundant in the ooze brought up between the Farüe Islands and Iceland, and between Iceland and Greenland; but they were almost entirely absent between Greenland and Labrador. They live at the bottom in great depths, and not near the surface; whence we may infer the gulf stream to be an under current in the localities named by Dr. Wallich. (pp. 19, 20.)

by the annual ice-drift, and by the invariable course of many whale-ships that have been beset in the ice; and by the drift of Sir James Ross's ships, of Lieutenant de Haven's, of the Resolute, after being abandoned by Captain Kellett, and, more recently, of the Fox, Captain M'Clintock. The last-named officer, however, says, that during his long and most remarkable winter's drift of eight months, from latitude $75\frac{1}{2}°$ N., down to the parallel of 65°, during which he was accompanied by several icebergs; he could detect neither surface nor ground current, and he therefore attributes the movement of the ice to the southward solely to the prevailing winds. But other observers and writers believe in the existence of a *southerly surface current flowing out of Davis' Strait.*

Bäer and Maury fully admitting the existence of this surface current, argue that there is a counter under-current setting into the polar basin to keep up the equilibrium; and the latter affirms that the warmer and salter under stream must rise to the surface somewhere in the north, and there produce a *polynia*,[*] or open sea of greater or less extent; such as that reached by Wrangell off the Kolyma in 1822; by Anjou off the Indigirka, and by Kane to the westward of Greenland in 1854.

Kennedy's Channel, as the latter piece of open water is named, is described by Mr. Morton, the only European of Dr. Kane's party who saw it, as being about thirty-five miles across. It was coasted northwards for fifty-five geographical miles; and from an elevation of 500 feet, at the limit of his journey, Mr. Morton, on the 24th of June, looked to the north-westward, over an expanse of water towards a "dark rain-cloud" on the distant horizon. From this height he saw only

[*] Spelt *poluinya* by Erman.

narrow strips of ice, with great intervening spaces of open water. A strong current was setting almost constantly to the south, but the tides in shore seemed to flow both north and south; the tide from the north ran seven hours, and there was no slack water. The wind at the time blew heavily down the channel from the open water, and had been freshening since the preceding day nearly to a gale; but it brought no ice with it.* Had there been ice-floes within a moderate distance to the northward, Mr. Morton would have readily recognized the "ice-blink" which attends them, so that twenty, thirty miles, or more, may be safely added to the extent of open water actually traced. Near Cape Jackson, at the south end of the open water, pieces of ice were observed moving northward in the channel at the rate of four miles an hour, and on the turn of the tide, returning southwards, at the same rate. At the south end of the channel the temperature on one trial was found to be +36° in the clear water of a rapid tideway close to the "ice-foot" or ledge of shore-ice; and on two other trials near the same place it was +40°, the last being of water drawn from the depth of five feet with the tide setting from the northward. The temperature of the air when the last-mentioned observation was made was +34° F. Dr. Kane states that the freezing point of sea-water in Rensselaër harbour was found to be 29° F. According to Dr. Walker, the temperature at which the surface begins to freeze in Baffin's Bay is 28½° F.†
Near Cape Independance, "many small pieces of willow, about an inch and a half in diameter, had drifted up the slope of the bay." The only willow, and indeed the only plants with a really woody stem in the high latitudes, approaching

* Kane's Arctic Expl., App. No. v., II. p. 378.   † Nat. Hist. Rev., Jan. 1860, p. 2.

80° N., are the *Salix arctica* and *Vaccinium uliginosum*, and if Mr. Morton means the diameter of the stem, and not the width of the bushy crown of branches, an inch and a half exceeds the diameter of any stem of these plants, even in much lower parallels of latitude. Willow is a common designation of slender twigs of any bush or tree, and may have been so applied by Mr. Morton. In any case these drift willows came from a distance, and did not grow near the bay in which they were found.* Much grass grew in this neighbourhood, and several flowering plants. Waterfowl abounded on the open water, the species being the same that frequent the Spitzbergen seas; very large flocks of eider ducks were swimming therein. To the southward of the open water of Kennedy Channel, a solid field of ice filled up about ninety miles of Smith's Sound, from side to side, for the two years that Dr. Kane remained shut up in Rensselaer harbour, but his chart is marked with arrows, shewing that a current running southward sets through it beneath the icy bridge.

Though the attempt made to carry the Advance northward by that opening was in accordance with the belief entertained by Lieutenant Maury, and many other cultivators of physical geography, of the existence of a polar polynia, which Franklin's ships were supposed to be traversing, Dr. Kane's remarks, part of which we quote, are made with true philosophical diffidence. "I am reluctant," he says, "to close my notice of this discovery of an open sea, without adding that the details of Mr. Morton's narrative harmonized with the observations of all our party." "It is impossible, in reviewing the facts which connect themselves with this discovery—the melted

* Kane's Arctic Explorations, II. Appendix, No. v.

snow upon the rocks, the crowds of marine birds, the limited but still advancing vegetable life, the rise of the thermometer in the water, not to be struck with their bearing on the question of a milder climate near the Pole." "There is no doubt on my mind, that at a time within historical, and even recent limits, the climate of this region was milder than it is now. I might base this opinion on the fact abundantly developed by our expedition, of a secular elevation of the coast-line. But independently of the ancient beaches and terraces, and other geological marks, which shew that the shore has risen, the stone huts of the natives are found scattered along the line of the bay, in spots now so fenced in by ice, as to preclude all possibility of the hunt, and of course of habitation, by men who rely on it for subsistence." "Tradition points to these as once favourite hunting-grounds, near open water." "I would respectfully suggest to those whose opportunities facilitate the inquiry, whether it may not be that the gulf stream, traced already to the coast of Novaya Zemlya, is deflected into the space around the Pole. It would require a change in the mean summer temperature, of only a few degrees, to develope the periodical recurrence of open water. The conditions which define the line of perpetual snow, and the limits of the glacier formation, may have certainly a proximate application to the problem of such water-spaces near the Pole."—(Kane, Arct. Expl. I. 308).

The open water of Kennedy's Channel, in the month of June, is not of greater extent than the spaces clear of ice that have occasionally been seen in summer by the whalers north of Spitzbergen. Supposing, as Dr. Kane suggests, that a current is deflected from the Spitzbergen seas round the north end of Greenland, and that in its course the warmer water rose to the surface, the temperature of 40°, observed by Mr. Morton, would at once be accounted for. The vegetation of Smith's Sound is very nearly the same as at Spitzbergen, but Dr. Kane procured seven more phenogamous species than have been enumerated in the Spitzbergen flora, towards which, the time and care he spent in collecting may have contributed.* The additional numbers are—*Ranunculaceæ*, 2 ; *Cruciferæ*, 3 ; *Rosaceæ*, 6 ; *Compositæ*, 1 ; *Ericaceæ*, 1 ; *Scrophularineæ*, 2 ; *Cyperaceæ*, 1—making sixteen additional species or varieties; and there is also an additional family—the *Empetreæ*, represented by *Empetrum nigrum*, a depressed shrub. *Vaccinium uliginosum* is another shrub of Smith's Sound not detected in Spitzbergen. A fern was also found in Smith's Sound.

The *Campanula rotundifolia*, though common enough in Greenland, was not seen by Dr. Kane so far north as Smith's Sound ; and there are three species, or perhaps only varieties, of *Caryophylleæ* more in the Spitzbergen list. Doubtless, further search would shew the two floras to be still more alike, especially if the lists of both were by the same botanist.

Mr. Morton, as mentioned above, found much grass beyond latitude 81°; and Dr. Hayes, in an excursion to the interior, eastward from Rensselaër Harbour, latitude $78\frac{1}{2}°$, to the distance of between forty and fifty miles, discovered a river flowing to the north-west, and a succession of terraced plains, generally covered with rich grass, with glaciers in the distance. He thought the vegetation much more luxuriant than on the immediate shores of Kennedy's Channel, the *Andromeda* being particularly vigorous and abundant. This militates

* See Spitzbergen Families of Plants, p. 210.

against Dr. Kane's opinion of the climate being rendered milder mainly by the warm currents of the sea.

As to the animals fed by this vegetation, they are, as far as has been ascertained, of the same species with those of Spitzbergen, with the addition of the musk-ox. This ruminant ranges over the islands north of Lancaster Strait and Melville Sound, and probably travels eastward to Smith's Sound, but it was not seen alive by Dr. Kane. He found, however, seven skeletons of the animal lying among fragments of limestone, imbedded in a paste of travertine, near the 79th parallel of latitude. The infiltration of the lime-water had begun to alter the structure of the bones; and a rein-deer skull, found in the same gorge, was completely fossilized.* No wolves were seen by Dr. Kane. Many rein-deer tracks were observed by Dr. Hayes during his journey into the interior.

The various facts which Dr. Kane has recorded respecting Kennedy Channel, lead to the conjecture that a current sets into it from the Polar basin, supplied from the east side of Spitzbergen; but as the superficial waters to the north and west of Spitzbergen flow from the north, the supply which comes to the surface in Kennedy Channel must be an undercurrent in the Spitzbergen seas, if it comes from thence, and not directly from the pole. A current or flood-tide, stronger than the ebb, comes also from the north, down the openings among the Parry Islands.

The Polar basin is supplied with water from the North Pacific and Bering's Sea, as well as from the North Atlantic. In Bering's Strait the current, according to general testimony, sets to the north. Commander Maguire, in making his way towards the north-east from Bering's Straits to Point

* Kane's Arct. Expl., I., pp. 95 and 456.

Barrow, found his progress to be greatly aided by that current during the prevalence of contrary winds; and in calms a "strong favourable current carried the ship past the grounded ice to the north-east at the rate of two miles an hour."* The same officer mentions that whalers, when making their way out of the straits with light favourable winds, were obliged to stem the current by using warps. Bering's Straits are neither wide nor deep at the present time. If the bottom has risen, in accordance with the secular movement of elevation, of which there are evidences on all the islands lying to the north of America, the diminishing influx of warmer water from the Pacific must have been gradually impairing the climate, and a corresponding loss of strength in the outflowing ice-bearing currents from the eastern openings must have taken place. South of St. Lawrence Island, at the southern entrance of Bering's Straits, counter-currents exist, either constantly or with certain winds. Lieutenant Hooper mentions that, in October 1848, the Plover was much delayed on approaching the straits by headwinds and strong currents, but that during the night after getting sight of St. Lawrence, the ship *drifted* between that island and the Asiatic coast, far to the north-west.†

Respecting the currents on the north coast of Siberia, Baron Wrangell says, that "between Svatoi-nos (longitude 140° E.) to Koliutchin Island (longitude 185° E., lying towards East Cape, and not far north of the arctic circle), during summer, the current is from east to west, or towards Bering's Straits, and in autumn from west to east. The prevalence of

* Proceedings of the Plover. Parliamentary Papers, Jan. 1855 (Blue Book), p. 905.

† The Tents of the Tuski, by Lieut. W. H. Hooper, 1853, p. 12.

north-west winds is doubtless the cause of the south-east current which we frequently observed in the spring. Our observations are confirmed by those of Liakhow in 1773, Schalarov in 1762, and Billings in 1787." . . . "The fur-hunters who visit New Siberia and Kotelnoi Island every year, and pass the summer there, have observed that the space between those islands and the continent (from sixty to one hundred and thirty miles) is never completely frozen over before the last days of October, although firm ice forms along the coast at a much earlier period. In spring, on the other hand, the coasts are quite free by the end of June, whereas at a greater distance from land the icy covering continues firm for a full month later. Throughout the summer the sea is covered with fields of ice of various sizes, drifted to and fro by the winds and currents."*

"The ice which the larger rivers bring down every year is never entirely melted the same season." "In the summer and autumn the ice breaks up into fields, and lanes of open water are met with near the land as well as towards the open sea. Winter hummocks (formed of pieces forced up over each other) are frequently one hundred feet in height. The thickness of the ice produced in a single winter is about nine feet and a half, and an exposure to a second winter will add about five feet more. Wherever the ice is formed from sea-water, and its surface is clear of snow, the salt may be found deposited in crystals. In the neighbourhood of the polynias, the layer of salt is often of considerable thickness."

"The *Great Polynia*, or that part of the polar ocean which is always open sea, is approached about twenty miles north of the islands Kotelnoi and New Siberia, and from thence in

* Wrangell, Polar Seas, Eng. tr., 502.

a more or less direct line to about the same distance from the coast of the continent between Cape Chelagskoi and Cape North (or between the 135th and 180th meridians). The shore ice extends some way farther from land at Cape North than at Cape Jakan" (eighty miles more to the west). The polynia was seen in 1811 by Tatarinow, in 1810 by Hedenström, in 1823 by Lieutenant Anjou, and in 1821 and 1822 by Baron Wrangell. This last observer adds—" Our frequent experience that north and north-west winds, and often north east winds also, are damp to a degree which was sufficient to wet our clothes, is also a corroboration of the existence of an open sea at no great distance in those directions."* "The inhabitants of the north coast of Siberia generally believe that the land is gaining on the sea. This belief is founded on the quantity of long-weathered drift-wood which exists on the *tundren* and in the valleys, more than thirty miles from the present sea-line, and decidedly above its level. In no circumstances of weather is either sea-water or ice now known to come so far inland. In Schalarov's map, Diomed Island is marked as separated from the mainland to the east of Svatoi-nos by a sea-channel, but no such strait now exists."†

In the navigable channel which bounds the American continent between Point Barrow, the estuary of the Great Fish River and the isthmus of Boothia, the tides are regular though (except in the straits) of small velocity, and producing little rise of water, rarely amounting to four feet; but in certain of the straits exceeding that rise. Some Arctic navigators have thought that they perceived a prevailing current setting to the eastward along the coast. Sir John Richardson found the flood-tide taking that direction between the Mackenzie

* Wrangell, Op. cit., p. 505. † Wrangell, Op. cit., p. 506.

and Coppermine Rivers, and in the Dolphin and Union Strait, both flood and ebb had so strong a current that it became advisable for the boats to lie by while the stream was adverse. A gale of wind, however, had a very decided effect in raising the water, three days of a strong north-wester being sufficient to flood for many miles the low lying meadows on the east of the Mackenzie, and to deposit long lines of drift timber a mile or two inland of the ordinary spring tides. At the distance of fifty miles to seaward off the Mackenzie, Captain Collinson experienced currents so strong that, with the boats towing a-head, he could not at times prevent the ship from being turned round. In Bellot's Strait, the first easterly outlet from the continental channel, Sir Leopold M'Clintock had to contend with tides running "like a mill-stream" at the rate of seven miles an hour. The flood came from the north-west, and the ebb flowed with nearly equal force. In Committee Bay, the bottom of the Gulf of Boothia, Dr. Rae ascertained a total rise of nine feet.

Opposite the eastern end of Bellot Strait, on the other side of Prince Regent Inlet, there is another strait leading through Eclipse Sound to Pond's Strait, and also communicating by northern channels with Lancaster Sound. The set of the tides in these straits and channels has not yet been determined. But in the Fury and Hecla Strait which bounds Cockburn Island on the south, and connects Foxe's Channel with the Gulf of Boothia, the rise of the tides was nine feet, and the stream came from the west during the twenty-four hours, with eddies in shore running in the opposite direction. The current from the west was at times as great as four miles an hour, and the observers thought that, in the summer season, it was so much stronger than in the winter time, as to mask the small stream of the ebb-tide that would have set westwards. In the frozen strait of Middleton the current of the flood or ebb-tide is so strong, according to the Eskimos, that polar bears, when compelled to take to the water, are often swept under the ice by the stream and drowned.

The channel further north has also a general set of current from the westward. In Prince of Wales' Strait Sir Robert M'Clure ascertained that the flood-tide came from the south, and that at spring-tides there was a rise and fall of three feet, with little if any rise at neaps. At Point Armstrong near Princess Royal Islands, in the Prince of Wales' Strait, a large quantity of drift-wood was seen by the same officer. It was all American pine, and, in the opinion of the carpenter, could not have been carried from its native forest above two years. As the Coppermine River brings down but a very small number of drift trees, and none at all descend the more easterly rivers, the drift trees of Prince of Wales' Strait come almost certainly from the Mackenzie, which annually rolls down vast numbers. Captain Collinson saw much drift-wood at the distance of fifty miles from the mouth of that river, and measured the trunk of a tree sixty-eight feet long, which must have grown to the southward of the Arctic circle.

In Banks' Strait (at the Bay of Mercy), a registry of ten months shewed a maximum rise of two feet, four tides after the full and change of the moon; Sir Robert M'Clure's observations coinciding with those made by Sir Edward Parry on the opposite side of the strait. There the ice-drift, whether impelled by currents or prevailing winds, coming round Prince Patrick Island, and down the passages between it and Melville Island, keeps the strait constantly filled with the pack. Sir Edward Parry remarks, that "the westerly and north-westerly winds were always found to produce the effect of clearing the southern shores of the North Georgian Islands (called on recent Admiralty charts the Parry Islands) of ice, while they always brought with them clear weather." He also notices the fact of his having sailed back from Winter Harbour to the entrance of Lancaster Sound in six days, a distance which took five weeks to traverse in the opposite direction.* While he remained in Winter Harbour of Melville Island, in the months of May, June, and part of July, the maximum rise of the tide was four feet two inches, and the minimum ten inches, the mean rise being rather more than two feet and a half. The highest tide was the fourth after full moon. From Dr. Sutherland's register of tides, kept near Cape Hotham, to the west of Wellington Channel, we learn that there the rise and fall varied from less than a foot to more than six feet. At Cape Beecher, on the north side of Wellington Channel, where it joins Queen's Channel, the tides, says Captain Penny, "flow regularly, but when strong winds blow from the north-north-west, they continue tide and half tide, the flood coming from the westward, and at a much greater rate;" that is to say, the flood tide continued nine hours and the ebb only three, the fall of water being rapid. In the autumn of the preceding year this gentleman, in conjunction with the American expedition, experienced a strong rush of water towards the north up Wellington Channel, caused, he states, by the long prevalence of south-east winds. In the straits formed in that channel by Baillie-Hamilton, and Dundas Islands, the tides are very rapid, the grinding of the ice on the beach producing a sound like thunder.† Farther

* First Voyage of Discovery, by Captain W. E. Parry. London, 1821, p. 299.
† Journal of a Voyage, etc., by Dr. P. C. Sutherland. London, 1852; II. pp. 152 and 161.

to the north, on the 77th parallel, beyond Grinnell Land, Sir Edward Belcher observed the *ebb* running strong to the eastward towards Jones' Sound.

That the general drain from Barrow's and Lancaster Straits is into Baffin's Bay, the preceding observations quoted from Sir Edward Parry, together with the drift of Sir James Ross's ships in 1849 from Port Leopold, that of the American Expedition in 1850, and of the Resolute in 1854, is sufficient to show. We have also seen that the Fury and Hecla Strait affords another outlet from the Polar basin, and that there, probably, is an intermediate one in Pond's Strait; between Spitzbergen also and Greenland a current comes from the north. In the contrary direction, there is the current setting northward through Bering's Strait, whose existence is fully established by observation, and one is surmised to flow between Spitzbergen and *Vostochnoi Severoi-nos*, but which has been actually traced no further than Novaya Zemlya. These are surface currents. Further experiments are needed to prove that there are under currents, though their existence has been inferred on theoretical grounds.

An able writer in the Natural History Review,* understood to be the Professor of Geology in the University of Dublin, gives the following theory of the Arctic tides:—"The great tidal wave enters the Polar Sea from the Atlantic by two distinct channels, separated from each other by the continent of Greenland. The first branch of the Atlantic tide, having swept past the British Islands and coasts of Norway, flows into the Polar Sea, past the islands of Spitzbergen, being assisted in its flow and retarded in its ebb by the remains of

* Natural History Review, April 1858, p. 65.

the Gulf-stream, whose heating effects are supposed to be felt even by the glaciers of Spitzbergen.

"Of the oscillations and movements of the Polar Sea itself north of Europe and Asia, we know but little, except the fact furnished to us by Von Wrangell, that its resultant on the north coast of Siberia is a current setting east by south, towards Bering's Strait; arrived at this point, the current is complicated in its action by the influx of the Pacific tide, whose movements are totally different in character. The combined Atlantic and Pacific tides (the latter predominating) flow and ebb in an east and west direction, along the coast of North America, with a preponderant set to the eastward, round Point Barrow, Cape Bathurst, through Dolphin and Union Strait, and Dease's Strait, and probably into Victoria Strait, as far as the bottom of Peel Sound and Bellot Strait, leading into Prince Regent's Inlet. It is highly probable, although it has not been distinctly proved, that off shore, both in Asia and North America, the Atlantic tide and Gulf-stream produce a resultant movement of the waters of the Polar Sea, which presses its loose pack-ice eastward and southward against the western and north-western shores of the Parry Islands, forming the great pack-ice observed by M'Clintock on the north-western shore of Prince Patrick's Island, and also the formidable double and triple floes to the west and north of Banks' Land, encountered by M'Clure. To the westward of Banks' Land, at some distance seaward from the American continent, is found the permanently ice-blocked sea, called by the Eskimos 'the land of the white bear.' This gigantic floe we believe to be formed by the continued eastern set of the deep tidal and oceanic currents of the Polar Sea east of Spitzbergen; and that it is prevented from permanently blocking up the coast line of the American continent only by the influence of the rapid tides which enter the Polar Sea through Bering's Strait."

"The second branch of the great Atlantic tidal wave, passing up to the westward of Greenland, fills Baffin's Bay, flows northward through Smith's Sound, and westward through Jones' Sound and Lancaster Sound, causing high water in succession in Prince Regent's Inlet, Wellington, Austin, and Byam Martin Channels. It finally meets the conjoined Pacific and Polar tides at the entrance of Banks' and Prince of Wales' Straits; the Pacific tide at Bellot's Strait, and the true Polar tide in the centre of Byam Martin Strait, in the space between Queen's and Wellington Channel, again in Cardigan Strait and Belcher Channel, and lastly at the ice-belt, dividing the open entrance of Smith's Sound from Kennedy's Channel. The limit of the Atlantic tide represents still water at all times of the tide, the currents flowing to and ebbing from the "head-line" of tide (in the manner well known in the Irish Sea and English Channel, forming slack water near the Isle of Man, and from Dover to Beechy Head). In a sea impeded by broken ice-floes, such a meeting of tidal streams will produce an almost permanent and immovable thickened floe." In another paper, the same author calculates the head of the tide, or point of meeting, to occur at ten or eleven o'clock Greenwich time. High water takes place three hours and a half sooner at the northern extremity of North Somerset than it does in Bridport Inlet of Melville Island.*

* Natural History Review, V. p. 123, July 1858. See also the preceding pages of the present work, where Professor Haughton's opinions are referred to.

CHAPTER XIII.

ICE.

Disruption of River and Lake Ice—Lapland—the Mackenzie—the Kolyma—Sea ice—Whale-fishers' bight—Effect of Drift Ice on the climate of Iceland—on Meta Incognita—Poles of cold—Thermic anomaly—Continental climate.

In treating of the oceanic currents in the preceding chapter some facts respecting the movements of ice in the Polar seas have been mentioned. The Siberian rivers Obi, Yenisei, Lena, Indigirka, and Kolyma, and the American Mackenzie, Coppermine, and Great Fish rivers, all rising far to the south of the Arctic circle, carry much ice into the Polar basin. Wahlenberg has remarked in his *Flora Lapponica* that the air must acquire a mean temperature of $39\frac{1}{2}°$ F. before the frozen rivers of Lapland break up completely. In the interior of subarctic Siberia and America, however, the spring is comparatively cloudless, and the direct rays of an unveiled sun have a manifest influence in hastening the epoch of the opening of the rivers, so that near the sources of the Mackenzie, for instance, about the 55th parallel, 36° F. is probably nearer the mean atmospheric temperature of the ten days which immediately precede the general disruption of the ice. As a matter of course, the upper or more southerly branches of these rivers break up first, and bearing down accumulations of water, ice, and drift-trees, the flood sooner or later is obstructed by a strong bridge of ice extending across the river. The water rapidly rises in the Mackenzie, often to the height of forty feet above its autumn level. Its pressure at length demolishes the obstructing bridge, and the flood sweeps over the islands and submerged banks, cutting down the trees as the grass falls before the mower's scythe. This operation is repeated more or less frequently before the debacle reaches the sea, and in some seasons much more destructively than in others. In the general thaw the land-floods, proceeding from the melting snow, break down the river banks in innumerable places, adding largely to the drift-trees; and in the Mackenzie the snags and sawyers are as common as in the mighty Mississippi. The Mackenzie usually breaks up where it crosses the Arctic circle, about the middle of May, or a few days earlier, the 23d of the month being unusually late. It takes about a fortnight for the flood to make its way from the Arctic circle to the delta of the river. The disruption of the river-ice is speedily followed by the arrival of geese; but the larger lakes in the same quarter are not navigable for a month or six weeks later.

The Oussa, which rises from the Ural mountains within the Arctic circle, and, running west-south-west, joins the Petchora near the 65th parallel of latitude, is frozen by the beginning of September, but firs, birches, alders, service trees, and willows grow on its banks, the forest being similar in character to that on the Mackenzie. The subsoil is permanently frozen, yet barley, rye, sheep, and cattle are products of the district.

On the Lower Kolyma the seasons appear to be more severe, and the spring later. Baron Wrangell, speaking of this province, says:—"At Nijnei Kolymsk (latitude $68\frac{1}{4}°$) the river freezes early in September; loaded horses can often cross the ice of the most northerly branch as early as the 20th of August, and the icy covering never melts before the beginning of June. When need is at the highest, suddenly large flights of birds arrive from the south, swans, geese, ducks, and snipes, and the general distress is at an end. At last, in June, the rivers open, and fish pour in abundantly ; but sometimes this season brings with it a new difficulty. The rivers cannot carry away sufficiently fast the masses of ice which are borne down by the current ; these ground in bays or shallows, and thus form a kind of dam, which impedes the course of the river, and causes it to overflow the banks ; in this way the meadows and villages are sometimes laid under water. These overflowings of the rivers take place more or less every year."*

It requires as much heat to melt a given quantity of ice as would raise twenty-eight times the mass of water one degree of Fahrenheit. Hence the drifting of a large quantity of ice down the rivers relieves the districts it leaves from the loss of heat that would have been consumed in melting it, and that which is carried down the rivers into the Polar basin produces a proportional deterioration of climate there.

Baron Wrangell's description of the sea-ice north of the Kolyma, already quoted, will apply generally to the ice in the sea north of America; but though there are high hummocks and ridges where currents or strong winds have pressed the floes and smaller pieces together, and caused them to override each other, there are no icebergs of any size in the Arctic American sea, from the absence of glaciers to furnish them, either on the continental shore or islands due north of it. The nearest approach to an iceberg on the American coast-line is a talus of drift-snow formed under a precipitous cliff washed

* Wrangell's Polar Sea, pp. 46 and 62.

by the sea, which breaks off by the action of the waves and sun after one or more summers. These are few and comparatively insignificant when contrasted with the mountainous bergs furnished by the vast glaciers of Greenland and Spitzbergen.

Sir Robert M'Clure and Captain Collinson, in the voyages from Bering's Straits to Banks' Island, obtained information of the fixed barrier of ice already noticed as distant from thirty to fifty miles from the continent. It is probable that this icebelt hangs on to a northern chain of islands. The Eskimos of Point Barrow have a tradition, reported by Mr. Simpson, surgeon of the Plover,[*] of some of their tribe having been carried to the north on ice broken up in a southerly gale, and arriving, after many nights, at a hilly country inhabited by people like themselves, speaking the Eskimo language, by whom they were well received. After a long stay, one spring in which the ice remained without movement they returned without mishap to their own country, and reported their adventures. Other Eskimos have since then been carried away on the ice, and are supposed to have reached the northern land, from whence they have not as yet returned. An obscure indication of land to the north was actually perceived from the mast head of the Plover when off Point Barrow.

In the latter quarter, lanes of water, in which the whale-hunt can be carried on, appear in some seasons by the end of April, though it is usually later in the year before much movement in the ice takes place. As early as the end of March, lanes of open water were seen between Beechey Island and Port Leopold by Captain Pullen in 1854, which he considered to be unusually early. But Sir Leopold M'Clintock,

[*] Blue Book on Arctic Matters, 1835, p. 939.

in the remarkable sledge journey of 105 days, employed in travelling round Prince Patrick's Island, and in surveying the adjacent shores, kept on the ice till the middle of July. Lancaster Strait and Melville Sound are seldom navigable for ships before the end of the month, and the harbours are often closed up till late in August, so that a ship that has wintered in one of them has only a fortnight or less time left for escape.

About the end of September fresh ice begins to encrust the surface of the sea, so as to terminate the general navigation for the season, and the history of Arctic enterprise in the earlier pages of this work shews that in certain localities, and in some seasons, the ice may be packed by prevailing winds and currents so as to obstruct the progress of ships for a whole summer, or even for several successive years.

The usual course of the whalers in Davis' Straits is to work to the northward along the coast of West Greenland early in the summer, and to cross over to the broken lands of *Meta Incognita* as soon as the "middle ice" has drifted far enough down the strait to allow them to pass round its northern end, or has become loose enough to let them sail through it.

On the east side of Greenland, a remarkable tongue of ice, mentioned by Dr. Scoresby, stretches abruptly to the north, between Ian Mayen and Bear or Cherie Island, and separates the west or *sealing district* from the east or *whale-fisher's bight;* this latter being the only pervious track to the northern fishing latitudes. Sometimes the bight is closed up on the north by ice, and ships are then prevented from approaching Spitzbergen, on which event the season is termed a *close* one, and is unproductive. The co-tidal curve indicated by Professor

Haughton, as stated in Chapter XII., would, if prolonged from Smith's Sound, take this direction, and this tongue of ice may occupy the area of comparatively slack water, interposed between the north-flowing current on the east of Spitzbergen, and the ice-bearing stream running southwards down the Greenland coast.

From some cause or concurrence of circumstances this tongue-like floe occasionally breaks up to an unusual extent, as was the case when Dr. Scoresby took advantage of the event to recommend 1818 as a favourable period for polar research. When the ice is drifted in extraordinary quantities upon the Iceland shores, it has deteriorated the climate of that island for a year or two, so as to produce famine from lack of pasture, as well as exposed the flocks to the ravages of the numerous polar bears that inhabit the ice; but at the same time the Icelanders obtain a supply of drift-wood which lasts for years. The year 1271 is mentioned in the Icelandic chronicles as one in which an extraordinary quantity of ice, multitudes of bears, and much wood, were cast on the coast by a north-west wind.*

The deterioration of climate by drift-ice, which is only occasional after intervals of many years at Iceland, recurs annually in *Meta Incognita*, and the adjoining corner of the American continent. The winds and currents conspire to fill the numerous intricate sounds and straits in that quarter with drift ice during the summer, by which the temperature of that season is kept much below the normal heat of other meridians in the same parallels of latitude. By the great extent of land, also, near the Arctic circle, as will be mentioned in Chapter

---

* "The shores of Iceland are visited by drift-ice only seven or eight times in a century."—Sir F. Leopold M'Lintock, in the *Engineer*, Dec. 21, 1860.

XV., the climate becomes what is called a continental one, and its winters more severe, so that both in summer and winter the temperature is kept low. Sir David Brewster explained these facts by supposing the existence of two poles of extreme cold in the northern hemisphere, one near latitude 80°, in longitude 92° west, the other in Siberia; round these centres, as poles, he represented the isothermal lines as circulating in lemniscate curves. Dovè gives a graphic exposition of the same facts under the designation of the *thermic anomaly*, by which he represents an area of abnormal cold as having a monthly progression, being inland on the parallel of 60° north at midwinter, and having moved north-east to *Meta Incognita* at midsummer. In like manner, the Siberian area of abnormal depression of temperature moves from the vicinity of Jakutsk, where it has its winter station, to Bering's Sea at midsummer. It is to be understood that these *isabnormal* areas denote merely the places where the temperature is lowest on their parallels of latitude, and not the coldest points in that hemisphere. They imply, of course, areas on the same parallels wherein the temperature exceeds the mean.*

* Distribution of Heat over the Surface of the Globe, by H. W. Dovè, London, 1853.

CHAPTER XIV.

WINDS.

Mr. Coffin's Theory—Lieutenant Maury's—Von Wrangell's Observations on the Winds of the Kolyma District—Winds at Fort Confidence—Teploi Weter—Repulse Bay—Baffin's Bay and Davis' Strait—Spitzbergen.

MR. J. H. COFFIN, in a treatise published in the sixth volume of the *Smithsonian Contributions to Knowledge* (1854), places a meteorological pole in latitude 84° N., longitude 105° W., and states that it is encircled by a zone twenty-three degrees and a half in breadth, of westerly or north-westerly winds, encompassed on the south between the parallels of 60° and 66° by a belt of easterly and north-east winds, as indicated by observations made at Great Bear Lake, Great Slave Lake, and Fort Enterprise, two stations in Greenland, and one at Reikiavik in Iceland. Lieutenant Maury's wind-chart marks the prevailing direction of the winds in the polar basin and northern seas as being westerly, but does not indicate the easterly encompassing belt of Coffin. These generalisations, though backed by references to observations, are partly founded on theoretical considerations.

Baron Wrangell, in treating of the winds of the Kolyma district in Arctic Siberia, says that the *north wind* is seldom fresh, or of long continuance; it is more frequent in summer, when it brings cold, than in winter, when it often brings mist and milder weather. The *north-east wind*, or more often the *east-north-east*, is seldom of long continuance, or violent. It usually clears the atmosphere from mist, and causes the thermometer to rise in summer and to fall in winter. The *south-east wind* drives away mist, and may be regarded as the prevailing wind in autumn and winter. There is a remarkable phenomenon called the *teploi weter* (the warm wind), which occurs sometimes in the middle of winter; it begins suddenly, when the sky is quite clear, with the wind blowing from the south-east by south, or south-east by east, and causes the temperature to rise from — 24°, or even — 47°, to + 32° or + 35° F., the barometer having in the preceding eight hours sunk four-tenths of an inch. The *south-south-east* winds do not influence either the barometer or thermometer. *South winds* seldom blow with much force. The *south-west wind* influences the temperature in summer little, but in winter it is the most piercing of all winds, and is called by the natives *schalonik*. The *west and north-west winds* prevail on the general average of the year; in winter the *south-east* prevails, in summer the *north-west*,—this latter wind blowing often in summer also; it is a cold wind in summer, and in winter brings snow and bad weather.*

Fort Confidence is situated on the north-east arm of Great Bear Lake, in latitude 66° 54′ N.; longitude, 118° 49′ W.; and consequently rather more than 80 degrees of longitude from the Kolyma. At this post, for seven winter months (October and April inclusive), in 1848-9, the wind was noted hourly, the total number of observations being 3430, of which 294 were calm. Direct *east winds* blew on 547 hours from the *barren grounds* towards the wooded valley of the Mackenzie;

* Wrangell, op. cit., pp. 49 and 513.

and on 286 hours west winds blew. The east and west direction of the arm of the lake, on which the house stood, had probably an influence on the frequency of these winds; excluding them from the calculation, we have 969 hours of north and north-easterly winds, and 348 of winds from the northerly and westerly quarters, or 1017 hours of winds coming more or less directly from the north. Of winds with southing, there were only 262 from the westerly points, and 718 with easting, or 980 hours of winds coming from any southerly points. The southing increased with the progress of the spring, and, had the summer months been included, would have predominated. The pressure of the atmosphere was greatest when the wind was south-east, decreased greatly when the wind came from any point to the north of east. Wrangell says that the south-east winds were preceded by a fall of the mercury on the Kolyma; and we may perhaps conclude that they occasioned a rise there, as they do at Fort Confidence, though he does not say so. At Fort Confidence the force of the winds was least in mid-winter, and from December to March, both months inclusive, calms were very frequent, but became rare in April. The sky was comparatively cloudy in October and November, and became remarkably clear in December and the succeeding four months. A storm of wind and snow always raised the temperature, which was uniformly low in a clear winter sky. In Arctic America, the phenomenon of warm winds (*teploi weter* of Wrangell) also occurs, and makes the month in which they happen, whether December, January, or February, warmer than the other two. The same warm wind was probably the cause of the rain which the Russian sailors observed in Spitzbergen in the month of January.

The observations of Dr. Rae made in the years 1846 and 1847 at Repulse Bay, in latitude 66° 32' N., and longitude 86° 56' W., or about 32 degrees of longitude east of Fort Confidence, on nearly the same parallel, furnish a convenient example of the winds on a different meridian. The period of observation embraces the entire year, except the last twenty days of August. The direct east and west winds were few at Repulse Bay, there being only 23 days of the former, and 22 of the latter. The days of north winds, and of northerly and easterly ones, were 130; of northerly and westerly ones, 261; or 391 days of winds having more or less northing. Of those having more or less southing, there were 82 days, viz., 52 with easting and 30 with westing, the directly east and west winds being excluded from this part of the numeration. The northerly and westerly winds were greatly in excess from December to April, Dr. Rae's experience agreeing with Sir Leopold M'Clintock's, who, in his winter drift down Davis' Straits, had almost constant northerly winds.

In the four months of May, June, July, and August, the northerly winds prevail in Baffin's Bay and Davis' Straits, being, according to Dr. Sutherland's record kept in 1850, 14 days of direct east wind, 4 of direct west; 54 with more or less northing, of which 43 belonged to the north-east quarter, and 11 to the north-west. Winds blew from the south-east quarter on 12 days, and from the south-west on 26; the total with southing being 38.

The effect of these prevailing northerly winds in bringing down ice from the Arctic basin, and filling the straits of the American north-eastern Archipelago in summer, is unquestionable, and, as has already been said, is probably the main cause of the abnormal depression of temperature in that quarter.* In the interior of the continent, when the ground is well clothed with snow, the influence of the winter winds on vegetation can be only very small; but, on the contrary, the southerly winds that prevail during summer in the valley of the Mackenzie must tend greatly to promote the growth of the flourishing forests which fringe the banks of that stream nearly down to the shores of the Arctic Sea.

Dr. Scoresby says of the Spitzbergen seas—" North-west and east winds bring with them the extreme cold of the icy regions immediately surrounding the pole, whilst a shift of wind to the south-west, south, or south-east, elevates the temperature to that of the surrounding seas." This is, of course, from his experience during the season of navigation, commencing in April, for he did not winter in Spitzbergen. A hard westerly gale with snow, occasions, he says, the greatest depression of the mercury in the barometer; and a light easterly wind, with dry weather, the greatest elevation—his experience agreeing in the latter respect with the observations made at Fort Confidence.

* See Chapters XIII. and XV.

CHAPTER XV.

TEMPERATURE.

Decreases with increase of Latitude—Effects of the Predominance of Land—Snow-line—Central Heat—Temperature of Soil—Epoch of Thaw—When the Rivers freeze again—First Appearance of Vegetation—Isothermal Lines—Table of Temperatures—Comparison of Latitude with Altitude.

TEMPERATURE performs an important part in the promotion of vegetation on the surface of the earth; and, though not the sole agent, is a principal one in the maintenance of that variety of plants exhibited by the various zones of the earth that succeed each other between the tropics and the poles. The sources of the heat are two. One existing in the centre of the earth has been clearly demonstrated, by direct thermometrical experiments in mines, shewing that the temperature increases with the depth to which the surface of the earth is penetrated. Geologists affirm that in ancient epochs of the earth's history, the mass of the earth was warmer than at present, and thus they account for the fossil plants of the older strata in northern or southern latitudes having more the character of tropical or subtropical productions, than the climates of the same latitudes will maintain in the present day. And as all bodies part with heat from their surfaces in every direction by radiation, it follows that the earth would be continually cooling did it not receive accessions of heat from the only body exterior to itself from whence it can come, namely from the sun.

The rays of the sun strike any one part of the earth only one half of the year, or about one half, for there is a slight difference in this respect between the northern and southern hemisphere; at the poles the day is six months long, and so is the night; while at the equator, where the day is also equal to the night, the length of each is only twelve hours. Between these extremes there are all intermediate stages of transition. The effect of the sun's rays lessens as their obliquity increases, and their thermal power ought consequently to diminish as the poles are approached. Philosophers, taking into account these elementary propositions, have endeavoured to elicit a rule by which the connection of the mean temperature of a place, with its latitude, may be calculated.

But other influences than mere distance from the equator contribute to produce the great variety of climates which experience has proved to exist, and Humboldt has shewn the importance of the irregular distribution of land and water, and of aerial and marine currents. As a graphic exposition of ascertained facts, he suggested the delineation of isothermal lines, and his idea has been ably acted on by Professor Dovè, whose charts, founded on an immense body of observations, collected from every available source, should be consulted by every one who is desirous of acquiring a comprehensive knowledge of the *distribution of heat on the surface of the globe.* He says that from 60° latitude to the pole the decrease of temperature is represented with much exactness by the following formula, in which t_x denotes the mean temperature of the year in degrees of Fahrenheit in the latitude x:—

$$t_x = + 3\cdot65° + 105\cdot75 \cos {}^2 x.$$

As far as lat. 80° the formula, he tells us, gives very approximate values, but at the pole there is a difference of about 1·35° of Fahrenheit.* Principal Forbes of St. Andrews, assuming it to be a fact that the temperature of the globe, on an average of all the meridians, reaches its maximum in latitude 6° 30' north, gives the following empirical formula, coincident with that of Kamtz, in which T is the mean annual temperature on Fahrenheit's scale of the parallel whose latitude is λ :—

$$T = 80\cdot 8° \cos.^2 (\lambda - 6° 30').$$

By this formula the temperature at the pole is + 1·0 Fahr., and on the Arctic circle about + 19·3 Fahr.

In the same paper, Principal Forbes states that the maximum proportion of land on any one parallel of latitude, being about six-tenths of the circumference, occurs almost exactly on the Arctic circle, and that in latitude 50° south, the entire circle of latitude passes through water, being the only portion of the known globe where this is the case. The effect of masses of land or continents is in every parallel to exaggerate the variation of temperature due to the seasons, and also to depress abnormally the mean annual heat beyond 45° of north latitude, and to raise it nearer the equator. In meridians which pass through one of the great oceans—the Atlantic for example—the decrement of temperature follows pretty nearly the formula of Sir David Brewster, or the simple cosine of the latitude; but when the continents are included, it is more accurately expressed by the square of the cosine, or the formula of Mayer. The great accumulation of

* Distribution of Heat, etc., by H. W. Dovè, printed for the British Association. London, 1853, p. 15. And Inquiries about Terrestrial Temperature, by James D. Forbes, F.R.S., etc., in the Trans. of the Royal Soc. of Edin., 1859.

land in Siberia sinks the temperature below the mean of the parallel.*

These quotations from the two works we have cited are all that we purpose to state on the general question of gradation of *mean temperature* with increase of latitude. The physical phenomena resulting from the diminishing temperature are more immediately our object, and one of the most obvious is the existence of perpetual snow on the summits and sides of hills at altitudes varying with the latitude, and also with other circumstances, which produce so many local modifications of the general law enounced by Professor Leslie, that we can be guided by actual observation only. This philosopher, starting with the erroneous assumption that the mean temperature of the atmosphere at the pole is $+ 32°$ or $+ 28°$ Fahr., tells us that the limit of perpetual congelation forms nearly the curve called the *companion of the cycloid*, bending gradually downwards from the high regions of the atmosphere as it recedes from the equator, reverting its flexure at the 45th parallel of latitude, and grazing the surface of the sea at the pole; the mean height of eternal frost under the equator, and at latitudes $30°$ and $60°$, being respectively 15,207 feet, 11,484, and 3818. He is probably correct in supposing that the lower limit of the snow line keeps near the sea in the Arctic polar regions, notwithstanding that the mean annual temperature of the air is thirty or forty degrees lower than that which he assigned to it; and the explanation of the fact may be sought for in the influence of direct radiation from the sun, reverberated from large tracts of land continuously for six summer months, compensating to a greater degree

* Inquiries about Terrestrial Temperatures, by James D. Forbes, D.C.L., F.R.S., etc. Trans. Roy. Soc. of Edin., xxii., pl. 1, p. 79.

than he had imagined for the obliquity of the sun's rays; in the effect of mild southerly winds, and, perhaps still more, in the existence of oceanic currents bringing warmer water and rafting off ice. The observations of Humboldt, Dr. Hooker, and others, shew how very much the height of the snow line on different sides of the same range of mountains is varied by conditions of aspect and of radiation from adjoining plains. That the reflection of the sun's rays from a snowy surface in a clear atmosphere has a most powerful effect on the thermometer, has been surmised by Professor James Forbes; and it will be found, doubtless, that between the upper and lower limits of perpetual snow within the Arctic circle there is a difference as great as on the sides of high mountain ranges.*

* It was not till the manuscript of this and the following chapters had been sent to the printer that I received Mr. L. W. Meech's paper on the *Intensity of the Heat and Light of the Sun upon different latitudes*, published among the *Smithsonian Contributions to Knowledge*, in 1856. This author's deductions from his elaborate mathematical investigations coincide with many of the statements given in the text as founded on observation; and his paper should be consulted by the reader who feels an interest in these matters. Room can be found here for only a few desultory extracts.

While the intensity (or thermal effect), at any one instant of time, decreases from the equator to the poles, and is *proportional to the cosine of the latitude*, the cumulative intensity during *twenty-four hours of polar day at the summer solstice is one-fourth greater than on the equator*. This the author states is owing evidently to the fact that daylight in the one place lasts but twelve hours out of twenty-four, while at the pole the sun shines on during the whole twenty-four. . . . The excess of thermal effect at the pole continues for eighty-five days, commencing on the 10th of May, ending on the 3d of August, and comprehending the whole summer season in the frigid zones. In the six winter months the intensity at the poles is 0. Let the number of days in a mean tropical year (365·24) represent the *thermal unit*, and the values of all the latitudes be converted in that proportion, then, while the thermal days in the year are 365·24 at the equator, they are 183·41 at the polar circles, and 151·59 at the poles, or five thermal months. Between 60° and 80° of latitude the height of the line of perpetual snow (or frost) descends 891 feet for every increase of five degrees of latitude, having an evident relation to the differences

It is certain that there is considerable phenogamous vegetation in the most northern lands that have been attained, and that lichens flourish on rocks rising far above the level of snow which continues to cover the ground from year to year. In the chapter on Spitzbergen it has been mentioned that almost all the valleys that have not a southern aspect are filled with snow or glaciers, yet Dr. Scoresby states that in climbing a mountain in King's Bay, of about 3000 feet in height, plants of *Saxifraga, Salix, Draba, Cochlearia,* and *Juncus,* which he had observed here and there for the first 2000 feet of elevation, did not disappear till he approached the summit. At the height of 3000 feet, the rays of the midnight sun caused streams of water to issue from the snow, and the temperature of the air in the shade was + 37° F. on the night of the 23d of July. He does not state what vegetation he saw on the summit of the hill, but it is probable that wherever the rocks were denuded of snow they supported crustaceous lichens, and that the upper limit of the snow-line about the 80th parallel of latitude, on the meridian of Spitzbergen, is elevated about 3000 feet. Within the Arctic circle, on the American continent, none of the mountain ridges are known to rise to the line of perpetual snow, though farther south the high peaks of the Rocky Mountains overtop it. Wrangell tells us that the thaw proceeds every summer at the Asiatic Liakhow Islands, disengaging the fossil bones of which the cliffs there are mostly composed; but on the Siberian continent no Arctic moun-

of the number of thermal days on the successive parallels. The intensities above mentioned represent the sun's effect at the summits of the atmosphere. " While passing through the atmosphere to the earth, the solar rays are subject to refraction, absorption, polarization, and radiation; also to the effects of evaporation, of winds, clouds, and storms." The thermometric heat at the surface of the earth being the resultant of a variety of causes.—(P. 21.)

tains are spoken of as clothed with perpetual snow, except on the promontory of Sieveroi Vostochnoi-nos, nor does Wrangell mention glaciers. There are, however, as has been already said, throughout the polar seas, scattered banks of snow, accumulated under cliffs with a northern aspect, which the summer heats have not wholly melted when the new snow begins to fall, and in certain localities packs of ice may remain for several summers, receiving winter additions equal to the summer's waste.

Innumerable observations have established the fact that the temperature of deep mines greatly exceeds that of the atmosphere at the surface of the earth; but the rate of increment, corresponding to the depth, varies with the locality, and is variously stated by experimenters at forty-five, fifty, sixty, and by some at one hundred feet of descent for each degree of Fahrenheit's scale of increased heat. The mean increase has not as yet been satisfactorily ascertained over any extensive district.*

In the nearly cloudless winters of the Arctic regions during the total absence of the sun, or in the nights of spring, the radiation into the dark blue depths of space produces the enormous depressions of temperature recorded by travellers in their thermometrical tables.

The heat thus parted with is replaced by the calorific rays of the sun when that luminary is above the horizon; but within the Arctic circle generally, the direct radiation of the sun during the whole spring of that region, or until after the sun has begun to decline from its greatest altitude, is employed in removing the snowy covering in which winter had clothed

* Adolph Erman found the increase of temperature in the mines of the Ural, about the 59th parallel of latitude, to be 1° R. (2°·25 F.) for every 112 feet of descent, or 1° F. for every 50 feet of descent.—*Travels in Siberia*, i. p. 238.

the earth; and the soil, though it thaws rapidly as the floods of melted snow pass over it, is exposed to the direct rays of the sun for only three months at most, and for a shorter time in the highest latitudes that have been reached by explorers. The two Arctic seasons of summer and winter are, therefore, of very unequal duration, being respectively of nine and three months, the latter including June, July, and August, being further restricted in the very high latitudes, and but little extended in the most favoured districts, such as on the Norwegian peninsula, and in the sheltered alluvial valley of the Mackenzie, or that of the Obi. It is in this short summer only that phenogamous vegetation can proceed; but the powerful effect of the sun's rays in May, and even in April, may promote the development of lichens growing on precipices where the snow cannot lie; or prepare trees rising above the snow for the ascent of the sap, which, as a general rule, does not flow freely till the snow is gone. Everywhere in Arctic America and Siberia the trees freeze to their centres in winter, and are not thawed till the end of March or beginning of April.

The thermal effects of the two seasons descend in waves through the soil, becoming gradually less and less distinct as the distance from the surface increases, and finally blending at depths which vary with the latitude and with local causes, but which Dovè states to be at about 100 feet below the surface. In severe polar climates the result of the comparative length and severity of the winter's cold is a permanently frozen substratum, whose southern limit coincides, according to Bäer, with the isothermal line of $+32°$ F., or the freezing point, its thickness increasing, of course, with the decrease of mean annual temperature calculated for a series of years.

The central heat of the earth sets bounds to the depth of the frozen soil, by lessening, as Dovè says, the extreme temperatures, without affecting the periods of variation on the surface; but borings within the Arctic circle have as yet been too few for the enunciation of any rule whereby the thickness of the permanently frozen bed can be correctly calculated. In Arctic Siberia and America the sun's rays thaw the surface-soil to the depth of from six inches to one or two feet, or more, under which the hard icy substratum presents an even surface, like a smooth bed of rock; and in woody districts, resembling rock, in the way that the roots of trees spread horizontally over it.

At Port Clarence in Bering's Strait, Mr. Berthold Seeman, in 1849, made several experiments to ascertain the depth of the summer thaw, and found that it varied from two feet, in some places, to four or five in others where the soil was sandy. On the northern coast-line of America, a number of pits were dug by Sir John Richardson, and nowhere did he find the frozen subsoil more remote than fourteen inches from the surface.

On the banks of Bellot Strait, in latitude 72°, Dr. Walker (surgeon of the Fox), sunk a brazen tube, two feet two inches long, into the soil, and placed therein a padded thermometer with a long stem. In the middle of September the loose pebbly surface-soil, six inches thick, was thawed, but immediately below it the subsoil, called "a yellowish mud," was firmly frozen. On the 15th of September the thermometer marked $+31°\cdot 2$ F. It was examined at intervals of a few days, throughout the winter, and shewed an invariable and tolerably regular decrease till the 10th of March, when it marked $+0°\cdot 5$. On the 28th of that month it had risen to $+0°\cdot 8$, and continued thenceforth to rise. Dr. Walker thinks, that had it been examined on the 16th of that month, it would probably have stood at zero, but he was then absent from the ship travelling. From the 28th March it rose, without any retrogression, to the 11th of July, when it indicated $+31°·8$ F. All the winter there was a covering of snow, deeper than the general thickness, over the place in which the thermometer was sunk; and this coating of snow increased from three inches, in the beginning of October, to eighty-four inches at the end of April, when it was thickest, and on the 1st of July it had melted away. Another thermometer, which was similarly sunk into gravelly soil, in the middle of January, in a place from whence the snow was constantly blown away, gave different results, and was less regular in its decrements and increments. When first sunk, it shewed (18th January) $-18°·7$. On February 26th, the sunken thermometer was at its minimum, $-25°·7$, the mean temperature of the atmosphere having been for ten previous days $-37°·4$. On the 16th of June, the sunken thermometer rose above the freezing point, the mean of the atmosphere for five days previously having been $+37°·4$. On the 11th of July, the sunken thermometer was $+37°·8$, the atmospheric mean, $+36°·9$, and on the 28th of July, the sunken thermometer shewed $+44°.8$, while the mean atmospheric heat for ten days was only $+42°·7$ F. The effect of the covering of snow, in preventing nocturnal radiation from the earth, and in moderating the direct influence of the sun's rays, is distinctly shewn by these two sets of experiments; as are also the different powers of a loose porous soil, and one retentive of moisture in transmitting heat.*

In the *Kamennaya-tundra* (or Stony-waste), exposed to

---

* These facts were kindly furnished by Captain Sir Leopold M'Clintock. The different thermometers used were compared, and the results reduced to one standard.

the action of the sun, in latitude 68° 42′ N., Wrangell observed that the summer thaw did not penetrate deeper than six or eight inches.*

At Iakutsk in Siberia (latitude 62¼° N.) a stratum of frozen soil, 382 feet in thickness, was pierced in digging a well, until water flowed from beneath it. At Fort Simpson, on the Mackenzie (in latitude 61° 51′ N.), the soil near the bank of the river thawed to the depth of eleven feet during the summer, beneath which there was a bed of frozen sandy earth of six feet thickness. At the depth of seventeen feet from the surface, the sand having no readily visible spiculæ of ice among it, had a temperature of 32° F.

At York Factory, on Hudson's Bay, five degrees south of Fort Simpson, but having nearly the same mean atmospheric temperature, the frozen stratum was not cut through until the shaft had been sunk to the depth of twenty feet and a half from the surface, or three feet and a half lower than at Fort Simpson; and the surface soil at the close of summer was thawed only for three feet. The soil at York Factory is alluvial, with a mossy surface, and is very retentive of water.

Near the 120th meridian of west longitude on the American continent, and on the verge of the Arctic circle, trees begin to thaw towards the end of March, and by the second week of April a decided softening of the snow occurs in bright sunshine. So much water flows from the melting snow during the first six days of May, that geese appear in favourable localities, and rivers that issue from lakes and shallow streams begin to flow, the mean temperature of the preceding ten days having reached 37° F. By the beginning of June (latitude 67°), the snow has gone, except where it had accu-

* *Wrangell*, French ed. Paris, 1843, ii. p. 150.

mulated in deep drifts, and thenceforth vegetation proceeds rapidly, until after a period of about a hundred days from the commencement of foliation, or by the 10th of September, the deciduous leaves are falling fast. Occasional snow showers occur before this date, and snow that falls towards the end of the month ceases to thaw; the soil too begins to freeze, and before the end of October the trees freeze likewise, though the frost does not reach their centres till the winter is further advanced.

On the 75th parallel, at Melville Island (Sir Edward Parry tells us), vegetation on the low grounds proceeds for seventy days, from June to September, but snow showers are not unfrequent in every summer month: and the snow remains for the winter from the first week in September. Patches of earth become visible on the 10th of June.

In 1853, three officers travelling over the ice, through the channels among the islands north of Melville Sound and Barrow Straits, had a mean temperature in June as follows:—

Captain M'Clintock, mean latitude, 77° temp. + 28·3° F.
„ Osborn, „ 76½ „ + 30·8
„ Richards, „ 76 „ 31:8

Though the mean temperatures of the first ten days of the month were only + 24° + 27° and + 26° respectively, the snow melted rapidly during that period, and before the end of the month had disappeared from the ice and from the low lands. Buds of Saxifrage were noticed among the melting snow on the first day of the month, and on the 21st that plant was gathered in flower. Brent geese were seen on the 1st, and an ivory gull was found sitting on its eggs on the 21st. The highest temperature in the shade recorded by any of these observers in the month was + 42° F., and the mean of the last decade of the month was from + 32° to + 37°.

TEMPERATURE.

On the 77th parallel Sir Edward Belcher observed the meadows to be partially denuded of snow early in June, and Dr. Kane mentions the first week in July as the time when patches of ground covered with flowering plants were seen in latitude 79°.

For the course of the isothermal lines for the mean of the year, and for each month, we must refer to *Dove's Tables of the Distribution of Heat* already cited. The following collection of temperatures within the Arctic circle, and in a few instances from places a little to the south of it, are from the best sources to which we have access:—

Observer.	Year.	Place and altitude above the sea-level in feet.	Lat. N.	Long.	Mean of three summer months.	Mean of year.
Parry	1827	Siptzbergen, Icy Sea, 0	$82\frac{1}{2}°$	20° E.	+ 33·0	
Forster	1827	„ Hecla Cove, 0	80°	$16\frac{3}{4}°$	+ 38·1	
Franklin	1818	„ At Sea, 0	80°	10°	+ 34·5	
		North Cape, Norway, 0	71°	$25\frac{1}{2}°$	+ 43·3	+ 32·0
(On the Muonio R.)	1802-6	Enontekis, Lapl., 1356	$68\frac{3}{4}°$	$39\frac{1}{2}°$	+ 54·9	+27·0
		Umeo, - - 0	$65\frac{1}{2}°$	$20\frac{1}{4}°$	+ 54·9	+33·3
		Uleo, - - 0	65°	$25\frac{1}{2}°$	+ 57·7	+ 35·1
		Hos. St. Gothard, 6390	$46\frac{1}{2}°$	$8\frac{1}{2}°$	+ 45·0	+ 30·4
(Siberia)	Many	Iakutsk, - - ?	62°	130° W.	+ 61·6	+14·0
(Greenland)		Godhaab, - - 0	$64\frac{1}{2}°$	$52\frac{1}{2}°$	+ 40·6	+26·8
Kane	1854	Smith's Sound, - 0	$78\frac{1}{2}°$	$70\frac{3}{4}°$	+ 33·0	— 3·2
Belcher	1852-3	Northumberland S., 0	77°	97°	+ 30·8	— 1·1
Rae (Greenl.)	1850-1	Wolstenholme S. 0	$76\frac{1}{2}°$	70°	+ 37·8	+ 4·5
Belcher	1853-4	Wellington Channel, 0	$75\frac{1}{2}°$	92°	+ 32·6	— 1·7
Sutherland	1850-1	Barrow St., Lanc. S., 0	$74\frac{3}{4}°$	94°	+ 35·9	+ 2·5
Parry	1819-20	Melville I., - 0	$74\frac{1}{4}°$	111°	+ 37·1	+ 1·4
M'Clure	1851-2	Banks' Land, - 0	74°	118°	+ 35·5	+ 1·8
Parry	1824-5	Port Bowen Reg. In., 0	$73\frac{1}{4}°$	89°	+ 36·9	+ 4·3
M'Clure	1850-1	Prince of Wales St., 0	73°	118°	+ 37·1	+ 1·1
Collinson	1851-2	Do. Do., - 0	$71\frac{1}{4}°$	$117\frac{1}{2}°$	+ 38·2	+ 7·9
Collinson	1853-4	Camden B., - 0	70°	$145\frac{1}{2}°$	+ 38·1	+ 6·2
Parry	1822-3	St. of Fury and Hecla, 0	$69\frac{3}{4}°$	$81\frac{1}{2}°$	+ 35·1	+ 5·8
Collinson	1852-3	Camb. B. Wollaston I., 0	69°	105°	+ 37·5	+ 4·4
Richardson	1848-9	Great Bear Lake, 500	67°	$118\frac{3}{4}°$	+ 49·0	+ 9·0
Parry	1821-2	Lyon Sound, - 0	$66\frac{1}{4}°$	83°	+ 35·1	+ 9·8
M'Murray	1846-7	Yukon R., - - 400 ?	66°	148°	+ 56·7	+ 14·6
Richardson	1825-6	Great Bear L., - 500	$65\frac{1}{4}°$	$123\frac{1}{4}°$	+ 50·4	+ 17·7
(Greenland)		Godhaab,	$64\frac{1}{4}°$	$52\frac{1}{2}°$	+ 40·6	+ 26·8

A comparison between the temperatures at the north cape of Norway and the Hospice de St. Gothard, given in the preceding table, shews that in Europe a difference of about twenty-four or twenty-five degrees of latitude is equal to between six and seven thousand feet of altitude, in depressing both the mean heat of the year and that of the three summer months during which alone vegetation can proceed at these places. Again, on comparing the places in Lapland with those on Great Bear Lake, we find that while the summer heats of the two countries are similar, the winters in Arctic continental America are much colder, and the mean heat of the year consequently greatly lower. This is doubtless due to the more continental character of the climate, the clearer winter atmosphere, and greater radiation from the earth in the vicinity of Great Bear Lake, than in Norway or Lapland, where the neighbourhood of the White Sea on one side, and of an open northern Atlantic on the other, agitated by the gulf stream, causes clouds and mists.

CHAPTER XVI.

VEGETATION.

Barren Grounds—*Tundren*—*Terræ damnatæ*—Line of Woods—+43-45° Summer heat—Line of Woods in America—And in Asia—Trees and their limits—Vegetation in Petchora-land—In Finmark—In Norwegian Lapland—In the American Barren Grounds—In the Valley of the Mackenzie—Peel's River—Kolyma—Aniui.

An Arctic circumpolar map shews three great chains of mountains, the Lulean, Ural, and Rocky Mountains, all running northward; those on the old continent having an inclination eastward, after entering the Arctic circle; and the main chain in America, as well as a second minor one terminating in Cape Barrow, in the Coronation Gulf, inclining westward. All of them lose in altitude as they approach the Polar Sea, are more abrupt on their western slopes, and have tracts of comparatively low lands spread out from their eastern bases. It is on these eastern levels that the "barren grounds" of America occur, and the "tundren" of Siberia.

In America, the barren ground district has its greatest extension near Hudson's Bay, where it descends to the 61st parallel, and in that direction may be said to include the north end of Labrador, bordering on Hudson's Straits, Meta Incognita, the whole of Greenland, and all the American islands of the Polar Sea. It is the absence of trees that has given name and character to "the barren grounds" of North America. The whole district is full of lakes, and it is traversed by one large river (the Great Fish River), and many smaller ones. Its surface is also varied by rocky hills of moderate altitude; and one ridge, alluded to above, named by Hearne the "Stony Mountains," runs from the Point Lake, and the bend of the Coppermine River, to terminate in Cape Barrow, a promontory of Coronation Gulf, which has an altitude of about 1500 feet. The district narrows greatly on the north of Great Bear Lake, and terminates at the delta of the Mackenzie. Greenland, though agreeing with the barren grounds in the absence of trees, differs in its lofty mountains and consequent presence of glaciers.

The winter winds sweep over this corner of America, rendering it uninhabitable in that season by the Red Indians, and the bulk of the Reindeer keep near its borders, so that they can retreat to the woods in storms. In places where the soil is moderately dry, it is densely clothed with the lichens, named *Corniculariæ*, which are mixed in moister spots with the Reindeer moss (*Cetraria*). Other plants also flourish where the soil is suitable, such as the Lapland rhododendron, the glaucous kalmia, the blueberry (*Vaccinium*), crowberry (*Empetrum*), the *Ledum*, bearberry (*Arctostaphylos*), the *Andromeda tetragona*, the cloudberry (*Rubus chamæmorus*), the *Rubus arcticus*, and various depressed willows. In favourable and sheltered meadows grasses and bents flourish in considerable variety, and on the banks of streams sometimes a growth of *Salix speciosa*, three feet high, or even more, may be seen. Also many flowering plants, of less note, but which serve to cheer the traveller, who traverses these wastes in the fleeting summer.

In character the Siberian tundren is very similar to the American ones. Thus Wrangell says,—" When one coming from the naked, frozen, moss-tundra reaches the valleys of the Aniui, which are sheltered by mountains from the prevailing cold winds, and where birches, poplars, willows, and low creeping junipers (*Juniperus prostratus*) grow, he thinks himself transported to Italy. In travelling across the wide tundra in dark nights, or when the vast plain is veiled in impenetrable mist, or when in storms or snow-tempests, the traveller is in danger of missing the sheltering hut, he will frequently owe his safety to a good dog, who will be sure to bring the sledge to the place where the hut lies deeply buried in the snow, and will suddenly stop, and indicate where his master must dig."

Even in the narrower country of Lapland there are districts which resemble the *tundren*. Linnæus calls them *terræ damnatæ*, and thus describes his experience of traversing one in the beginning of June, when the melting snow had flooded the country :—" We had next to pass a marshy tract (in Lapmark), where at every step we were knee-deep in water, and if we thought to find a sure footing on some grassy tuft, it proved treacherous, and only sunk us lower. Our half-boots were filled with the coldest water, as the frost in some places still remained in the ground. I wondered how I escaped with life, though certainly not without excessive fatigue, and loss of strength." The guide who had been despatched to seek assistance returned. " He was accompanied by a person whose appearance was such, that I did not know whether I beheld a man or woman. Her stature was very diminutive ; her face of the darkest brown, from the effects of smoke, her eyes dark and sparkling, her eye-brows black. Her pitchy coloured hair hung loose about her head, and on it she wore a flat red cap. She had a grey petticoat, and from her neck, which resembled the skin of a frog, were suspended a pair of large loose breasts of the same brown complexion, but encompassed by way of ornament with brass rings. She addressed me with mingled pity and reserve in the following words:—" O thou poor man! what hard destiny can have brought thee hither, to a place never visited by any one before? This is the first time I ever beheld a stranger. Thou miserable creature! how didst thou come, and whither wilt thou go?"

The *northern termination of the woods*, co-incident with the south-west borders of the barren-grounds and tundren, though partly dependent on soil and on contiguity of the sea, yet furnishes an approximate measure of the climate of various meridians, as well as of the elevation of the country. It oscillates nearly on the line of mean temperature of the three summer months, or between the *isotherals* of + 43° + 45° Fahr. In America, this boundary line of the woods, rising with an increase of westerly longitude, passes the 106th meridian in the neighbourhood of Artillery or Peshew Lake, between the 63d and 64th parallels of latitude; strikes the Coppermine River at Point Lake, runs northwards some way on its banks, then cuts the Arctic circle, and passes a little beyond the 67th parallel on the north side of Great Bear Lake. In this part of the country the woods are confined to the valleys, and after skirting the *Beghoola-dessy*, a stream of considerable magnitude, for an indetermined distance to the north, they attain the 69th parallel, on the delta of the Mackenzie.

On the left bank of that river the northern end of the Rocky Mountain chain comes within ten or twelve miles of the sea-coast. To the westward of this chain there is a barren district whose limits have not been ascertained, but the woody banks of the Yukon touch the Arctic circle, and running west under the name of the *Kwichpack*, that river falls into the sea some way to the south of Norton Sound. Forests of white spruce occur on the *Noatak*, a river which falls into Eschscholtz Bay on the Arctic circle.

In Northern Asia the *line of woods*, as traced by Baron Wrangell, commencing near the Bay of the Holy Cross, at the head of the Gulf of Anadyr, rises from the Arctic circle with considerable undulations, in its course eastward through 50° of longitude, until it reaches the 71st parallel of latitude on the deltas of the Iana and Lena. The great north-eastern promontory of Asia is probably wholly destitute of trees, and three *tundren* on the lower Petchora are specially named and described by Count Keyserling.*

In giving a very brief sketch of the range of trees and their kinds within the Arctic circle on different meridians, and of some other phenomena of vegetation, it is convenient to begin with Europe where they have been most fully explored. No corn is grown in the lower Petchora district.

On the 28th meridian east from London which passes from the Gulf of Finland through the extensive sheet of water in Lapland, named *Enara træsk*, the spruce (*Abies excelsa*), ceases at the 68th parallel, and the Scotch fir (*Pinus sylvestris*), at the 69th. In Swedish Lapland, a little more to the east, Von Buch and Martius traced the spruce a quarter of a degree further to the north.

In Norway the trees advance still more northwards, probably owing to the vicinity of the gulf stream. The forests of Altenfjord yield Scotch firs sixty feet high, and birches

* Reise en das Petchora Land. St. Petersburg, 1846.

which average forty-five. On the northern slope of the mountain Kjolen, in that valley, the fir ascends 800 feet, and in a dwarf and isolated condition to twice that height. Near Kistrand, on Porsanger Fjord, in latitude 70° 28' N., the Scotch fir was seen by Lund, but the spruce fails a degree or more further south. At Hammerfest, in latitude 70¾° N., there are dwarf alders and aspens, bird cherries (*Prunus padus*), rasps and currants. On the extreme island of Mageröe, to which the North Cape of Europe belongs, and which reaches 71° 11' N., there are among other ligneous plants *Salix glauca* and *lapponum, Betula pubescens* or *nana*, and the common juniper.

Von Buch, as quoted by Malte Brun, gives the following tabular view, calculated for the 70th parallel of latitude in Norwegian Lapland or Finmark. Limit of the red pines 730 feet of altitude; of the birch, 1483; of the whortleberry (*Vaccinium myrtillus*, 1908; of the dwarf birch, 2576; of the *Salix myrsinites*, 2908; of the *Salix lanata*, 3100; and of perpetual snow, 3300. Mr. William Dawson Hooker says, that at Hammerfest he observed an attempt at a garden behind one or two of the houses, where a few radishes, turnips, lettuces, and parsley plants struggled to elevate their starveling heads into an ungenial atmosphere. About a dozen stalks of immature rye were raised as a curiosity but were not expected to ripen.* Barley is cultivated as far north on the Scandinavian peninsula as the 70th parallel, and oats up to the 65th, in sheltered valleys whose rocky cliffs reflect the sun's rays with much power.

It must be attributed mainly to the constant presence of ice drifting from the north that Iceland, Greenland, with its inland glaciers, and the barren *Meta incognita* islands that form

* Hooker, *lib. cit.*, p. 18.

the western shores of Davis' Straits, present such a contrast in their treeless desolation to woody Norway.

On Melville Island and the neighbouring shores lying north of Lancaster Strait and Melville Sound, seventy-seven phenogamous plants have been detected, of which fifty-seven are dicotyledinous, and only one has a ligneous stem, the prostrate *Salix arctica*. The *Andromeda tetragona* also occurs there, but its stem is a mere thread, although the whorled and withered leaves adhere to it for successive winters.

At Repulse Bay, on the Arctic circle, it was on this Andromeda that Dr. Rae depended for fuel during the two winters he passed there, though the barren grounds nourish other shrubby plants, such as roses, rasps, *Andromeda polifolia*, and *calyculata, Arctostaphylos uva-ursi* and *vitis-idœa, Rhododendron lapponicum, Ledum palustre, Azalea procumbens,* and various *Salices,* but these are local, and on exposed situations rare. The neighbourhood of the "frozen strait" which Captain Middleton and Sir George Back found to be impenetrable in the years of their voyages, is probably the reason of the extreme barrenness of Repulse Bay.

In the valley of the Mackenzie, on the 135th meridian, the spruce fir (*Abies alba*) is the most northern tree that forms a forest, reaching to a much higher latitude than the pines, contrary to what occurs in Norway, where the pines are the most northern. In latitude 68° 55′ N., the trees, which up to this parallel cover the immediate banks of the river and the islands of the delta, terminate suddenly in an even line, probably cut off by the sea-blasts. Beyond this line a few stunted spruces and scrubby canoe-birches straggle up the acclivities, struggling for existence, and clinging to the earth. The forest is formed by the spruces, but among them there are many canoe-birches of much slenderer growth, their stems not exceeding five inches in diameter. The *Populus balsamifera* and *Alnus viridis* grow to the height of twenty feet, and the *Salix speciosa* to that of twelve near the termination of the woods. The hills skirting the river in these latitudes are nearly bare, supporting only a few scattered depressed trees. The *Rosa blanda*, and seven other shrubby Rosaceæ, nine or ten dwarf prostrate *Ericaceæ*, the common juniper, the trailing form of the *Juniperus virginianus*, the *Betula pumila* and *nana*, the *Eleagnus argentea* and the *Shepherdia canadensis*, together with a number of willows, comprise the ligneous plants which accompany the spruce to its northern limit on the Mackenzie, some of them going beyond it. *Pinus banksiana*, the most northern American member of the genus, has not been traced far within the Arctic circle, and the *Pinus resinosa* does not go beyond 57°.

Wheat has not been raised within the Arctic circle in America, nor indeed within six degrees of latitude of it. It requires a summer heat of 120 days, but is said to be cultivated up to the 62d or 64th parallel on the west side of the Scandinavian peninsula. Barley ripens well at Fort Norman on the 65th parallel, in the valley of the Mackenzie, after the lapse of 92 days from the time of its being sown. All attempts to cultivate it at old Fort Good Hope, two degrees further north, have failed. Sixty-six degrees of latitude may therefore be considered as the extreme limit of the *cerealia* in America, which is four degrees short of the northern extreme of barley in Norway. Oats do not succeed so far north as barley or bere.

At Fort Good Hope, on the Mackenzie (the new fort), in latitude $66\frac{1}{4}°$ N., a few turnips and radishes, and some other culinary vegetables, are raised in a sheltered corner, which receives the reflection of the sun's rays from the walls of the house, but none of the *cerealia* will grow, and potatoes do not repay the labour.

On Peel's River (67° 35′ N.) the trials made to raise esculent vegetables failed; nothing grew except a few cresses. Turnips and cabbages came up about an inch above the ground, but withered in the sun, and were blighted by the early August frosts.

The general character of the *tundren* of the east of Siberia is like that of the American barren grounds.

On the Lower Kolyma, Wrangell observes that "the severity of the climate may be attributed as much or more to the unfavourable physical position as to its high latitude. To the west there is the extensive barren tundra, and to the north a sea covered with perpetual ice; so that the cold north-west wind which blows almost without intermission meets with no impediment; it brings with it violent snow-storms, not only in winter, but frequently in summer. The vegetation of summer is scarcely more than a struggle for existence. In the latter end of May the stunted willow-bushes put out wrinkled leaves, and the banks which face the south become clothed with a semi-verdant hue, which an icy blast from the sea suffices to destroy. At Nijnei Kolymsk, in latitude $68\frac{1}{2}°$ N., the neighbourhood is especially poor. It is a low marsh, with a thin layer of vegetable earth on the surface, intermixed with ice that never thaws; it supports a few stunted larches, whose roots, being unable to penetrate the frozen subsoil, extend along its surface. A few small-leaved willows grow on banks facing the south. The nearer we approach the sea the more rare become the bushes, and on the left bank of the Kolyma they cease entirely, twenty miles to the north of Nijnei Kolymsk, or near the 69th parallel. On the right bank of the river, where the soil is drier, they extend further north than on the dreary icy moor of the other side. On the right bank there are patches of good grass, wild thyme, wormwood, wild rose, and forget-me-not. The currant, the black and white whortleberry, the cloud berry, and *Rubus arcticus* bloom there, and in favourable seasons bear fruit. No cultivation is attempted, though at Shredne-Kolymsk, which is two degrees more to the south, I have seen radishes, and even cabbages, but the latter formed no heads."* Here the larch is mentioned as the most northern tree. It is dwarfed, and disappears on the Mackenzie long before the same latitude is reached. We have already quoted a passage from Wrangell in which the valleys of the Aniui are, mentioned as supporting birches, poplars, willows, and creeping junipers, but the forests seem less flourishing in this quarter of Siberia than on the Mackenzie, where they are formed of white spruce. Nevertheless, the line of woods is represented on Wrangell's map as crossing the Kolyma below Nijnei Kolymsk, and as rising to the 71st parallel of latitude on the Iana. With him it may mean the utmost limit of the isolated depressed trees. In America these are met with here and there in the barren grounds, and convey to the traveller the impression of the forests having in former times extended further north. The same idea crossed Baron Wrangell's mind in regard to the Siberian woods.

At Obdorsk, on the estuary of the Obi, nearly on the Arctic circle, a pit sunk into the frozen soil to the depth of seventeen feet from the surface, had a temperature of $+ 30\cdot25°$ F. On the Obdorsk range of mountains, Erman observed single

* Wrangell, Polar Sea, etc., p. 51.

straggling larches at the height of six hundred feet above the alluvial valley of the Khanami. These mountains rise nearly 5000 feet above Obdorsk. The stone pine *(Pinus cembra)* is on this meridian the most northern tree of the family, and especially a prostrate variety of it. The birch *(B. alba)* was not seen at Obdorsk, but it was a conspicious object twenty or thirty miles further south, in latitude 66°. At Beresov, in latitude 64°, rye and barley thrive well. Erman states that the condition assigned for the cultivation of barley is that the mean temperature of any one of the three summer months shall not fall below $+ 7°$ R $= 47°\cdot75$ F. The mean temperature of the three summer months at Beresov is actually $+ 65°$ F., and of none of the months more than three degrees lower.*

Taken as a whole, there is a close similarity in the vegetation of the different meridians within the Arctic circle. Nearly the same genera are repeated on all, and the majority of species are alike. The trees of the old continent, however, and of America, are for the most part specifically distinct. In the higher latitudes, which the trees do not reach, there is very little difference in the phenogamous plants of one meridian from those of another, the mosses are nearly identical, and only two or three lichens are peculiar to Arctic America.

* Erman, Travels in Siberia, i., 474.

CHAPTER XVII.

ZOOLOGY.

Rein-deer—Musk Ox—Polar Hare—Marmots—Lemmings—Arctic Fox—Wolverine—Polar Bear—Brown Bear—Black Bear—Argali—Goat-Antelope — Birds — Geese — Water-fowl — Raven — Owls — Snow-Bunting—Lapland Finch—Lesser Redpole—Marine Mammals—Fishes—Herrings—Muksun—White Fish—Tchiir—Nelma—Beghula—Kundsha—Golzy—Lenok—American Trouts—Kolyma Sturgeon.

THE most important land animal within the Arctic circle is the rein-deer, or *rennthiere* of the Germans, so named doubtless because of its fleetness. This animal, common to the Arctic coasts of Europe, Asia, and America, frequents the most northern islands that man has reached. It is comparatively abundant in Spitzbergen, and some small herds remain all the year in the extremes of arctic Greenland and on the islands north of Melville Sound. But the bulk of the species retire from the Arctic coasts and barren grounds in September, October, and the beginning of November, to the vicinity of the woods, where they assemble during the rutting season in large and very numerous bands. The passes among the mountains and lakes which the deer frequent in their migrations southwards are known to the natives, and sought by them for securing a winter's supply of venison. Indeed the movements of the rein-deer regulate those of the northern Indians, and of the families of Eskimos who inhabit the continental shores of the Arctic sea. On the approach of the milder weather in the spring months of April and May, the female reins travel north again, and drop their young on the coast; the males taking the same route in separate bands. As early as the 1st of March, Dr. Rae observed the rein-deer migrating steadily northwards at Repulse Bay, in latitude $66\frac{1}{2}°$ N., and some bands were seen a week previously. In October the migration southwards was nearly over in the same quarter, though a few stray bands were seen in November. At the time when the northerly movement is at its height, the lichens of the barren grounds, *Cornicularia tristis*, *divergens* and *ochrileuca*, *Cetraria nivalis*, *cucullata* and *islandica*, and *Cenomyce rangiferina* are softened by the melting snow, and furnish an excellent food to the deer. The grasses too, and the bents whose vegetation was suddenly arrested at the beginning of winter, are shedding their seeds in the spring as the snow disappears, and their culms, not entirely deprived of sap, are at that season good hay. In the beginning of June, Captain Osborn, when travelling over the ice between the islands north of Melville Sound, observed numerous seeds of plants, among which he recognised those of the poppy, willow, and saxifrage, travelling over the smooth floes before the wind. The northern islands are thus supplied with seed in seasons when the plants growing far north have not heat enough to bring their own fruit to maturity.

The migrations of the rein-deer are as constant in Siberia as in America. "About the end of May," says Baron Wrangell, "these animals leave the forests in large herds, and seek the northern plains nearer the sea. The hunt is not so successful in this season as in the autumn, since, the rivers being frozen over, the hunters have not the same opportunity of intercepting them. The true harvest is in August or September, when the rein-deer are returning from the tundren to the forests. In good years the migrating body of rein-deer on the Aniui consists of many thousands, and though they are divided into bands of two or three hundred each, yet the herds keep so near together as to form only one immense herd of from thirty to sixty miles in breadth. They always follow the same route, and cross the river at the same place." In another passage the same author says, that the rein-deer remain almost immovable during the severe storms of winter.

Erman mentions similar periodical migrations of this animal over the Samoyed lands on both sides of the Obi. The rein-deer, like other ruminants, is fond of salt, and Erman states, that the appetite of the animal for the alkaline urine of man, is one of the chief means by which it is tamed. The half-tamed deer follow a sledge-driver with eagerness, to lick his urine from the snow.

In Lapland, also, the wild rein-deer go to the northern coasts in summer, as Linnæus long ago told us. In Siberia and Europe this animal is domesticated, but the only domestic animal in Arctic America is the dog.

The musk ox (*Ovibos moschatus*), another native of the Arctic regions, is much more limited in its distribution than the rein-deer, which it resembles in its habits. It belongs properly to the eastern barren ground district, and does not range south of the Arctic circle, but frequents in small bands all the islands north of Melville Sound. Dr. Kane, as has been mentioned, found a number of its bones in Smith's Sound; but we have no account of more than a solitary skull having been found in Greenland. In the ice-cliffs of Eschscholtz Bay, its bones, not to be distinguished from those of the existing species, are associated in great numbers with the bones of elephants and other extinct animals, neither of them mineralized, and none of them having sustained much loss of animal matter. The musk ox is not known to live at the present time in that corner of the continent. This animal was contemporary with the extinct mammoths and rhinoceri of the ice-cliffs of Eschscholtz Bay, and it is probable that the antiquity of the rein-deer is as great.

The polar hare, some marmots and lemmings, also dwell on the Parry Islands, and the beasts of prey are the arctic or stone fox, the wolf and wolverine. The polar bear, which preys specially on seals, belongs rather to the sea. The brown barren ground bear, not as yet distinguished specifically from the Norwegian bear (*ursus arctos*) has not been seen beyond the continental shores of the Arctic Sea. All these animals are common to the Old World and the New, except perhaps the marmot and polar hare, which are American; yet even they have representatives in Siberia, which may, on further examination, prove to be the same species. The black bear, which is a proper American species, enters the Arctic circle, but only in the woody tracks that run northward along the great rivers. The argali or big-horn, which is confined to the Rocky Mountain range, and on it reaches the 68th or 69th parallel of latitude, has a near ally in the Siberian argali. The goat-antelope (*Aploceros montanus*), which also inhabits the Rocky Mountains, but is not known to pass far within the Arctic circle, has not, as far as we know, a representative in the Old World. Bones of the species exist in the Eschscholtz ice-cliffs along with those of the mammoth.

As far as quadrupeds are concerned, the lands within the Arctic circle form but one zoological province.

The polar region, excluding merely the points where the woods cross the Arctic circle, presents an uniformity in its native birds on all meridians. All the birds that frequent the high latitudes are natives, and though their stay at the breeding places does not exceed three months, they are to be considered as merely visitors in the southern regions, which they traverse in going and coming during the remaining nine months of the year. The brents and snow-geese breed in the most northern lands on all meridians, as do also the various birds enumerated in the chapter on Spitzbergen. The dovekie (*Alca alle*) keeps the sea in the high latitudes all the winter wherever open water exists, but numbers of the species migrate southward. Of land birds the ptarmigan is a winter resident in the north, and the raven and snowy owl, though scarce, may be seen in winter wherever food is to be obtained. Altogether, about fifty species frequent the Parry Islands in summer, among which the snow bunting and Lapland finch are graminivorous. These also visit Spitzbergen. Further south, within the Arctic circle, where shrubs exist, the lesser redpole is also a winter resident. The laughing goose (*Anser albifrons*) is common to the Arctic coasts on the old and new continents, but the Eskimo goose (*Anser Hutchinsii*), which also goes north to the shores of the Arctic sea to breed, is confined to America. The Canada goose follows the larger rivers in their course through the Arctic regions, and is properly an American species, though it is not unfrequent in Europe. Whether it frequents the northern Siberian shores or not we do not know. The arrival of the summer birds, and especially of the ducks and geese, is to the northern nations of Siberia and the Red Indians and Eskimos of America not only an unequivocal sign of spring, and as such welcomed with joy, but also the commencement of a period of plenty succeeding one of privation or famine, for the months of March and April are those in which food is most difficult to be procured by the uncivilized natives of Arctic countries. The geese arrive in the north as soon as patches of ground become visible. Vegetation is so rapid, that simultaneously with the melting of the snow, the development of leaves and flowers occurs. The geese, on their first arrival, feed on the crowns of the roots of *Eriophori* and *Carices*, which, just before the leaves show, have some sweetness; and for a few days after the snow has gone on the barren grounds, both the birds and bears feed with relish on the berries of *Empetrum* and *Vaccinium*, which are then laid bare. The flight of a goose being forty or fifty miles an hour, or more, with a favouring gale, these birds may breed in the most barren northern solitudes, where they find safety, and in a few hours, on a fall of deep autumn snow, convey themselves, by their swiftness of wing, to better feeding grounds.

The seals, walruses, and whales being the marine beasts on which the Eskimos and coast Samoyeds depend mainly for subsistence, must not be overlooked in the enumeration of Arctic animals. The blubber which these beasts yield is a substance of absolute necessity to the northern Eskimos, who subsist wholly on animal matters, supplying them with a kind of food essential to their health in winter, as well as with fuel ; and all their domestic furniture, much of their clothing, their boats, and fishing gear, are made from the skins and bones of marine mammals, —the rein-deer skins completing the articles of Eskimo dress. The pursuit of the seals and cetaceans by the civilized races of Europe and America, carried on with systematic perseverance and the appliances of science, must eventually drive these animals from their more accessible haunts to the extreme recesses of the polar area. Already the formerly highly productive seas of Spitzbergen and Davis' Straits are almost fished out, and the inlets and channels in the *Meta Incognita* Archipelago have been invaded by the whalers. At present the *Meta Incognita* islands are among the most densely peopled Eskimo areas; but should the seals and whales, which are eagerly pursued there by Europeans, be driven to remove elsewhere, the natives, after much suffering, must also abandon their ancient seats. If the fishes of the polar seas were well known, we should probably find the species much the same on every meridian; the temperature of the sea, when ice is present, varying but little near the surface. Dr. Sutherland found the mean in Davis' Straits and Lancaster Sound to be a little below $+ 33°$ F., the range in these months being less than three degrees; but in September the surface water became a degree or two colder. In the Spitzbergen seas the mean was $+ 35\frac{1}{2}°$ during summer, according to Sir John Franklin's and Sir Edward Parry's trials. At considerable depths the temperature is greater and more uniform, and a comparative uniformity of animal life is likely to prevail there. There is, however, one remarkable difference in the presence of sturgeons in the rivers of Asia into which they ascend from the icy sea, whereas no sturgeons have as yet been detected in the American rivers which fall into the polar basin, not even in the Mackenzie, whose sources are situated so far south as to interlock with those of rivers that fall into the Pacific and Hudson's Bay, and abound in fine sturgeons of several species.

The only moderately full list of the northern fishes is that of Fabricius in his Greenland Fauna, with the additions made by Professor Rheinhard ; but that applies to stations in West Greenland, situated south of the Arctic circle, and differs greatly, both generically and specifically, from a list of the fishes known to frequent the Kamtschatdale coast and Bering's Sea. Within the Arctic circle, however, there is reason for believing that the diversity of species on different meridians is comparatively small. Two flat fish (*Pleuronectes glacialis* and *scaber* of *Pallas*) extend eastward from Bering's Sea to Coronation Gulf, and the *Acanthocottus quadricornis* ranges from Spitzbergen westward to the same Coronation Gulf.

A small variety of the common herring makes the entire circuit of the Arctic circle. This is noted by Pallas as a well-known and abundant inhabitant of the White Sea and Siberian Ocean, as well as of the coasts of Kamtschatka and Sea of Ochotsk. It was taken by Sir John Franklin's party in Coronation Gulf, and it occurs in the Greenland seas, where it is known to the Eskimos by the name of *Kapiselik*, but is said to be rare in that quarter, though Fabricius saw some sculls of it in the Fjord of Frederikshaab in the month of July. In Kamtschatka, Pallas informs us that it seeks the innermost bays and even fresh-water lakes at midsummer in innumerable sculls for the purpose of spawning. Sometimes the exit from these fresh or brackish lakes is obstructed by banks of gravel thrown up by storms, and the herrings and their young remain shut in until liberated by the breaking up of the barrier next summer. The Kamtschatdales avail themselves of these occasions to take vast numbers by cutting a gap in the bank and placing nets in the opening.* Erman states that sculls of it ascend the Obi, and, passing into the tributaries of that river,

* Pallas, Zoog. Rossica.

winter there.* Baron Wrangell, in his account of the Kolyma district, says—" In September the sculls of herrings begin to ascend the rivers, and almost all the population hasten to the favourable spots for catching them. The numbers in good years are so enormous that 3000 may be taken at a draught, and 40,000 in a few days with a single net. The largest herrings come into the Kolyma, those of the Alaseia are smaller, and those in the Jana and Indigirka are still less."† Pallas mentions ten inches as the ordinary length of the Arctic herrings. No comparison has been made, as far as we know, between these Arctic herrings and the true herring of the British Islands and German Ocean since exact ichthyology has been cultivated, and *Coregoni* often pass among the vulgar for herrings.

Baron Wrangell speaks of three other species of fish as of much importance to the inhabitants of Arctic Siberia. The *muksun*, a *Coregonus* with a strongly arched or humped back, has a prominent snout, and is said to resemble the Gwiniad of the Welsh lakes. It may be said to represent the white fish (*Coregonus sapidus* of Agassiz), so important to all the natives of Rupert's Land between the great Canada lakes and the Arctic Sea. No fish in any country or sea excels the white fish in flavour and wholesomeness, and it is the most beneficial article of diet to the Red Indians near the Arctic circle, being obtained with more certainty than the rein-deer, and with less change of residence in summer and winter. Another Siberian species, named by the Russians *tchiir* and the Samoyeds *chy-calle*, is the *Coregonus nasutus* of Pallas, and resembles the last, but has a lower back and a short blunt

---

* Erman enumerates three kinds of salmon of the Obi, viz., the *Siurok* (*S. vimba*), the *Stchokur* (*S. tchokur*), and *Padyan* (*S. polkur*). *Travels in Siberia*, II. p. 57.   † Wrangell, Polar Sea, p. 67.

snout somewhat like that of the white fish of Rupert's Land. Both these Siberian *coregoni* ascend the rivers from the Arctic Sea. The white fish of Rupert's Land occurs in great perfection in inland lakes, to which its ascent from the sea is cut off by cascades, but it and several other species of the same and of allied genera exist in the estuaries of rivers falling into the Arctic Sea, more especially in some of the outlets of the Mackenzie, or of the river to the eastward of it named the Beghula-dessy, where the Eskimos take numbers and value them much as articles of diet.

A third Siberian species, the *Nelma* or *Salmo leucichthys* of Pallas, ascends the Obi, Lana, Covyma, Indigirka, Kolyma, Aniui, and Tchukotsky rivers. It reaches the size of six feet in length, and thirty or forty pounds in weight. It is called by Wrangell, a large kind of salmon-trout; and it has its representative in the *Stenodus Mackenzii*, which attains a similar size, and ascends the river after which it is named as high as Great Slave Lake. Its Indian name is *Beghula*, and it abounds in the Beghula-dessy.

Trout of various kinds and of large size inhabit the rivers that fall into the Arctic Sea, and on the American coast near the mouth of the Coppermine River, a species closely resembling the sea-trout of England, was abundant in the shallows. The true *Salmo salar* is said to be rare in Siberia, but others of the genus, such as the *Kundsha*, the *Golzy*, and the *Lenok*, are common. In America too, four or five great trouts and chars, like those of Greenland, people all the large lakes in the Arctic regions. The natives, when furred animals fail, know how to make clothing of the skins of large fishes of the salmon kind. The skin of the Burbot (*Lota*) is used on the Obi in place of glass for windows.

In America, though sturgeon abound in Hudson's Bay, and in the rivers that fall into the Northern Pacific, none exist in the Mackenzie or in any river that falls into the Arctic Sea, but in Siberia this fish is said by Pallas merely to be less frequent in these north-flowing rivers, and Wrangell mentions the sturgeon as one of the fishes of the Kolyma district. Sturgeons also enter the Obi, and like other anadromous fishes of that river, ascend the tributary streams that flow from the mountains, and therein they pass the winter. Such of them as remain in the main stream, are said to die of convulsions in the month of January; but it is said also, that these fish crowd together in deep holes of the river during the severe season, and remain there at rest.*

Fish can be preserved all the winter in Arctic climates in a frozen state, but even when taken towards the close of the summer months and hung up in the open air, it keeps in an eatable state, though not without some taint, till the following spring. It is therefore a viand of vital importance to the native inhabitants and fur-traders residing near large rivers or lakes.

* Erman, l. c. I. p. 402.

CHAPTER XVIII.

GEOLOGY.

Primitive, Silurian, Devonian, Carboniferous, Liassic and Tertiary Formations—Pseudo-volcanic Products—Wood-hills—Fossil Elephant and Rhinoceros—Liakhow Islands—Bison—Ice-cliffs of Eschscholtz Bay—Drifts—Spread of the Tundren.

The Geology of the Polar Regions is too wide a subject, and one too imperfectly known for a full view of it to be attempted in a compilation like the present one, and it will suffice to say, that the following formations have been recognised—primitive granites and gneiss; silurian deposits; devonian; true coal-fields; liassic beds; tertiary deposits, including thick beds of coal or lignite, and also newer alluvial and drift beds. The primitive igneous rocks prevail in the eastern barren grounds of America, and in the moss tundren of Siberia. They form high hills in Spitzbergen, and constitute the most northern islets of that Archipelago. They also occupy much of Greenland. Cape Barrow, on the east side of Coronation Gulf, a lofty granitic promontory, at least 1500 feet high, is the end of the Laurentinian range proceeding from Point Lake and Fort Enterprise, situated on the western edge of the barren grounds; and most of the islets in Coronation Gulf are knolls of granite. Granite exists also in North Cornwall, in the neighbourhood of liassic fossils.

The paleozoic limestones, and other more recent fossiliferous deposits occurring in the high latitudes, and containing the remains of animal forms which resemble those living only in warmer climates in the present age, are interesting as indications of the changes of temperature, as well as of surface that have taken place on the earth.

Silurian limestone occupies much of the area within the Arctic circle, at least on the American continent and islands. It forms the whole continental coast from Cape Parry (longitude 124° W.), to Cape Krusenstern (longitude 114° W.) : also the north ends of the islands bounding the entrances of Navy Board, Admiralty, and Prince Regent's Inlets. North Somerset and Boothia consist mostly of the same ; and it occurs on King William's and Prince of Wales' Islands ; also on both sides of Wellington Channel up to the 77th parallel, where newer deposits come in. Mr. Salter observes, that from the eastern borders of Europe to the Rocky Mountains of America, and from the southern states of America to the Polar regions, there is a general similarity in the fossil contents of these old rocks, and some of the most common species are the same. But even over this wide area, the seas of which must certainly have communicated with each other, there are great local differences, and the northern districts are wanting in many characteristic southern forms.* In the arctic parts of America, the silurian deposits are in general as little disturbed from their horizontal position, as in the United States and Rupert's Land. These horizontal limestones split into thin slates and fragments by the action of alternate frosts and thaws, and on the shores of the Arctic seas form the most barren of all surfaces, identical probably with the "stoney tundren" of Siberia. The continual disintegration prevents all vegetation, and

* Sutherland's Journal, etc., ii. p. ccxviii.

except where sandstone beds and crevices occur, even lichens are scarcely to be met with. The "barren grounds" or mossy tundren of the primitive igneous districts, are much more fertile in grasses and other food for herbivorous animals.

Carboniferous limestone exists on the north-western coast of Banks' Island, on Melville and Bathurst Islands. At Village Point, in latitude 76° 50′ N., and longitude 97° W.; at Depot Point, Grinnell Land (of Belcher), latitude 77° 5′ N., and at various other places in the carboniferous limestone tract there are coal beds. These coal beds are considered by Professor Houghton to be very low down in the carboniferous series.*

A liassic basin extends on the 77th parallel of latitude from the 95th meridian to the 120th, or for a distance of 270 geographical miles. On Exmouth Island, at the height of 570 feet above the sea, bones of a species of Ichthyosaurus were found by Sir Edward Belcher in limestone resting on sandstone, and, as said above, in the vicinity of out-crops of granite. This formation exists near Cape York on the east side of Baffin's Bay, and at Capes Horsburgh and Warrender, on the west side of the same.

A tertiary coal in workable beds comes to the surface at Disco, in Greenland. The Garry Islands, lying off the Mackenzie, contain beds of a tertiary coal which takes fire spontaneously on exposure to the atmosphere. Higher up the Mackenzie, at the junction of Bear Lake River, on the 65th parallel of latitude, there is a tertiary coal deposit of considerable extent, which yields hand specimens entirely similar to Garry Island ones. The forms of trunks of trees, lying confusedly in a horizontal or nearly horizontal position, are preserved in some of

* Nat. Hist. Rev., Jan. 1858, p. 45, and July 1860, p. 360.

these beds, which are much iron-shot. In others, composed of glance coal, the wood-like structure is lost, and pieces taken from any of the beds split into small rhomboidal fragments, no longer presenting the grain or layers of wood. Layers of pipe-clay, with minute grains of amber, and plastic clay interposed among the beds of lignite, contain delicate impressions of leaves belonging to plants of the yew tribe (*Taxites*), of a plant resembling *Vaccinium*, of one similar to a maple, of others like the mulberry, the lime (*Tilia*), and the hazel—in short, an assemblage of plants such as a climate like that of the northern United States would support. The lignite, examined carefully by Mr. Bowerbank with the microscope, was considered by him to be coniferous, but it offered much difficulty to microscopic investigation. These lignite beds are constantly on fire where they meet the atmosphere, and the interposed clays are burned like bricks, producing many pseudo-volcanic products. A precisely similar formation, but with less of the coal exposed, exists near Cape Bathurst, and has been erroneously called volcanic in some of the recent arctic narratives. The true coal beds of the Arctic seas must have been deposited when the climatal conditions of the earth were so totally different from those of the present epoch as scarcely to afford materials for comparison; but the lignite beds were evidently accumulated when the configuration of the surface departed much less from that which at present exists. The leaf beds were undoubtedly formed of vast layers of leaves quietly deposited along with a fine mud in still water. Had these leaves been transported from any considerable distance, their nerves, margins, and hairs, could scarcely have remained in the perfect condition in which impressions in the matrix shew them to have been when they slowly subsided in the turbid but still water. They are leaves of deciduous trees belonging to genera which do not in the present day come so far north on the American continent by ten or twelve degrees of latitude.

A still newer ligneous deposit exists in several localities on the Arctic Sea. On the flat alluvial shores at the mouth of the Mackenzie there are conical hills, which rise a hundred feet or more above the general level. Where these hills are escarped by the action of the water, they are seen to consist of sand of various colours, in which vast quantities of drift timber are imbedded, the whole mound being coated with a black vegetable earth, which, in that climate, must have been ages in forming. At the present time, the highest floods reach only to the bases of these mounds, on which they strew a line of newly-drifted spruce fir trees. Sand blown by the winds among these logs will slowly increase the extent of the formation.*

In a valley of Banks' Island, some distance from the coast,

* Franklin's Second Overland Journey, 4to, 1828, p. 209.

There is a prospect of the formations on the Mackenzie, one of the localities of most interest to the geologist, being investigated by competent observers, since there are now able naturalists employed in that district and in various parts of Rupert's Land by the Smithsonian Institution. Prof. H. Y. Hind received two *Ammonites* from Mackenzie's River resembling Jurassic species, but which he thinks may belong to the *Cretaceous* epoch.

The same geologist remarks that tertiary coal or lignite occurs in tertiary and cretaceous formations in more or less continuous areas along the flanks of the Rocky Mountains from Mexico to the Arctic Sea. Of these deposits he enumerates the following. The great Missouri Lignite Basin, extending 600 miles on that river and 300 up the Yellowstone. It possesses the mixed character of a fresh water and estuary deposit, and cannot, Dr. Hayden thinks, be older than the Miocene period. Lignite has also been traced from the Coulées of the Mouse (Little Souris) to the head waters of the Milk River, a distance of 500 miles.

Dr. Hector traced lignite strata 211 miles along the north branch of the

and 300 feet above the sea-level, Captain M'Clure, accompanied by Dr. Armstrong, visited a formation similar to that just described. "The ends of trunks and branches of trees," says the last-named officer, "were seen protruding through the rich loamy soil in which they were imbedded. On excavating to some extent, we found the entire hill to be a ligneous formation, being composed of the trunks and branches of trees—some of them dark and softened, in a state of semi-carbonization; others quite fresh, with a woody structure perfect, but hard and dense. In a few situations the wood, from its flatness and the pressure to which it had been for ages subjected, presented a laminated structure with traces of coal. The trunk of one tree was twenty-six inches in diameter. Other pieces, though still preserving the woody structure, sunk in water. Numerous pine cones and a few *acorns* were found in a state of incipient silification. Many of the trunks crumbled when struck with the pickaxe, some approached lignite in character, and as far as our excavations penetrated, nothing but the trees and the loamy soil in which they were imbedded was met, though in some places the decay of the wood seemed to form its own soil. Many portions of the branches of trees were found silicified on the surface of the hill and on the neighbouring heights. Some were impregnated with iron, and had a metallic tinkle when struck. Rills of water, impregnated with iron and sulphur,

Saskatchewan. They disappeared four miles below Edmonton, when the cretaceous rocks came to the surface. The same gentleman saw an extensive deposit of tertiary coal on Red-Deer River in lat. 52° 12′ N. and long. 113° W.

The lignite beds which are now worked at Nanino, Vancouver's Island, were ascertained by Dr. Hector to be of the Cretaceous age. Dr. Evans some years since, found Tertiary coal in Oregon and British Columbia.—See *Assiniboine and Saskatchewan Expedition* in 1858, by Henry Youle Hind, M.A., F.R.G.S.

flowed over the surface. "On several of the neighbouring hills I observed distinct stratifications of wood running horizontally in a circular course, formed by the protruding ends of the trunks to which the bark adhered."*

This description of Dr. Armstrong's would apply, in a great part, to the lignite formation on the Mackenzie, at the mouth of Great Bear Lake River, though none of the lignite in the latter situation is so little changed as some of the trees on Banks' Land. The wood from the latter quarter was considered by Dr. Joseph Hooker to be the white spruce (*Abies alba*), and Dr. Harvey pronounced one of the fossilized cones to belong to the same species. The white spruce is the principal forest tree on the Mackenzie, and extends the farthest north; but the *acorns* are remarkable things to be found in such a deposit, as no oaks grow on the banks of any American river that falls into the Arctic Sea, nor approach within many miles of the dividing water-shed.

Malte Brun mentions similar lignite formations as occurring in Iceland. "Besides the fossil bituminous wood, another kind is also found in the earth, which has only undergone a change of colour, odour, and solidity; sometimes merely a flattening, but with no appearance of mineralization. This wood is met with in argillaceous and sandy ground, at the height of some fathoms above the present level of the ocean, while the beds of turf and bituminous wood most generally commence twenty-five or even a hundred fathoms below this level.

Great masses of wood have also been deposited in Siberia at elevations which the sea never reaches in the present day. The ground at Jakutsk on the Lena, lying 270 feet above the

---

* Armstrong, *lib. cit.* p. 395. Nat. Hist. Review, April 1858, p. 73.

level of the sea, and 8° of latitude removed from the mouth of the river, is thus described by M. Erman:—" The internal constitution of the ground was learned in sinking M. Shergin's well to consist, to the depth of one hundred feet at least, of loam, fine sand, and magnetic sand. They have been deposited from waters which at one time, it may be presumed, suddenly overflowed the whole country as far as the Polar Sea. In these deepest strata are found twigs, roots, and leaves of trees, of the birch and willow kinds; and even the most unbiassed observers would at once explain this condition of the soil by comparing it to the annual formation of new banks and islands by the floods of the Lena; for these consist of similar muddy deposits and spoils of willow banks, but they lie about 110 feet lower than the ground which was covered by the ancient floods. Everywhere throughout these immense alluvial deposits are now lying the bones of antediluvian quadrupeds, along with vegetable remains. It cannot escape notice that as we go nearer to the coast, the deposits of wood below the earth, and also the deposit of bones which accompanies the wood, increase in extent and frequency. Beneath the soil of Jakutsk, the trunks of birch trees lie scattered only singly; but on the other hand, they form such great and well-stored strata, under the tundren between the Jana and the Indigirka, that the Jukahirs never think of using any other fuel than fossil wood. They obtain it on the shores of lakes, which are continually throwing up trunks of trees from the bottom. The search for ivory also grows continually more certain and productive along the coast of the icy sea." *

These phenomena attain the greatest development on the

* Erman's Travels in Siberia, ii. p. 279.

Liakhow islands, which lie on the 75th parallel of latitude north of the Indigirka. "The wood-hills of New Siberia," observes Hedenström, "can be seen at the distance of seventy miles. They consist of horizontal beds of sandstone, alternating with bituminous beams or trunks of trees, to the height of 180 feet. On ascending the hill, fossilized charcoal is everywhere met with, encrusted with an ash-coloured matter, which is so hard that it can scarcely be scraped off with a knife. On the summit there is a long row of beams resembling the former, but fixed perpendicularly in the sandstone. The ends, which project from seven to ten inches, are for the most part broken. The whole has the appearance of a ruinous dyke."*

These vertical stumps were probably set up by man, as the custom is with the Eskimos of the present day. On the Mackenzie there are precipitous cliffs, apparently of hard stone, but, in fact, composed of incoherent sand, fixed in a matrix of ice, which cements the whole into a rocky cliff. At the close of summer, the surface thaws deep enough to support a stake driven into it.

Lieutenant Anjou describes the wood hills as extending for about three miles and a half along the southern coast of New Siberia, and rising abruptly from the sea to the height of 120 feet. It consists, he says, of earth, in which planks are imbedded in groups of more or fewer than fifty, with the ends cropping out, the thickest being two inches and a half in diameter. The wood was brittle, semi-hard, black, faintly shining, imperfectly combustible, and burnt with a pitchy smell, glimmering in the fire without flame. In another passage, Lieutenant Anjou remarks, that the trees, though generally

† Wrangell's Siberia.

horizontal, are very irregularly disposed, and that the largest had a diameter of about ten inches.*

Buried trees of a similar description were found by Hedenström on the Moss-Steppe Tundra, east of the Jana, remote from the present line of the forest. The inhabitants designate them as subterranean trees of Adam's time. They glow when lighted, but emit no flame.

From an early period of the Russian explorations of Siberia, the tusks of the fossil elephant or mammoth have been sought for on the shores of the Polar Sea as a valuable article of commerce, and they have been found in greater or smaller numbers in various localities, from the Taimur River to Bering's Straits. Even in the first quarter of the present century, when Erman visited the Gulf of Obi, large quantities of mammoth tusks were collected there. But the great deposit of these bones was discovered in 1773, on the Liakhow islands, by the merchant whose name they bear, and by his associate Protodiakonow. The soil consists (these adventurers said) of sand and ice, with such quantities of mammoth bones, that they seemed to form the chief substance of the island; the skull and horns of a bovine animal, and bones of a rhinoceros (*Rhinoceros tichorinus*) were also found there; and its toes, mistaken by the hunters for the claws of an enormous bird, were by them used in the structure of their bows. Heads of deer, with antlers differing somewhat from those of the rein-deer, also form part of the bony deposit. For eighty years after the discovery of Liakhow islands, the tusk-hunters worked every summer at the cliffs without producing any sensible diminution of the stock. The solidly frozen matrix, in which the bones lie, thaws to a certain

* *Wrangell*, translated by Mrs. Sabine, pp. 383, 487, 492, and 499.

extent annually, allowing the tusks to drop out, or to be quarried, by the Jukahirs. In 1821, 20,000 lbs. of fossil ivory were procured from the island of New Siberia, some of the tusks weighing 480 lbs. The skull, flesh, and skin of the *Rhinoceros tichorinus* have been procured there. And at the mouth of the Lena, the entire carcass of a mammoth was discovered so fresh that the dogs ate the flesh for two summers. The skeleton is preserved at St. Petersburgh, and specimens of the woolly hair, with which the skin was covered, exist in England. The history of this remarkable discovery has been repeatedly given in popular works.

Elephants' teeth are also abundant on the American side of Bering's Strait, and have long been articles of Eskimo traffic with the Asiatic Tchutche. In the zoology of the voyage of the Herald there is an account of the fossiliferous ice-cliffs of Eschscholtz Bay, first discovered by Admiral Kotzebue of the Russian navy; and Mr. Berthold Seeman's botany of the Herald's voyage contains a view of the cliffs. These cliffs are described by some who visited them as composed of pure ice; by others, as being merely faced with ice of considerable thickness. The fossils lie upon the top of the ice, imbedded in, and more or less completely covered with, boggy or sandy soil. The bones had lost little of their animal matter; and those of the mammoth (or elephant), when the earthy substance was removed by acids, shewed the fine membranes of the sieversian canals in great beauty. Hair was dug up along with the elephant skulls, and the whole deposit had a strong charnel-house smell. The species found were, the— 1, mammoth (*Elephas primigenius*); 2, the horse (*Equus fossilis*); 3, the moose-deer (*Cervus alces*); 4, the fossil reindeer (*Tarandus*); 5, fossil musk-ox (*Ovibos*); 6, a musk-ox, of greater size than any living one (*Ovibos maximus*); 7, Arctic fossil bison (*Bison latifrons*, Fischer ? *B. Americanus*, Leidy??); 8, the heavy horned fossil bison (*B. crassicornis*, Rich ; *B. antiquus*, Leidy ?) At least fifteen different individual mammoths must have contributed the bones collected by Admirals Kotzebue and Beechey, and Captain Kellett, in Eschscholtz Bay. Other parts of the coast near Bering's Strait, as far eastward as Point Barrow, yield mammoth teeth ; and an entire skeleton of this species was discovered by the Indians inland, on the elevated country near the sources of the Yukon. No mastodon remains have been found in America further north than the south side of the Saskatchewan Valley, about latitude 51°. It is foreign to the plan of this compilation to enter into speculations on the manner in which such accumulations of bones could be formed, but it may be stated as probable that many, or all, the animals were migratory, like the quadrupeds now frequenting the same districts.

The shells of many mollusks, of the same species with those now inhabiting the surrounding seas, are scattered over the Arctic islands, and a general opinion prevails among the coast inhabitants of Siberia, as well as with voyagers to the Arctic American seas, that the shores and islands are rising. Professor Haughton thinks that there is evidence of the islands of Lancaster Strait and Melville Sound having risen 500 feet within a comparatively recent geological period. An opinion also exists among the inhabitants of both continents that the tundren and barren grounds are encroaching on the forests. Solitary outlying dwarf spruces cling to the ground many miles from the edge of the woods, but the traveller meets with no seedling trees, nor even with young ones, straggling out in the same way.

Facts of a similar kind have been observed in Norway. "I was much struck with the evidence that presented itself of pretty large trees having formerly existed upon Qualoën (Whale Island, on which Hammerfest stands), where nothing now but a few stunted birches can be seen. Dead stumps of considerable size of this kind of timber still stand erect, some of them with branches bearing twigs, even as small as my little finger, with the bark sufficiently recent to tell that the decayed trunks it encompasses belong to the genus *Betula*, thus indicating a comparatively recent date of destruction. The air of Qualoën possesses a peculiarly drying anti-putrescent quality, so that I doubt not but these trees, or rather these remains of trees, may have existed in this state for perhaps centuries, as it is not in the memory of man that living trees, of such magnitude, grew on the island; but tradition says that Qualoën was formerly covered with fir-timber of great magnitude.* The same intelligent writer mentions a report of the entire skeleton of a whale lying on the summit of *Fugle-oe*, an island that rises four or five hundred feet above the sea. Time did not permit him to ascertain what truth there was in report, by ascending the hill, which, on the 14th of July, was still covered with snow.

* *Notes on Norway*, by William Dawson Hooker. Glasgow, 1837, p. 19.

CHAPTER XIX.

INHABITANTS.

Greenlanders, Skrællings, Eskimos, or Inuit—Name—Area—Native Names—Physical Aspect—Habits—Dress—Sledges—Bows—Arrow-Points—Kajaks—*Umiak*—Tents—*Iglut* or Winter Huts—*Kashim*—*Kollek*—Mode of Living—Eaters of Raw Flesh—Whale Hunts—Seal Hunts—Traffic with Asiatics—*Arimaspi*—Temper—Kashim among the *Kuskutchewuk*—Superstitions—Shamauism—*Innuæ* or Souls—Creation—Society—*Tchukche*—Sedentary Tchukche or Namollos—Liakhow Islands—Ancient Buildings of the Eskimo type.

THE sea-coasts of Arctic America, including Greenland, are inhabited solely by one nation, called by the Scandinavians of the tenth century *Skrællingar*.* The seamen of the Hudson's Bay Ships who trade annually with the natives of Northern Labrador, and the *Meta Incognita* islands on the opposite or northern side of Hudson's Strait, have for long called them "Seymos," or "Suckèmos," appellations evidently derived from the vociferous cries of Seymō or Teymō, with which the poor people greet the arrival of the ships. French writers call them Eskimaux, which name Charlevoix conjectures may have been derived from the Abenaki word *Esquimantsic*, signifying "eaters of raw flesh."† English authors, in adopting this term, have most generally written it "Esquimaux,"

---

* Arngrimi Jonæ *Gröenlandia*, Kiobenhavn, 1732. *Grönlandia antiqua*, Thormodo Torfæo, Hafniæ, 1706. *Skräl* fem. pl. in Danish signifies "Screamers." *Skrællingar*, in Swedish, "wretches."

† The Abenaki inhabited the area at present occupied by New England, and are not likely to have come into contact with the Eskimos since Cabot's time, though in days of Scandinavian discovery the Skrællings occupied the coast line as far south as Vermont.

but Dr. Latham, and other recent ethnologists, write it "Eskimos," after the Danish orthography, a practice which is followed in this compilation.

The Eskimos are essentially a littoral people, able to gain subsistence on an icy sea exclusively, with a facility which no other nation has attained to, and occupying a larger extent of continuous sea coast than has ever been held by any other primitive race. At present, they retain possession of the shores of the continent from the lower parts of Labrador, along Hudson's Strait, and down the east side of Hudson's Bay nearly to James' Bay; (from thence round the west side of Hudson's Bay up to the 60th parallel of latitude, the *Eithinyuwuk* and *Tinné* have seized the coast-line, but more to the north) on the shores of the Welcome, and throughout the whole northern sea-board of the continent, round to Bering's Strait and Kotzebue Sound, the Eskimos are the sole maritime inhabitants. To them also belong the entire Greenland coasts, up to the 81st degree of latitude, or as high as Europeans have as yet penetrated; also the large group of *Meta Incognita* islands, and all the islands of the American Polar Sea, whereon traces of their recent or ancient dwellings have been everywhere discovered by the recent searching expeditions.

The autochthonal designation of the nation at large is *Inuit*, the vowels having the same sounds as in modern Italian, and consequently the word, if written in English according to its pronunciation, would be "Eenoo-eet." It signifies men (of their own race), and has a singular *Inuk.**

* Mr. Simpson, surgeon of the Plover, for the two years that that ship was stationed at Point Barrow, writes the national appellation according to the pronunciation of the Western Eskimos, *En-yu-in*, and its singular *Enyuk;* the *n* of the plural being substituted by the western tribes for the terminal *t* of the eastern ones.

The Greenlanders give themselves the distinguishing epithet of *Kălălik* (in the singular *Kalalek*), while the natives of Repulse Bay call themselves *Ahăknan-helik;* those that frequent Back's Great Fish River bear the designation of *Utku-hikalik* (stone-kettle Eskimos), or as Augustus, Sir John Franklin's interpreter named them, *Utku-hikaling-mëut*. Further to the westward, near Cape Alexander, the *Kang-or-mëut* (Snow Goose Eskimos) possess the coast; in the vicinity of the Coppermine River the *Naggëuktor-mëut* (deer-horns) dwell; and the eastern outlets of the Mackenzie and the Rein-deer Mountains are frequented by the bold and numerous *Kittegarëut*, The coast to the west of the Mackenzie, as far as Barter Reef, is occupied by the *Kangmali-enyüin*, and the Point Barrow tribe bear the distinctive appellation of *Nuwung-më-un*. *Nu-wuk* being the native name of Point Barrow. The *Nuna-tangmë-un* inhabit the country traversed by the *Nunatok*, a river which falls into Kotzebue Sound.

In Greenland the natives term a Dane *Kablunak* (plural *Kablunet*), and the same word is recognised as denoting a white man or European along the American coast, as far west as Barter Reef. But as the Eskimos are very observant of peculiarities of features, dress, or gesture, they readily invent epithets to denote either people or individuals; thus in Greenland, the Dutch, who at one time traded a good deal thither, have a proper designation, and at Point Barrow the natives termed the crew of the Plover sometimes *Shakenatana-mëun*, "people from under the sun;" or *Emakh-lin*, "seamen;" or *Ingaland-mëun*, "men of England;" but most commonly *Nelluäng-mëun*, "unknown people." They have also distinctive names for the Red Indians, of their several vicinities, and several expressions for stranger Eskimos.* Throughout the long lines of coast which this people inhabit, they are generally scattered in small bodies of five or six families together, or even fewer in situations so remote from their enemies, the Red Indians, that they are not apprehensive of attack; and only in places that are favourably situated for hunting deer and marine animals, and, it may be added, for commerce, do they congregate in large numbers, such as Hudson Strait, the delta of the Mackenzie, the banks of the Colville, of the Nutawok, and Point Barrow. Yet, from Labrador to the northern extremity of Smith's Sound, including both sides of Greenland, and along the whole northern coast of America, the variations of dialect are small and unimportant. Mr. Miertsching, who learnt the language on the Labrador coast, understood, and could make himself intelligible to, the Eskimos in the vicinity of the Mackenzie and in Camden Bay; and the native Eskimo interpreters from Hudson's Bay employed by Sir John Franklin, Sir John Richardson, and Dr. Rae, had still greater facility of conversation with the north coast tribes. The language is similar in its grammatical construction to the other native American tongues, but differs widely from all of them in its vocabulary.

The Eskimos are remarkably uniform in physical appearance throughout their far-stretching area, there being, perhaps,

* The language is copious; thus, in addition to the national epithet of *Inuit* "people," they have in the eastern dialects *Ang-ut* (sing.), (dual) *Ang-utek*; plural *Ang-utit*, "man, relations, or stock." *Seksariak* means an unprotected man; and *Tunnisuga*, "my nation." All the relations in which a man stands to other men have distinct words to express them. In the western dialects outside of Bering's Strait, *Tatchu* signifies "man," *Tagut* and *Yugut* "Eskimos," corresponding to which, on the other side of the continent, the Labrador Eskimos call the image of a man in a glass, or his shadow, *Tatchak*. The looking-glass itself *Tatchartut*, "it reflects images."

no other nation in the world so unmixed in blood. Frobisher's people were struck with their resemblance in features and general aspect to the Samoyeds, and their physiognomy has been held by all ethnologists to be of the Mongolian or Tartar type. Dr. Latham calls the Samoyeds Hyperborean Mongolidæ, and the Eskimos he ranges among the American Mongolidæ, embracing in the latter group all the native races of the New World. The Mongol type of countenance is, however, more strongly reproduced in the Eskimos than in the Red Indians—the conterminous *Tinnè* tribes differing greatly in their features, and the more remote Indians still more.

Generally the Eskimos have broadly egg-shaped faces with considerable prominence of the rounded cheeks, caused by the arching of the cheek-bones, but few or no angular projections even in the old people, whose features are always much weather-beaten and furrowed. The greatest breadth of the face is just below the eyes, the forehead tapers upwards, ending narrowly but not acutely, and in like manner the chin is a blunt cone; both the forehead and chin recede, the egg-outline shewing in profile, though not so strongly, as in a front view. The nose is broad and depressed, but not in all, some individuals having prominent noses, yet almost all have wider nostrils than Europeans. The eyes have small and oblique apertures like the Chinese, and from frequent attacks of ophthalmia, and the effect of lamp-smoke in their winter habitations, adults of both sexes are disfigured by excoriated or ulcerated eyelids. The sight of these people is, from its constant exercise, extremely keen, and the habit of bringing the eyelids nearly together when looking at distant objects, has, in all the grown males, produced a striking cluster of furrows radiating from the outer corner of each eye over the temples.

The complexions of the Eskimos, when relieved from smoke and dirt, are nearly white, and shew little of the copper-colour of the Red Indians. Infants have a good deal of red on the cheeks, and when by chance their faces are tolerably clean, are much like European children, the national peculiarities of countenance being slighter at an early age. Many of the young women appear even pretty from the liveliness and good nature that beams in their countenances. The old women are frightfully ugly, the discomforts which age entails in savage life, especially on the weaker sex, spoiling the temper and the lineaments of the countenance, in a people who are not restrained from giving expression to their emotions by any conventionalities of refinement. To this must be added the deteriorating influences on the complexion of a close atmosphere and high temperature in the winter huts, alternating with sudden exposure to the blasts of arctic snow storms, followed by the scorching effects of the rays of a spring sun reflected from the snow, all of which concur in giving a harshness to the countenance. Now and then a benevolent-looking old man is met with but not frequently, and a cheerful and pleasant looking old woman is rare indeed among them.

The young men have little beard, but some of the old ones have a tolerable shew of long gray hairs on the upper lip and chin, which the Red Indians never have, as they eradicate all stray hairs. The Eskimo beard, however, is in no instance so dense as a European one.

The hair of the head is black and coarse; the lips thickish; and the teeth of the young people white and regular, but the sand that, through want of cleanliness, mixes with their food, wears the teeth down at an early age almost to the level of the gums, so that the incisors often have broad crowns like the molars.

The average stature of the Eskimos is below the English standard, but they cannot be said to be a dwarfish race. The men vary in height from about five feet to five feet ten inches, or even more. They are a broad-shouldered race, and when seated in their kayaks, look tall and muscular, but when standing, lose their apparent height by a seemingly disproportionate shortness of the lower extremities. This want of symmetry may arise from the dress, as the proportions of various parts of the body have not been tested by accurate measurements. The hands and feet are delicately small and well formed. Mr. Simpson* observed an undue shortness of the thumb in the western Eskimos, which, if it exists further to the east, was not noted by the members of the searching expeditions. From exercise in the occupations of hunting the seal and walrus, the muscles of the arms and back are much developed in the men, who are moreover powerful wrestlers.

Ablution or attention to personal cleanliness is little practised by the Eskimos of any tribe. Water is a scarce article for eight months of the year, and it does not appear that it is a practice to grease the skin with marrow fat as the younger Red Indians are accustomed to do. Egede does mention that the Greenland females occasionally wash their hair and faces with their own urine, the odour of which is agreeable to both sexes, and they are well accustomed to it, as this liquor is kept in tubs in the porches of their huts for use in dressing the deer and seal skins. The men, he says, moisten their

* Blue Book, 1855.

fingers with saliva and therewith rub off the salt that the spray of the sea may have left on their faces. All the Eskimos have a great dislike to wet their limbs in salt water, believing it to be injurious, and the men do not willingly step into the sea unless in water-tight boots. The tongue of the mother is the towel used for cleansing the child, and it is the same handy instrument wherewith the scum is licked off a piece of meat by the woman who cooks, before she presents it to her husband or to a stranger.

The men do not dress their hair in any peculiar fashion, but merely shorten it on the crown, and allow it to hang loose on the cheeks and neck. The women turn theirs up in a large ornamental bow on the top of the head as was once the fashion with European ladies, and the side locks are plaited or tied together with strings of beads, and allowed to hang down in a club to the shoulder. The patterns in which the women tie their hair vary in distant localities, but there is a general similarity of mode, and it is everywhere totally different from that of the neighbouring Red Indians, whose women seldom attempt the conversion of their hair into an ornament, though the young Indian males do.

In Greenland and throughout American Eskimo-land the women tatoo their faces in blue lines produced by making stitches with a fine needle and thread, smeared with lamp-black. Every tribe has a recognised form of tattooing. Westward of the Mackenzie the men cut a hole in the lower lip near each corner of the mouth, which they fill with a labret of bone, stone, or metal; small green pebbles obtained at the mouth of the Mackenzie are used for this purpose, and are neatly set in wood or bone. This unsightly fashion is practised by the Wakash nation of Vancouver's Island, and may have passed northwards from them to the Eskimos of the west coast. The same inconvenient custom of inserting disks of wood or other material into the lips and ears exists among the aborigines of the Brazils.

The dress of the Eskimos consists of a pair of drawers, over which they wear breeches, which come below the knee. The body is clothed in a close jacket like a Guernsey frock or jumper, with long sleeves, a hood, and a hole to pass the head through, but no side openings. This comes down to the haunches, and has a peak in front and another, generally a longer one, behind. These peaked tails are longer and broader in the women's jackets, but there is no other difference in the dresses of the sexes, except that married women have a larger hood, in which they carry their infants. Both sexes wear boots with wide tops which come up over the hips, and are used by the women especially as pockets, in which they occasionally deposit their children on the outside of the thigh, or any other article that they will hold. The boots used in summer are of seal-skin, and quite water-tight; in winter rein-deer skin, dressed with the fur, being warmer, is more employed.

The material of which the dresses are made varies with circumstances. The most prized for the two main parts of the clothing, the frock and trousers, is rein-deer skin, which the women know how to dress with the hair so as to render the skin thin, soft, and pliable, like shamoy leather. The finest dresses are made of the skins of unborn deer, which, after being properly prepared, are doubled and worn with their hair both on the inside and outside of the dress. The boots are of seal-skin dressed in a different way, but with equal skill, so as to be water-tight. The females are very superior needlewomen, and are chiefly occupied in winter in the making of garments. In defect of deer-skins, the skins of birds, particularly of the northern divers, are used to make jackets and breeches, and even the skins of fishes are similarly employed. Polar hare-skins are used to ornament the dresses, and for socks during winter.

For summer use, when seated in their kayacks, the men are provided with water-tight shirts formed of the intestines of the whale, or of the skins of young seals, which are so well drawn round the aperture in which they sit as completely to exclude the water; and even should the kayack be upset, the man knows how to right himself by the action of the paddle without allowing water to enter the cavity of his small but elegant vessel.

Seal-skins, blown up like bladders, are used as buoys for the harpoons, being adroitly stripped from the animal, so that all the natural apertures are easily made air-tight by wooden plugs. With equal industry and skill almost every part of the land and marine animals which are objects of the Eskimo chace, are economised. Of the horns and bones of the deer, knives, spear-points, and fish-hooks are made, or they are used in the framing of sledges. The bones of the whale are employed in roofing huts, or in the construction of sledges in situations where drift-timber is scarce. Strong cord is made from strips of seal-skin hide, and the sinews of musk oxen and deer furnish bow-strings or cord to make nets or snares. The Eskimo bow, a most powerful weapon, is artistically formed of three pieces of spruce fir, carefully split with the grain, the two end pieces having a curve in the opposite direction to that of the central one. Along the back fifteen or twenty nicely twisted sinews are laid and bound down at intervals, giving great strength to the weapon.* A strong arm is required, as well as much address, to bend an Eskimo bow. In the hands of a native hunter it will propel an arrow with sufficient force to pierce the heart of a musk ox, or to break the leg of a reindeer. Iron obtained by barter or from wrecks is employed to point weapons or to make flenching knives, but among the *Kittegarëut* native copper is extensively used for that purpose and for making ice-chisels. The more northern Eskimos are compelled to resort to the antlers of the deer for the construction of the indispensable ice-chisels. Flint or chert, obtained from Silurian limestone, is chipped to make arrow-heads, precisely similar to the flint weapons so commonly found in the soil of various parts of Europe, and even now frequently fashioned by the natives of Australia. The nature of the material has caused the form of the weapon to be alike in all these distant localities.†

The kajak, which is shaped like a weaver's shuttle, pointed at both ends, is framed of wood, bone, or whale-bone, according to the locality in which it is built, and is flattish above, and convex in the bottom. It is covered with seal-skin, a

---

* A very powerful bow also made of fir is in use by the natives dwelling on the northern Obi, and is stated to be the peculiar manufacture of the Kasuimski. The bow is strengthened by thin slices of the horn of the fossil *rhinoceros tichorrhinus*, very neatly joined to the fir by fish-glue, and requires great dexterity to bend it fully. The Kasuimski are inhabitants of the banks of the rivers Kas and Suim. Erman, lib. cit., I. p. 431.

† Before the area occupied by the Eskimos of the Mackenzie was abridged by the encroachments of the *Tinnè*, the Eskimos used to ascend the stream to the "rapid," three hundred miles from the sea, in quest of chert for their arrows. About fifty years ago some Eskimos ventured up the river to seek this material at the rapid above Fort Good Hope, and were repelled by the Dog-ribs, who, being armed with guns, shot one of the party. Even at this day the Indians far up the Mackenzie are in continual dread of the appearance of Eskimo enemies.

circular hole being left in the centre for the sitter. No Red Indian has invented a boat similar, either in material or form, none of the birch-bark canoes of the *Tinnè* or *Eithinyuwick*, being decked or constructed for navigating a stormy ocean. The mixed races that inhabit the peninsula and islands of Alaska, and the neighbouring shores of the continent, have decked kajaks with holes for two sitters.

The umiäk, or "women's boat," is a large rectangular open boat, roomy enough to hold ten or twelve people, with benches for the women who row or paddle. It is buoyant enough to float only a few inches deep, with a full load, and notwithstanding its clumsy appearance is easily propelled, and can be readily transported over ice or land on a sledge.

The harpoons and lances used in killing whales or seals, have long shafts of wood or of the narwhal's tooth, and the barbed point is so constructed, that when the blow takes effect, it is left sticking in the body of the animal, while the shaft attached to it by a string is disengaged from the socket, and becomes a buoy of wood.

The barb of the whale spear has also the line of a seal-skin buoy attached to it, which, with the other weapons, is kept ready for use on the deck of the kajak.

The sledges of the Eskimos are totally unlike those of the Indians, being much larger, and placed on runners. They are framed of drift-wood firmly joined with thongs, and in default of wood, the bones of the whale are employed, in pieces fitted to each other with neatness, and firmly sewed together. To make the runners slide smoothly, a coating of ice is given to them, and this is done also in Siberia. "Every evening," says Baron Wrangell, "the sledges were turned over, and water poured on the runners to produce a thin crust of ice, which glides with incredible ease over firm snow. Chopped moss is sometimes added, and as the ice freezes on the runner it is polished with the palm of the naked hand, however low may be the temperature of the air."*

In the months of June, July, August, and part of September, the Eskimos dwell in conical tents covered with deer-skin leather, framed by poles which meet at the top. In October, in default of tents which the well-to-do Eskimos alone possess, a temporary residence is built of large slabs of ice, and during winter journeys or when hunting seals on the ice in spring, hemispherical huts are admirably built of blocks of snow. No other people has devised such a use for this beautiful light, translucent building material. The Red Indians when overtaken by a storm, at a distance from the woods, burrow in the snow, but have not ingenuity enough to build huts of it, though, had they known how to do so, parties that have perished in attempting to cross lakes in bad weather, might have saved their lives.

In most places that lie conveniently for the prosecution of the whale-fishery, winter residences (*iglut*) are built. The site is generally near the edge of a bank, and the soil is excavated down to the frozen subsoil. The building is then raised with walls, floor, and roof, of drift-timber, the latter being supported by ridge-poles; and four upright pillars of wood are added when the house is large enough to require them. The entrance is on the side by a partly or wholly subterranean passage. The whole building is covered with earth to the thickness of a foot or more, and in a few years it becomes overgrown with grass, looking from a short distance like a small tumulus. Sometimes two or three families join

* For a view of these sledges, see Sir Edward Parry's second voyage.

in constructing a dwelling; each family having a recess appropriated to itself, while the centre is reserved for the cooking lamps.

In populous villages a larger house, termed a *Kashim*, is generally built by the joint labour of the community, with more care, the floor and inside walls being formed of split or dressed logs. This building is used on festal occasions. Most of the houses have an air-hole in the roof, that can be occasionally opened. There is no chimney, wood being rarely used for fuel. The lamp (*kollek*) fed by whale-oil, and trimmed with wicks of moss, is used for cooking, and supplies heat enough for warming the huts.

When the family is assembled in the winter-house, and the door is blocked with a slab of snow, the interior atmosphere soon becomes stifling, and to the unaccustomed European, unendurable. The inmates of both sexes strip off their upper garments in the hut, retaining only the breeches, and pass much of their time, if not compelled to go abroad to seek provisions, in air tainted with putrefying odours; for the scrapings of skins, stinking fish, and much other strong-smelling offal, is cast into the door-way or into the centre of the hut.

The ordinary routine of Eskimo life in most localities is as follows:—In the month of September, the band, consisting of perhaps five or six families, moves to some well-known pass, generally some narrow neck of land between two lakes, and there await the southerly migration of the rein-deer. When these animals approach the vicinity, some of the young men go out and gradually drive them towards the pass, where they are met by other hunters, who kill as many as they can with the bow and arrow. The bulk of the herd is forced into the lake, and there the liers-in-wait in the kajaks spear them at leisure. Hunting in this way day after day as long as the deer are passing, a large stock of venison is generally procured. As the country abounds in natural ice-cellars, or at least everywhere affords great facilities for constructing them in the frozen subsoil, the venison might be kept sweet until the hard frost sets in, and so preserved throughout the winter, but the Eskimos take little trouble in this matter. If more deer are killed in summer than can be then consumed, part of the flesh is dried, but later in the season it is merely laid up in some cool cleft of a rock, where wild animals cannot reach it, and should it become considerably tainted before the cold weather comes on, it is only the more agreeable to the Eskimo palate. When made very tender by keeping, it is consumed raw, or after very little cooking. In the autumn also, the migratory flocks of geese and other birds are laid under contribution, and salmon-trout and fish of various kinds are taken. In this way a winter stock of provision is procured, and not a little is required, as the Eskimos being consumers of animal food only, get through a surprising quantity. In the autumn, the berries of the *empetrum nigrum, vaccinium uliginosum,* and *vitis-idea, rubus chamæmorus,* and *arcticus,* and of a few other arctic fruit-bearing plants are eaten, and the half-digested lichens in the paunch of the rein-deer are considered to be a treat; but in other seasons this people never tastes vegetables, and even in summer animal food is alone deemed essential. Carbon is supplied to the system by the use of much oil and fat in the diet, and draughts of warm blood from a newly killed animal, are considered as contributing greatly to preserve the hunter in health. No part of the entrails is rejected as unfit for food, little cleanliness is shewn in the preparation of the intestines, and when they are rendered crisp by frost, they are eaten as delicacies without further cooking.

On parts of the coast where whales are common, August and September are devoted to the pursuit of these animals, deer-hunting being also attended to at intervals. The killing of a right whale or of a sufficient number of the *killeluak* (*Beluga albicans*) secures winter feasts and abundance of oil for the lamps of a whole village, and there is great rejoicing. On the return of light, the winter-houses are abandoned for the seal-hunt on the ice, sooner or later, according to the state of the larder. The party then moves off seaward, being guided in discovering the breathing-holes of the seal or walrus by their dogs. At this time of the year huts are built of snow for the residence of the band, and at no season is the hunter's skill more tested, the seal being a very wary animal, with acute sight, smell, and hearing. It is no match, however, for the Eskimo hunter, who, sheltered from the keen blast by a semi-circular wall of snow, will sit motionless for hours, watching for the bubble of air that warns him of the seal coming up to breathe. And scarcely has the animal raised its nostrils to the surface before the hunter's harpoon is deeply buried in its body. This sport is not without the danger that adds to the excitement of success. The line attached to the point of the harpoon is passed in a loop round the hunter's loins, and should the animal he has struck be a large seal or walrus, woe betide him if he does not instantly plant his feet in the notch cut for the purpose in the ice, and throw himself in such a position that the strain on the line is as nearly as possible brought into the direction of the length of the spine of his back and axis of his lower limbs. A transverse pull from one of these powerful beasts would double him up across the air-hole, and perhaps break his back, or, if the opening be large, as it often is when the spring is advanced, he would be dragged under water and drowned. Accidents of this kind are but too common. When the seals come out on the ice to bask in the powerful rays of a spring sun, the Eskimo hunter knows how to approach them by imitating their forms and motions so perfectly that the poor animals take him for one of their own species, and are not undeceived until he comes near enough to thrust his lance into one. The principal seal-fishery ends by the disruption of the ice, and then the reindeer are again numerous on the shores of the Arctic Sea, the birds are breeding in great flocks, and the annual routine of occupation, which has been briefly sketched, commences anew.

On the continental line of coast west of the Mackenzie part of the tribes pass the summer in traffic, hunting of course on the way as opportunities offer. Mr. Simpson, who, from his position at Point Barrow, was able to learn more correctly than any other person the course of this traffic, describes it as follows:—" Having cleared out the furniture from the *iglu* (winter hut), the *umiāk* (woman's boat) is put upon a sledge (*uniëk*), secured with thongs, and stowed with the summer tent and all the baggage of the family, also with the *kayaks*, the children and old people, making a very considerable weight to drag. On a low sledge (*kamōtik*), of stouter structure, seal-skins filled with oil for barter are placed. Three grown people are appointed to drag the large sledge, and one is allotted to the *kamōtik*, assisted by dogs distributed according to the load. Fourteen parties, with as many boats (in the aggregate seventy-four souls), passed the ship on their way eastward during the third day of July 1854. On the fourth day after leaving Point Barrow they arrive at Dease Inlet, which is then a sheet of water, and the mode of transport is reversed, the sledge being now carried in the umiak, and the small boats are towed. In favourable seasons the voyage is continued by paddling and tracking the boat along the shore, between which and the ice there is generally a narrow lane of water. At Smith's Bay the sledges are left, and the parties pass through a chain of lakes to Harrison's bay, whence they return by a river to the sea. In these lakes abundance of large fish are taken by nets, a few birds are obtained, and occasionally deer. About the eleventh day they reach a small island where the various groups assemble to enter the Colville River in a body. They take the deepest of its four channels, and on the west bank find the *Nunatang-mëun* assembled waiting for them. About ten days are spent in bartering, dancing, and revelry on the flat ground between the rows of tents of each party, pitched a bow-shot apart. The time is one of pleasant excitement, and is passed nearly without sleep.

" About the 20th of July this friendly meeting is dissolved, the *Nunatang-mëun* ascend the Colville on their way homewards, and the others descend to the sea to pursue their voyage eastward. Spending much time in hunting, they take four or five days to reach Point Berens, twenty miles east of the Colville; and after sleeping by the way four nights more, they leave the women and children at Boulder Island. The men and two women then embark in three boats, and reach Barter Reef after a fifth sleep. On approaching the eastern Eskimos, being the weaker party, they are very wary, and take up a position near a small island, to which they can retreat on any alarm, and advance cautiously, making signs of friendship. They say that formerly great distrust was manifested on both sides, but of late years more women go, and they have dancing and amusements, though they never remain long enough to sleep there.

"Some more details of this traffic may not prove uninteresting. At the Colville, the *Nunatang-mëun* offer goods procured at Kotzebue Sound from the Asiatics in the previous summer, consisting of iron and copper kettles, women's knives, double-edged knives, tobacco, beads, and tin for making pipes; and also articles procured from the Eskimos on the river Kowak, such as stones for labrets, whetstones, arrow-heads, and plumbago. Besides these, deer-skins, the skins and horns of the argali, furs, feathers for arrows and head-dresses, are among the articles of trade brought by the *Nunatang-mëun*. In exchange for the *Nunatang-mëun* articles, the Point Barrow people give the goods they procured to the eastward the year before, and the produce of their own sea-hunts, namely, whale or seal-oil, whalebone, walrus-tusks, thongs of walrus-hide, seal-skins, etc., and proceed with their purchases to Point Barter, as stated above. There they obtain in traffic from the Western Eskimos, wolverine, wolf, argali, and white-dolphin skins, thongs of deer-skin, stone-lamps, English knives, small white beads, and lately, guns and ammunition.*

"The Point Barrow people are acquainted with about 600 miles of coast, between Point Hope and Barter Reef. Beyond the latter they know, by report, the tribes which have been named above, as far as the makers of stone-lamps or kettles

---

* The Asiatic Tchukche find this trade so important, that a settlement of 200 people has been formed on the small but lofty rocky island Ukiwok, in Bering Strait, for carrying on the trade. The people of St. Lawrence Island are also engaged in it; and a body of skilful factors dwelling on Sledge Island, are entrusted by the Tchukche with tobacco, clothing, and other articles to exchange for furs, fossil, ivory, and other articles collected on the banks of the Kwichpak.

at the mouth of the Great Fish River. More remote still than the *Utkuhikalig,* there exists, they say, according to rumour, men who have two faces, one in front and the other at the back of the head. Each face has one large eye in the centre of the forehead, and a large mouth armed with formidable teeth. The dogs of these people, which constantly attend them, have also but a single eye each."

It is curious to discover in this fable the story of the Hyperborean Cyclopes, or Arimaspi, mentioned by most of the ancient geographers on the authority of Aristeas of Proconnesus, himself a myth, or as Herodotus calls him, "ghost of a man."*

> The Arimaspian troops, whose frowning foreheads
> Glare with one blazing eye.
> *Prometh. vinet.* (Potter's tr.)

Articles of Russian manufacture find their way from tribe to tribe along the American coast, eastward to Repulse Bay, Mr. Simpson having recognised the double-edged knife seen at Winter Island in the possession of the Eskimos by Sir Edward Parry, as precisely similar to knives exported from Siberia to Hotham Inlet of Kotzebue Sound; and noticed in the latter locality stone-lamps or kettles from Back's Great Fish River. Sir John Richardson also saw a sword-blade at Point Atkinson, east of the Mackenzie, of Russian fabric. Love of barter is an almost incontrollable passion with the Eskimos, who encounter many dangers in gratifying it. Von Bäer compares them in this respect to the Phœnicians; and by Baron Wrangell, we are informed that the commodities with which fossil, ivory, furs, etc., are purchased from the Eskimos of

---

* Herod. iv., ch. 14 and 15, Rawl. tr. *Arima,* one; *spu,* the eye, in Scythic. Id. iv., app. p. 197.

the Island of Kadyak and of Kotzebue Sound, are obtained by the Tchukche at the fair held annually at Ostrownoie near the Kolyma, in exchange for the ivory and other things they had transported over Bering's Strait in the preceding year.

During the long confinement to their hovels in the dark winter months, the Eskimo men execute some very fair figures in bone, and in walrus or fossil ivory, besides making fish-hooks, knife-handles, and other instruments neatly of these materials, or of metal or wood. Some of the bone articles purchased from the Eskimos are used in games, resembling the European one of cup and ball, or in other contrivances for passing the time. Imitations of the human figure are common, and also of canoes, sledges, and other instruments of their menage, or of the animals known to them; but there is no reason to believe that any of the figures they make are ever worshipped as gods. They part with any of them freely in barter.

Generally the Eskimos are good-natured and cheerful, and their social disposition is distinctly shewn in the arrangement of their winter dwellings, sometimes with recesses for the accommodation of two or three families under one roof, the centre being common to all; sometimes by placing the houses side by side with a narrow dividing lane, into which the doors open, the lane being easily converted in winter into a porch of communication by roofing it over and shutting up the ends with slabs of snow. The *kashim*, or council-house, is an erection larger than a common dwelling-house, constructed in the larger villages as a place of assembly for the community, where the men feast, and whereto both sexes are admitted to dance, according to the customs of the northern and eastern

Eskimos.* Bäer, in describing the manners of the *Kuskutchewuk*, an Eskimo tribe living outside of Bering's Strait, on the banks of a river which falls into the sea on the 60th parallel of latitude, states that with them the *kashim* is the sleeping apartment of all the adult, able-bodied males of the village, who retire to it at sunset ; while the old men, the women, and children, with the *shaman*, sleep in the ordinary dwellings. Early in the morning the shaman goes to the kashim with his tambourine, and performs some ceremony, such as his fancy prompts, for he is bound by no established precedent. Feasts are held in the kashim, particularly a great festival at the close of the hunting season, in which the success of each hunter is proclaimed, and the liberality of the contributors to the feast applauded. The only women admitted on these festive occasions into the kashim are those who have been initiated with certain formalities. In the ordinary domestic life of the Eskimo tribes, the men eat first and the women afterwards, but the woman cooks, and it is her duty and privilege to lick the gravy from the meat with her tongue before she presents it to her husband.

Among the Eskimos that inhabit the coast of the Welcome, south of Chesterfield Inlet, there were, according to the information of Augustus, Sir John Franklin's interpreter, sixteen men and three women who were acquainted with the mysteries of Shamanism, the women practising the art on their own sex only. When the shaman was employed to cure a sick person, he shut himself up in a tent along with his patient,

---

* We have no account of the existence of these council-houses on the Labrador coast, nor does Egede mention them as a Greenland institution, but in the Labrador vocabulary the following words occur :—*Kashim-iüt*, "an assemblage of men in council ;" *Kashimin-wikhak*, " a place of assembly for council."

and sung over him for several days, abstaining from food all the time. Blowing on the affected part is an approved remedy with the Shamans, and Europeans are often requested to blow on the faces, eyes, or ears of Eskimos who are ailing. The Angekoks employ ventriloquism, swallow knives, extract stones from various parts of their bodies, and use other deceptions to impress their countrymen with a high opinion of their supernatural powers.* Certain women, Egede says, by living strictly according to rule, acquire the pretended power of stilling the wind, causing the rain to cease, etc. They are called *Arnak-aglarpok* (women who abstain at certain times). Similar powers were formerly (and perhaps still are) claimed by certain Lapland witches, that were regularly propitiated by English seamen trading to Arkhangel, who made it a point to land and buy a wind from these poor creatures. A spurious kind of witches called *Illiseersut* are said to be feared and hated by the Greenlanders, and often destroyed without mercy.

There is always much difficulty in obtaining a correct notion of the religious belief of a heathen people. The language must first be mastered, and by the time that that is accomplished, the priests, or shamans, who are the sole expounders of the heathen superstitions, have become jealous of the superior attainments of the white man, and shrink from exposing their practices to his ridicule. In the main, however, the account of the religion of the Greenlanders furnished by the first missionary, Hans Egede, who founded the modern colonies in 1721, agrees with information collected by the Moravian Brethren of the present day.

* *Angekok* (in the plural *Angekut*) is the Greenland, Labrador, and Mackenzie River name for the Shaman. Among the Kuskutchewuk the name is *Analh-tuk* or *Tungalh-tuk*.

These heathens believe, say the missionaries, in two greater spirits and many lesser ones. One named *Torngarsuk** is stated by Egede to be the being who is consulted on all occasions by the Angekoks or Shamans, but is known to the common people only by name. Even the initiated vary in their opinions, some saying that he is devoid of form, others affirming that he has the shape of a bear, or that he has a large body with only one arm, or that he is no bigger than a man's finger. He abides, say they, in the interior of the earth, or under the waters, where there is continual fine sunshiny weather, good water, deer and fowls in abundance. From him the Angekoks learn their art. Torngarsuk is supposed to be of the male gender, and to be generally friendly to man. There is another great spirit of the other sex, having no proper name, but considered to be of a very bad and envious disposition. Neither to the one or other is worship or honour paid, all intercourse with them being left to the Angekoks. These boast of a close intimacy with Torngarsuk, and that it is from him they obtain their familiar, or *Torngak*, who accompanies them on their journeys when they go to ask advice of Torngarsuk about the curing of disease, procuring good weather, or dissolving the charms by which land and marine beasts of chase have been secluded from the hunters. †

The inhabitants of Cumberland Sound name the evil spirit *Torngak*, and call the good spirit *Sanak* or *Sana*, saying that they implore his aid when in trouble or in want, and then *Takak* (the moon) provides for their necessities, giving them rein-deer, seals, or other benefits. Sanak once lived on

---

* Torngak (dual, Torngek, pl.—*ait*) a devil, with the nominal affix *arsuk*, " a great devil," or great Spirit.

† Egede.

earth, they said, but had retired to the moon and was still there. Among this section of the nation Brother Warmow discovered an imperfect knowledge of the deluge, which led him to think that in former times they had obtained some acquaintance with Scripture history, of which only a few vague traditions remain among them now.*

The Labrador Eskimos believe that a very old woman, named *Supperguksoak* rules the rein-deer, and selects those which the *Innuit* need; and to *Torngarsuk*, they attribute the office of herding the whales and seals, his employment being that of the Grecian Proteus. On him the Eskimos call when they need seal's flesh;† and in the interior of the country Supperguksoak assembles the souls of the deceased to hunt rein-deer. The condition of the soul is generally supposed to be better after death than during life, and its happiness is by most believed to consist in the possession of abundance of fowls, fish, seals, rein-deer, and other corporeal enjoyments. Its abode is judged by some to be under the sea, from whence the Eskimos obtain their best and most abundant food; others think that the departed spirits resort to the upper sky over the rainbow, there being no uniform opinion on these subjects.

Brother Warmow visited a dying man in Cumberland Inlet. He seemed to have no fear of death, but to rejoice in the prospect of going to his children, who had died in past years. When asked where they were, he said he did not know, but thought that they must be in another world where they were happy. When intelligence was brought to Warmow of the death of this man, he hastened to his hut, and found the

---

* Brother Warmow's Journal of his residence in Cumberland Inlet in the winter of 1857-8. Missions Blatt. March 1859.

† Missions in Labrador der Evangelischer Bruder. Gnadau, 1831.

Eskimos engaged in dragging the corpse, tied up in skins, over the snow to a distance from their habitations, carrying also the weapons used by the deceased in the chase to be deposited with the body. When the party arrived at the appointed place, Warmow commenced digging a grave, and asked the Eskimos to assist, and to gather stones to cover it, that the dogs might be prevented from devouring the body; but they replied the poor animals are hungry, let them eat it if they like. They also told him that if the grave was covered at all it must be done by the Angekok, or one of his pupils. On Warmow taking from among the weapons one or two very neat bows and arrows, telling the bystanders that this was done at the request of the captain of the ship, who wished to have them as memorials, they expressed great surprise, and said that the deceased would come and demand his property. The widow, and a boy who helped to make the grave, had to sit mourning in the house for three days, during which time no Eskimos dared to go near them. When any one is supposed to be dying, the relatives carry all the property out of the house that they wish to preserve, as all that is left with him at his death must be deposited beside the corpse. The near relatives of the deceased seclude themselves for a time after his death, and abstain from food. If the body has been dragged to the place of sepulture on a sledge, the sledge is left at the grave. Among the Eskimos of Point Barrow, the family of the dead mourn him for five days, according to Mr. Simpson; while among the Greenlanders the mourning and loud daily lamentations are kept up for some weeks.

Egede mentions a minor kind of spirits or *Innuæ,** from

---

* Innuk (plural *Innuit*), a man. *Innusek*, a life. *Innue*, a track. *Innujok* a portrait of a man. *Innulivok*, he heals the sick.

among which Torngarsuk selects the Torngak or familiar of the Angekok. Some Angekoks have their deceased parent for a Torngak. The *Kongcuscrokit* are marine Innuæ, that feed on fox-tails. The *Ingnersoit* inhabit rocks on the shore, and are very desirous of the company of the Greenlanders, whom they carry away for that purpose. The *Tunnersoit* are alpine phantoms. The *Innuarolit* are pigmies that live on the eastern shores of Greenland ; and the *Erkiglit* who reside on the same coasts are people of a monstrous size, with snouts like dogs. *Sillagiksertok* is a spirit who makes fair weather, and lives upon the ice-mountains—*Silla* means air or wind, also the world and reason. To the air the Greenlanders ascribe some sort of divinity, and lest they should offend it, they are unwilling to go out after dark. *Nerrim-Innua* is the ruler of diet.

Concerning the creation of the world, Egede reports that the Greenlanders have little to say, except that, as things are now, they believe they have always been ; yet various and contradictory fables respecting the origin of things may be gathered from them.

The Eskimos have neither magistrates nor laws, yet they are orderly in their conduct towards each other. The constitution of their society is patriarchal, the head of the family ruling as long as he has vigour enough to secure success in hunting. When age impairs his strength and powers of mind, he sinks in the social scale, associates with the women, and takes his seat in their boat. Individuals of greater judgment and ability take, of course, a lead in such a community ; and Mr. Simpson says, that among the western Eskimos, certain wealthy heads of families (*Omalik*), have great influence, their wealth being the exponent of their superior prudence and management. Among the Eskimos that frequent the Hudson's Bay posts of East Main and Churchill, there are leading men, it being the policy of the Hudson Bay Company to institute native chiefs, who may direct the motions of the tribe, and act for it at the trading-posts; but there are no recognised chiefs among the Eskimos, who have no intercourse with Europeans, and it is left to any individual who is aggrieved, to seek redress for an injury by his own might; or, should a murder be committed, for the relations of the deceased to retaliate, or, if they prefer it, to exact blood-money. Both the Eskimos, and their neighbours the Tinnè, believe that the Shamans of either nation can cause calamity or death at a distance by their arts. Hence on the occurrence of an epidemic or other unforeseen disaster, the one people is always ready to accuse the other of the mischief, and retaliation is made by slaying one or more of the other nation. In this way the feud is maintained, until the desire for barter overcomes the passion for blood, and then the matter is compromised by the people who have killed most men, paying blood-money for the surplus. In the conflicts that have ensued on such occasions, the Tinnè, being armed with muskets, have had that advantage counterbalanced, in a measure, by the greater daring of the Eskimos, who are a bold and resolute race of people when their rights or hunting-grounds are menaced.

In their demeanour with strangers the Eskimos do not affect the solemnity of the Red Indians, but are lively, talkative, and noisy, especially when the women are of the party. On the first approach of boats to the huts of a band of Eskimos, who have not seen white people before, the men assemble on the shore with their bows and arrows, and throw themselves into extraordinary attitudes, standing on one leg, leaping making hideous grimaces, and uttering loud shouts. In some such way, the stories of people with one leg, or with one eye, or two faces, told by Aristæus and Pliny, may have originated. But as soon as the Eskimos are convinced of the peaceable intentions of the new comers, their hostile demonstrations cease at once, and they hasten with ardour to exchange such things as they possess for those offered to them.

In their intercourse with strangers, the Eskimos are inveterate thieves, and steal with a dexterity that could not be acquired without long practice; yet among themselves the rights of property would appear to be sacred. Mr. Simpson alone, of writers who have had an opportunity of studying the Eskimo character, mentions stealing among themselves as a vice of the Point Barrow Eskimos. This people also lie, as it were, naturally, as often giving a false as a true response to questions; yet such is their communicative disposition, that with patient listening a true version of any story may be elicited from them by one skilled in their language and habits.

Quarrels among themselves would appear to be rare, and when they do occur, they are settled by boxing, the parties sitting down and striking blows alternately, until one of them gives in. The women are treated sometimes with indifference, seldom with harshness, and they have much to say in bargains, and in all public transactions. The children are affectionately and indulgently cared for. Marriages are made, pursuant to betrothals, when the females are very young; but the wives are rarely mothers before their twentieth year, and the children are commonly suckled until they are about four years of age. The number of children of each marriage are consequently few; and frequent famines, with the diseases and accidents of savage life in so severe a climate, keep down the population, so that the numbers of the nation are most probably on the decrease, though this is a point on which correct data are wanting. The number of deserted encamping places, of very ancient date, scattered over the Arctic islands, are no certain evidence of a decreasing population, as many circumstances might lead to extensive wanderings by a people who can make very long summer voyages in their boats, and are able to subsist wherever animals are to be found.

In ancient times the Eskimos ranged down the Atlantic coasts of America, as far as Vermont, if the evidence of the Scandinavians is to be confided in, and they probably once occupied much more of the Pacific coast of America than they do at present.

Von Bäer establishes, on the authority of Russian voyagers and traders, the fact that the Eskimo race predominates down to the peninsula of Alaska, Tchugatch Bay or Prince William's Sound, and the island of Kadjak (or Kodiak) on the 58th parallel of latitude. The language is, however, not pure in that quarter, but is altered by an admixture of words derived from the *Kolyutschin* tribes of the *Tinnè* stock, with which they have intermarried. The similarity of the timber houses of Nootka Sound to the Eskimo winter-huts, so different from any dwellings of other Red Indians, and by no means rendered necessary, or even advisable, by severity of climate in British Columbia, give some reason for inferring that the Eskimos, in former times, held the coast down to the Strait of Da Fuca, and that their houses were occupied and imitated by the Wakash Indians who drove them away. The natives of Oonalashka, figured by Webber in the atlas of Cook's third voyage, have the Eskimo features strongly marked, and are very unlike to either the Nootkians of Vancouver's Island, or to the Tchukche of north-eastern Asia, drawn by the same artist.

The rein-deer, Tchukche, of the eastern extremity of Siberia, north of Kamtschatka, now occupy a considerable portion of the area within the Arctic circle, but as they are evidently an intrusive race who have driven away or enslaved the former inhabitants of that coast, it is not our purpose to describe them here. M. Matiuschkin, in Von Wrangell's narrative, gives an account of their summer menage, when they come to trade at Ostrownoie; and Mr. Hooper, in his recent work, "Ten Months in the Tents of the Tuski," supplies many details of their hospitality and domestic habits in their winter residences. This officer mentions that there evidently appeared to be two distinct races in their villages.

The subdued race is named by Dr. Latham, NAMOLLOS, and from the vocabularies in Klaproth's Asia Polyglotta, are considered to be undoubted Eskimos. They do not possess rein-deer like their present masters the Tchukche, and as they dwell in fishing villages, and have not that facility of travelling great distances over land which the domestication of the rein-deer gives, they are sometimes called by way of distinction *Sedentary Tchukche*. Their present residence as a distinct people is, according to Dr. Latham, from Cape Tchukotski westward round the shores of the Gulf of Anadyr to the mouth of the river of that name, embracing nine or ten degrees of longitude, but habitations such as they and the American Eskimos construct, still exist in the more northern country occupied by the rein-deer Tchukche and the Eskimos that are enslaved or incorporated with them, also in various localities along the northern shores of Siberia, as far as the Kolyma. When Andrejew, in 1763, visited the Liakhow Islands, he found everywhere ruined yourtes constructed of earth. One of these which he describes is so much like what a large Eskimo *Iglu* would be after much earth had fallen into the interior, and decay had done its work, after a long lapse of years, that we can scarcely hesitate in ascribing its construction to people of that race. Andrejew's account of it is as follows :—

" From the northern side of this island (*Kotelnoi*) a sand-bank runs into the sea to a distance of eleven fathoms from the shore. The greater part is covered at high water, but it happened to be dry when we saw it. On this sand-bank is a soft sandstone rock, forming at the height of three fathoms a terrace, on which a kind of fortress has been erected. To increase the breadth of the terrace, the trunks of ten strong larch trees have been placed with their roots upwards, forming a support to a fabric of beams, of the usual form of the Russian *lobassy* or store-houses, the whole resembling very much a bird's nest. The floor in the interior is formed of the trunks of larch trees, covered by a layer of earth seven inches deep. In the inside, at a little distance from the outer wall, a second wall about three feet six inches in height is formed of split trunks or rough boards, and the intervening space filled with earth. The walls without are likewise protected with earth and sods of moss. The roof consists of pine and larch branches irregularly thrown upon each other, and formerly probably covered likewise with earth, but at present the greater part of the covering has fallen in. To hold the building together, cross beams have been let into the cornice, and fastened with thongs of leather. The beams appear to have been hewn, not with an iron axe, but with one of stone or wood, and look almost as if they had been gnawed by teeth. The building is at present four and a half fathoms in length, and four fathoms in breadth, but appears to have been originally six fathoms square. From this yourte a path leads to the shore, and another to the summit of the rock under which it stands. The fortress has been built with great care and owing to the height and narrowness of its site, must have required great labour. By what nation it has been built cannot now be determined. It is not Russian."*

We may remark that the American Eskimos are accustomed to construct stages on posts formed by planting drift trees in the sand with their roots up. Numbers of these stages used for placing skins and food beyond the reach of dogs are erected on the sea-beaches close to the villages, though not generally touching the houses; that position in the case of the stage on Kotelnoi island being evidently due to the narrowness of the terrace selected for a building site. M. Andrejew says nothing of a subterranean entrance to the house, but the bank through which it had been worked might have crumbled away.

* Wrangell's Polar Sea, p. 462.

CHAPTER XX.

SAMOYEDS.

Their Country — Carpini — Heberstein — *Slata Baba* — *Peschchori* — Erman — Fletcher — Richard Johnson — *Waigatch* — Shamanism — Steven Burrough — Samoyed notions of a God — *Koedesniks, Tadebes* or Shamans — Samoyed Possessions — *Chaya* or *Khayodeya*, the Holy Island — Character of the Samoyeds — Eaters of Raw Flesh — Stature — Physical Aspect — Nomades occupying a Diminishing Area — OMOKI — *Yukahirs*.

THE SAMOYEDS, according to Dr. Latham, come nearest to the Eskimos in their physical appearance, and they resemble likewise the Lapps who form the third circumpolar division of mankind. With the latter, indeed, they are supposed to have an ethnological relationship, both being considered by some authors to be members of the Yugrian Mongolidæ, though Dr. Latham places the Samoyeds in a separate division of Mongolidæ, which he names Hyperborean.

Samoyed, according to Adolph Erman, means an inhabitant of a swampy land (*Sama*), and is identical with *Samolain* the name assumed by the indigenous Finnlanders. It is, therefore, a Finnish word, and not one of Russian origin as has been supposed. The national name in the Samoyed tongue is *Nyenech* or *Khasovo*, both of these terms being equivalent to "men or people." As early as the year A.D. 1096, they were designated in the Russian Chancellerie, as *Siragnezi*, "eaters of raw flesh," a practice which they still pursue.*

---

* Erman repudiates the supposed Russian origin of the term Samoyed, and the meaning attributed to it of "self-eaters or cannibals." Tr. in Sib., 1, p. 12.

The northern Samoyeds occupy exclusively the great northwest promontory of Asia which is terminated by the *Sieveroi rostochnoi Nos*, and seek their subsistence for part of the year on the shores of the icy sea. They are the nearest people to the north pole, except the Eskimos of Smith's Sound. The nation more or less intruded on by other races, also, inhabits the district west of the Ural mountains between the mouths of the Petchora and Obi, which they name their proper home or *Arkya* the "Great Land."* They also range westward round the shores of the *Kanin nos*, to the river Mozine on the 44th meridian. The island of *Waigatz* or *Kolguev*, is one of their holy places, and is termed in their own tongue, *Khayodeya*.

Joannes de Plana Carpina, ambassador to Turkey in 1246, mentions the *Samoyedi* as a people that were seen by the Tartars on their campaign against Russia in the vicinity of the frozen ocean. Between them, however, and the actual shore of the icy sea, a people were reported to dwell who had stag's feet and dog's faces.†

Sigismund Von Heberstein, ambassador to Russia from Germany in 1517-1526, mentions his cognisance of Russian itineraries into the Samoyed country lying to the west of the Urals, but seems to have had no personal acquaintance with the people. He repeats, however, a story, which in some form or other had been current for six hundred years or more, of a golden old woman or *Slata Baba* situated on the further bank of the Obi. This idol he represents to be an old woman holding her son in her lap, and he states further that recently another infant had been seen which is her grandson. This

---

* Erman, l. cit. II., 86.

† Hakl. I. p. 30. See the preceding chapter (p. 324) for an Eskimo legend of people with dog's faces, called *Erkiglit*.

fable is probably of ancient origin, and may be one version of the gold-guarding Griffons of Herodotus.

On this subject Erman says that the Samoyed country in the vicinity of Obdorsk, at the mouth of the Obi, "remained unvisited by strangers till the last century, while the western districts of Timansk and Kaminsk were frequented by Novo-gorod merchants before Rurik's time. The Samoyeds of that day related to these traders that men of unknown origin and unintelligible language were living in subterranean dwellings in the high insulated mountains which rise in the district of Timansk. In later times, up to the beginning of the present century, both Russians and Samoyeds have found deserted caverns of this kind (called in Russian *peschchori*) so frequently that it has been conjectured that the name usually given to the river had its origin in this circumstance. The metal utensils and fire-places in these caves leave no doubt that they were inhabited in ancient times by itinerant metal finders, of whom similar traces are found further south in the Ural and Vogul country.

"It is manifest that Greek information respecting the gold-seeking Arimasps in the northern Ural referred in reality to some temporary dwellers in the western parts of the Samoyed country. The obscurest portion of the narrative of Aristæus of Proconnesus, in which he tells us that the Arimasps, seeking metals in the extreme north of Europe, drew forth the gold from under the Griffons, will be found at this moment, in one sense, literally true, if we only bear in mind the erroneous zoological language of the inhabitants of the Siberian tundren. All these people believe that the compressed and sword-shaped horns of the fossil *Rhinoceros tichorhinus* are birds' claws, and some of these tribes, the Yukagirs in particular, find the head of this mysterious bird in the peculiarly vaulted cranium of the rhinoceros, its quills in the leg-bones of other fossil pachyderms; and they state that their forefathers saw and fought wondrous battles with this bird, just as the Samoyeds preserve to this day the tradition that the mammoth still haunts the sea-shore, dwelling in the recesses of the mountain and feeding on the dead.

"Now, if it be not denied that this northern tradition presents to us the prototype of the Greek story of the Griffons, it must be allowed to be strictly true that the metal-finders of the northern Ural drew the gold from under the Griffons, for gold-sand lying under the formations of earth and peat, which are filled with these fossil remains, is at the present day a very common phenomenon."*

Dr. Giles Fletcher, who wrote of the Samoyeds towards the close of the sixteenth century, after having visited their country, says—"As for the storie of *Slata Baba*, of which I have read in some mappes of these countries, I found it to be but a verye fable. Onlie near to the mouth of the great river Obba there is a rocke which, being somewhat helped by the imagination, may seeme to beare the shape of a ragged woman with a childe in her armes, where the Obdorian Samoites use much to resort by reason of the commoditie of the place for fishing; and there sometime conceive and practice their sorceries and ominous conjecturings about the good or bad speed of their journeys, fishings, huntings, and such like." †

Dr. Fletcher says further that they acknowledge one God, but represent him by such things as they have more use and good by. Their leader in every company is their *papa* or

---

* Erman, Travels in Sib., II. p. 88.
† Hakl Soc. Pub. The Russe Commonwealth, Lond. 1591, p. 99.

priest (Shaman). They are all black-haired and naturally beardless, and therefore the men are hardly to be discerned from the women by their looks, save that the women wear a lock of hair down along both their ears. They are clad in seal-skins with the hairy side outwards, with their breeches and netherstocks of the same, both men and women (Fletcher, l. c.).

Richard Johnson, who in 1556 accompanied Steven Burrowe (or Burrough) in his voyage to Waigatz in the Serchthrift, gives an account of the idolatrous offerings of these people which we slightly abridge—" And the sayde Samoüds which are about the bankes of Pechere (Petchora) make sacrifices in the manner following. Euerie kinred doeth sacrifice in their owne tent, and hee that is most ancient is their priest. First the priest doeth beginne to playe upon a thing like a great sieue with a skinne on one ende like a drumme, and the sticke that he playeth with is about a spanne long, and one end is round like a ball, covered with the skinne of a harte. Also the priest hath upon his head a thing white like a garlande, and his face is covered with a piece of shirt of maile, with manie small ribbes, and teeth of fishes, and of wilde beastes hanging upon the same maile.* Then he singeth or shouteth, and the rest of the company answere him with *Igha, Igha, Igha,* and then the priest replyeth again with his *voyces*. In the ende he becometh as it were madde, and falling downe as he were dead, hauing nothing on him, but lying upon his backe, I might perceive him to breathe. And when he had lyen still a little while, they cried three times together, *Oghao,* and as they use these three calles, he riseth with his head and lieth downe againe, and then he rose up and sang with like voices as hee

* This face mask seems from the description to resemble closely one figured in Cook's third voyage.

did before, and his audience answered him *Igha, Igha, Igha.* Then he commanded them to kill five *Olens* or great Deere, and continued singing both hee and they as before. Then hee tooke a sworde of a cubite and a spanne long (I did mete it my selfe), and put it into his bellie halfeway and sometime lesse, but no wounde was to bee scene, they continuing their song. Then he put the sword into the fire till it was warme, and so thrust it into the slitte of his shirte and thrust it through his bodie, as I thought, in at his navill and out at his fundament; the poynt beeing out at his shirt behinde, I layde my finger upon it, then he pulled out the sworde and sat downe.

"This being done, they set a kettle over the fire to heate, and when the water doeth seethe, the priest beginneth to sing againe, they answering him. Then they made a thing being foure square, and in height and squareness of a chaire, and covered with a gown very close the forepart thereof, for the hinder part stood to the tent's side. The water still seething on the fire, and this square seate being ready, the priest put off his shirt, and the garland from his head, with those things which covered his face; and he had on yet, all this time, a paire of hosen of deere's skins, with the haire on, which came up to his buttocks. So he went into the square seat, and sate down like a tailour, and sang with a strong voyce of halowing. Then they tooke a small line, made of deere's skinnes, of four fathoms long, and with a small knotte the priest made it fast about his necke, and under his left arme, and gave it to two men standing on both sides of him, which held the ends together. Then the kettle of hote water was set before him in the square seat, and the seat covered over againe with a cloth. Then the two men did draw the ends of the line till it became stiffe, and I hearde a thing fall into the kettle of water, whereupon I asked them that sate by me what it was that fell into the water, and they answered me that it was his head, his shoulder, and his left arme, which the line had cut off, I meane the knot which I sawe afterwarde drawen hard together. Then I rose up, and would have looked whether it were so or not, but they laid hold on me, and said if they should see him with their bodily eyes they shoulde live no longer. Then they begane to hallow these words, *Oghaoo, Oghaoo, Oghaoo,* many times together; and when they were outcalling I saw a thing like the finger of a man thrust through the gowne from the priest, but they said it was a beast, and not his finger, for he was yet dead. And I looked upon the gowne, and there was no hole to be seene; and then at the last the priest lifted up his head, with his shoulder, and arme, and all his bodie, and came forth to the fire. This, their service, I saw during the space of certain houres."*

Steven Borrough describes the Samoyed idols which he saw at Waigatz in these terms:—" Hee brought me to a heap of the Samoëd's idols, which were in number above three hundred, the worst and most unartifical worke that ever I saw; the eyes and mouthes of sundrie of them were bloodie, they had the shape of men, women, and children, very grosly wrought, and that which they had made for other parts was also sprinkled with blood. Some of their idols were an olde sticke, with two or three notches made with a knife in it. Loshak tolde me that these Samoëds have no houses, but onely tents made of deers' skins, which they underproppe with poles; their boates are made of deers' skins, and when they

* Hakluyt, I., pp. 284-5.

come on shore they cary their boates with them on their backes."*

A later writer† states that the Samoyeds admit the existence of a Supreme Being, creator of all things, eminently good and beneficent, but who takes no interest in the affairs of men, and requires no worship. They believe, however, in a very powerful evil being, from whom come all the misfortunes that befall them, yet neither do they worship him though they fear him greatly. Such reliance as they have in the counsels of their *Koedesniks* or *Tadebes* is founded on the influence these Shamans are supposed to have with this malevolent spirit, otherwise they submit with apathy to every untoward occurrence. It is through the intervention of the sun and moon that the favours of the Supreme Being reach them, but they never worship them as they do the idols which they carry about with them by the advice of the *Koedesniks*. They appear to care very little about these idols, and wear them only because it was the custom of their ancestors to do so. They entertain some notion of the transmigration of souls, and inter with the deceased all his weapons, and other property, which they think he will need in another state of existence, but they have no clear conceptions of what the future life is. The whole administration of the Shamans is confined to giving advice to the idols, and to the hunters when they are sick, or more than commonly unsuccessful in hunting. On the whole, the Samoyeds think their condition too happy a one to be desirous of changing it.

They have no laws and no government among themselves, but hold the marriage tie to be binding; they do not marry a

---

* Hakluyt, I., p. 281.

† *Hist. Gen. des voy.* xxiv., p. 66. Trans. in Pinkerton's collection.

female of their own tribe, and few crimes are perpetrated in the community to which they belong; yet they do not consider it to be forbidden to take the wives or daughters of others by force;—if they abstain from doing so, they are not deterred by principle.

Their senses and faculties are in relation to their mode of life as nomads and hunters. They have a piercing eye, very delicate hearing, and a steady hand; they shoot an arrow with great accuracy, and are exceedingly swift in running. On the other hand, they have a gross taste, weak smell, and dull feeling; and submit willingly to the *yassak*, or fur-tribute, imposed by the Russians, because their fathers did so. In other respects they are individually independent, deferring only to the senior of each family, or to the Shamans, but not holding themselves bound to obey even them. They purchase their wives, and some girls have cost a hundred or a hundred and fifty rein-deer. In the possession of herds of rein-deer consists Samoyed wealth, and P. von Krusenstern in 1845 calculated the numbers owned by the Samoyeds of the lower Petchora, near Pustoserk, at 40,000 head; a much inferior number to what they formerly had, owing to a succession of misfortunes. They wander over the tundren with their deer, fish in the lakes, and visit Waigatz, their Holy Island, called, in their language, *Chaja* (*Khaya*), for the purpose of making offerings on the mountain Chaissé. On that island, also, they procure many foxes, much fish in the bays, and also belugas (*Beluga albicans*).

The Samoyeds are said to eat the flesh of the rein-deer always raw, and to consider it as a luxury to drink the warm blood of these animals, similarity of climate having given to them like tastes with the Eskimos, and according to the author of the memoir* from which we here quote, they say that the blood preserves them from scurvy. They consume their fish also raw,† but other kinds of food they cook, and every member of the family eats at pleasure, there being no fixed hours for their meals. Some animals they refuse to eat, such as dogs, cats, ermines, and the squirrel, from which they abstain for some unknown reason.

Our author describes the Samoyed men as being below the middle size, but he never saw any under four feet, while some were above middle height, and individuals exceeded six feet. They are of a nervous make, broad and square built, with short legs and small feet. The neck very short, and the head large in proportion to the body; the face flat, and the eyes black and tolerably open; the nose so much flattened that the end is nearly upon a level with the bone of the upper jaw, which is strong and greatly elevated. The mouth is large. Their jet black hair, hard and strong, hangs from their shoulders, and is very sleek. Their complexion is a yellow-brown, and their ears are elevated. The men have little or no beard, and little or no hair anywhere but on the head.

The physiognomy of the women exactly resembles that of the men, except that their features are rather more delicate, their body more slender, the leg shorter, and the foot still smaller; otherwise it is difficult to distinguish the sexes,

---

* *Memoire sur les Samoiëdes*, Konnigsberg, 1762.

† Baron Wrangell says that during his residence in Siberia he learnt to relish *strugannia*, or thin flakes of frozen fish, without cooking, and fresh raw rein-deer marrow; and even after his return to civilized life, he continued to prefer *strugannia* before it thaws, seasoned with salt and pepper, to dressed fish. The Swedes eat pickled salmon and herrings raw; and the Canadian voyagers consider frozen trout to be a delicacy.

especially as their dress is very nearly the same; the chief distinction in the clothing of the sexes being that the women trim their skin-garments with some scraps of coloured cloth and the youngest sometimes separate their hair into two or three tresses, which hang down behind.

Dr. Latham gives an engraved portrait of a Samoyed man, copied from Von Middendorf, which might pass for an Eskimo, though it is uglier than any of the latter race that we have seen. The thick lips, high cheek bones, and other features, are generally alike, but there seems to be more prominence of the jaws. The superstitions of the two people are in many things similar; and their modes of living and economic contrivances are very nearly identical.

The Samoyeds are on the whole little calculated, by the low condition of their intellect, to contend with other nations, and they have accordingly experienced a considerable contraction of their area, by the encroachments of nations of the Finnic or Yugrian stock, themselves pressed forwards by the emigrating waves of the more populous or more powerful Celts, Teutons, and Turks. In the most ancient times, probably, the Samoyeds divided the polar coasts with the Eskimos, the latter having by far the most extensive range. Dr. Latham mentions a southern division of Samoyeds named *Soiot*, with which, as being far to the south of the Arctic circle, we have no concern in this work.

The *Yeniscians* and *Yukahirs* are ranged by the same ethnologist in his group of Hyperborean Mongolidæ, and are considered by others as tribes of the Yugrian or Finnic race. These people are evidently intruders on the polar area, not being to the manor born, as the Eskimos and Samoyeds are. The Yukahirs are active traders, more shrewd bargainers than the simple Samoyeds, and may have sought the north primarily in pursuit of furs and other articles of commerce. Heeren conjectures that Greeks of Pontus traded with the inhabitants of the Ural or Riphæan Mountains for furs, their caravans passing through the country of a tribe of Scyths who had revolted and fled from the rule of the dominant tribe of their own nation. One readily agrees with Baron Wrangell in the sentiments he expresses in the following terms, when speaking of the present inhabitants of the Kolyma district:—

"Nomade races," he says, "under milder skies, wander from one fruitful region to another, gradually forget the land of their birth, and prefer a new home. But here is nothing to invite. Endless snows and ice-covered rocks bound the horizon. Nature lies shrouded in almost perpetual winter. Life is a continual conflict with privation, and with the terrors of cold and hunger. What led men to forsake more favoured lands for this grave of nature, which contains only the bones of an earlier world? It is in vain to ask the question of the inhabitants, who are incessantly occupied with the necessities of the present hour, and amongst whom no traditions preserve the memory of the past. Nothing definite is known concerning the natives, even at the not very remote epoch of the conquest of Siberia, by the Russians."

"I have indeed heard an obscure saying, that there were once more hearths of the OMOKI on the shores of the Kolyma, than there are stars in a clear sky; there are also the remains of forts formed of the trunks of trees, and tumuli, the latter especially near the Indigirka; both may be supposed to have belonged to these Omoki, who have now disappeared. From the little that I could gather on the subject, it would seem, that the Omoki were a numerous and powerful people, that they were not nomades, but lived in settlements along the rivers, and supported themselves by fishing and hunting, Another numerous tribe, the Tukotschi or Tchuche, appear to have wandered over the *tundren* with their herds of rein-deer, and to have left names to the features of the country. Both races have disappeared, the Omoki have probably perished from want and sickness, and the Tchuche have left the district for the north-east corner of the continent, where they are still to be found (as rein-deer nomades), or may have been in part absorbed by newer arrivals, forming a portion of the present scanty population of the country."* We may remark that the term *Omoki*, has a strong analogy with some Eskimo words, and the instance of *Omalik*, "a wealthy person," occurring in a preceding page, may be mentioned.†

Dr. Latham classes the Omoki as a tribe of Yukahirs. The evidence on which they can be affiliated to any race is small, but the buildings ascribed to them, log-huts, called in Siberia as in Rupert's Land, "forts," are probably Eskimo erections; and the Omoki may have been a western branch of the Asiatic Eskimos, separated from those on the Gulf of Anadyr by the invasion of the Tchuche, and finally becoming extinct, either by the attacks of the Tchuche, or as Baron Wrangell suggests, by the spread of some epidemic disease. A people living in villages, supported by the chace either wholly or in conjunction with fisheries, could not be so numerous as report makes them to have been, in a land where the chief beasts of chace are migratory.

* Wrangell, Polar Sea, p 52.   † Chapter xiv., p. 324.

CHAPTER XXI.

LAPLANDERS OF YUGRIAN ORIGIN.

Finnish Runots—*Same* Nomades—Sledges—Snow-shoes—Dress—Reindeer—Boats—Regnard—Maupertuis—Knud Leems—W. Dawson Hooker.

THE LAPPS are the rudest, or one of the rudest people of the Yugrian Branch of the Turanian stock, following Dr. Latham's ethnological arrangement. By many authors the Yugrians are called Fins, but this term is of Gothic or perhaps Celtic origin,* and is not applied by any of the Yugrian people to themselves. The Yugrians or Lapps are supposed by Humboldt and others to have dwelt primitively on the Ural Mountains, and they are supposed to have occupied the north and west of Europe before the advent of the Indo-European Celts, Teutons, or Litho-Slaves. To a very early people of this stock have been attributed flint arrow-heads, and stone tools found in Norway, precisely alike in character to others dug up in abundance in Ireland.† But this argument cannot be safely relied on to establish the identity of a race, as flint weapons and stone adzes collected in Australia and North and North-west

---

* *Phinn*, the Celtic for "giant," has been already alluded to as perhaps bearing in Ireland some relation to the ancient intercourse of the Phœnicians with that country. What was the origin of the word *Fin*, applied to the Yugrians by the Norsemen, we have seen nowhere mentioned. The *Skrithi-finni*, were probably more clamorous than the others. (*Skrige*, to scream).

† Saturday Review, Dec. 31, 1859.

America, differ little or not at all in form, from those found in the soil of Europe.*

The Finnish Runots are surmised to have been composed before the Yugrians entered Western Europe, and the *Kalevala* is thought to embody the older traditions of a section of the race. In this poem, the Blacksmith Ilmarinen, who wrought the heavens of blue steel, has a conspicuous part assigned to him, betraying an acquaintance with one of the most refractory metals at the time of its composition.

The Lapps term themselves *Same* or *Sabome.*† By their Norwegian invaders they were considered to be dwarfs, skilled in extracting metals from the bowels of the earth, and possessing great power as magicians. It is to be observed, however, that the older the burrows in Lapland are, they contain less iron, and more stone weapons, associated with skulls of the Yugrian type.

It is an Yugrian custom to abstain from intermarrying in their own tribe, and as it is interesting to note similar observances in distant countries, it is worth mentioning that the

---

* Rask proved that the Finnic language had once been spoken in the most northern extremities of Europe, and that allied languages extended like a girdle over the north of Asia, Europe, and America. He maintains that the Eskimo is a scion of the Scythic or Turanian language, spreading its branches over the north of America, and indicating the antediluvian bridge between the continent of Europe and America. According to his views, therefore, the Scythian is a primary language over an area reaching from the White Sea to the valleys of the Caucasus, and extending laterally in Western Europe, as far as Britain, Gaul, and Spain; and in America from Greenland westward, as well as southward. This original substratum he supposes to have been broken up first by Celtic inroads, secondly by Gothic, and thirdly by Sclavonic immigrations.— (*Archæol. of United States,* wherein the passage is given as a quotation from Max Müller's results of Turanian research.

† Latham, Var. of Man, p. 105. *Same* according to Erman, refers to the swampy nature of the country they inhabit.

American Eskimo tribes on the coast of Bering's Sea, follow a like rule.

The Lapps, like other possessors of rein-deer, are necessarily nomadic to a greater or lesser extent. We do not intend to enter at length into the manners of a people living so near to the civilization of Europe, and doubtless at the present time greatly influenced by it, and by the Christianity they have embraced, imperfectly as its precepts have been taught them; but a few extracts from the account of Regnard's journey, made in 1681, will serve to complete what we have to say on the ethnology of the Polar Regions.

"The Lapland sledge is called *pulea*, and is raised in front like a small boat, for keeping off the snow. The prow consists of a single plank, and the body is composed of several pieces sewed together with strong thread of rein-deer sinew, without a single nail;* this is joined to another piece of about four fingers breadth, which goes beyond the rest of the structure, and is like the keel of a ship. It is on this that the sledge runs, and from its narrowness, constantly rolls from side to side. The traveller sits inside as in a coffin, with half his body covered, and is there tied in immovably, with the exception of his hands, one of which holds the reins, while with the other he supports himself when falling. He is obliged to balance himself carefully, least he should lose his life, as the sledge descends the steepest rocks with horrible swiftness.

"The Lapland snow-shoes or snow-skeuts, made of two narrow deals, a shorter one for one foot, and a longer for the other, extending to eight feet or more, are peculiar to the

---

* Nails in a climate where there is a great range of temperature, soon become loose, and are much inferior to the fastenings of sinew used by the Lapps and Eskimos.

country. (The American Indian snow-shoes are a net-work of sinew, on a frame. The Eskimos do not use them.)

"No other weapons are used by the Lapps in hunting than the bow or cross-bow. The former is employed in killing the larger beasts, as the boar, the wolf, and the wild rein-deer; and the smaller animals are knocked down by aid of the cross-bow. These people are so skilful that they never fail in striking the object. Some of their arrows are pointed with the bone of a fish or with iron; others are round like a ball cut through the middle. The inhabitants are ignorant of the use of corn; fish-bones ground with the bark of trees, are used instead of bread in the north.

"We regarded the first Laplanders we saw very attentively, they are made quite differently from other men. The tallest of them is not more than three cubits high, and I know not any figure more truly laughable. They have large heads, broad and flat faces, level noses, small eyes, large mouths, and *thick beards* descending to their stomachs. All their limbs are proportioned to the littleness of body; their legs are thin, their arms long, and the whole of this little machine seems to move on springs.

"Their winter dress consists of the skin of a rein-deer, descending like a sack to the knees, and tied round the thighs with a sash of leather; the shoes, boots, and gloves of the same stuff. A purse, made of the entrails of a rein-deer, hangs on the breast, and contains a spoon. A lighter summer dress is made of the skins of birds. Their cap is made of the skin of a loom (*Colymbus*), so placed that the bird's head falls over their brow, and its wings cover their ears.

"When a Lapp marries he gives his services for a year to his father-in-law, after which period he removes with his wife and all her property.

"One meets with very few old men who are not blind, owing to the smoke of the huts and glare of the snow. No sooner is a man dead than the family abandon the house lest the soul of the deceased should do them an injury, and they even demolish the house. The coffin consists of a hollow tree, or of the sledge, and in it all that the deceased had of value, as his bow, lance, etc., are placed that he may continue to exercise his former profession on his return to life.*

"The Laplander nourishes no other domestic animal than the rein-deer, but in this creature they find all that they require. They throw away no part of the animal, but make use of the hair, the skin, the flesh, the marrow, the bones, the blood, and the nerves. The skin serves them for clothing, dressed for winter use with the hair, or for the summer with the hair removed. Its bones are of the greatest utility, for making their bows and cross-bows, arming their arrows, making their spoons, and for adorning every thing they make. The rein-deer tongue and marrow are their greatest of delicacies. They frequently drink its blood, but more commonly preserve it in a bladder and allow it to freeze. From this ball they cut off as much as they desire to boil with their fish Their thread is drawn from the sinews of this animal, using the finest to sew their clothes and the coarsest to join the planks of their vessels. The milk of the rein-deer is the only beverage they possess, being mixed with an equal quantity of water owing to its richness. They draw a gallon of milk daily from the best rein-deer, which yield it only when they have a fawn. Of this milk they make very nutritious cheeses, which are fat, and have a very strong smell."

The Laplanders rear the deer used in travelling from a

* This is identical with the Eskimo practice on a death in a family.

wild male and a tame female, and these strong deer are said to be able to carry the sledge three times in a day beyond the horizon. The swiftest and strongest, when hard pushed on firmly frozen snow, can travel six French leagues in an hour, but cannot support this toil for many hours.

Maupertuis describes the Lapland boats as being skiffs formed of a very few thin deals, so thin and flexible that when borne by violent rapids in the rivers against rocks, they sustain the shock without injury. It affords, he says, a sight terrible to strangers and astonishing, to behold this frail machine in the midst of a deafening cataract, sometimes borne up aloft, at others lost amid the waves, which threaten to overwhelm it.

Knud Leems, professor of the Lappish language, wrote a detailed account of the manners, mode of living, religion, and superstitions of the Danish Laplanders in 1767, which is republished in Pinkerton's Collection. As his account has reference to a later state of society, modified greatly by intercourse with the missionaries and the more civilized Fins of the Gulf of Bothnia, no part of the treatise is transcribed here, but the reader who is desirous of knowing more of this people is advised to consult it.

In our first chapter, Ohther is quoted as mentioning the Finnic Queenes, the inland inhabitants of Norway, who occasionally crossed the mountains to make raids on the Norman occupiers of the western coast. The name still survives as the following extract shews. It relates to the working of the copper-mines of Kaafiord near Hammerfest. "The workmen are chiefly *Quäns*, with a few Norwegians. These two races are so perfectly distinct, as not to be easily confounded with one another. The former are a dull heavy-looking tribe, broad shouldered, their faces flat and square, with high cheek bones and sallow complexions; they came originally from the Gulf of Torneä, but have for a considerable time been settled in Finmark, for agricultural and other purposes; they are industrious, tolerably steady, and generally make good workmen. The Norwegians, on the other hand, who are the original denizens and proprietors of the soil, are tall, well built, compactly formed, and sinewy, with fair complexions, longish faces, and sharp features; they are more talented than the Quäns, and look down upon their more mercenary neighbours as interlopers and intruders on their territories. What the Quäns, however, want in intellect, they make up by superior industry, steadiness, and perseverance; for the Norwegian peasant, more especially the miner, is sadly addicted to drunkenness, making it almost a point to get intoxicated every Saturday, which here, as it is unfortunately in England, is the pay day.*

* Notes on Norway, by William Dawson Hooker. Glasgow, 1837; p. 32.

PART SECOND.

CHAPTER I.

A.D. 1576-1840.—ANTARCTIC POLAR REGIONS.

Terra Australis incognita—Juan Fernandez—*Tierra ferme*—Saloman Islands—Mendana—New Hebrides—De Quiros—Santa Cruz—*Tierra Austral*—*Australia de l'Espiritu Santo*—Silver ore—De Torres—Torres' Strait—Australia—Cook—Enters the Antarctic circle—Low temperature—South Georgia—Sandwich Islands—Southern Thule—Cockburn Island—Its vegetation—Bellinghausen—Weddell—Biscoe—Weddell—Balleny—Dumont d'Urville—Cote Clairie—Wilkes.

THIS portion of the work will necessarily be short. It refers to an area wholly within the snow-line, uninhabited by man, without land animals, and only in a few instances traversed by navigators. When the ancients had, on mathematical grounds, admitted the globular form of the earth, and Parmenides, as Strabo* says, indicated five zones or climates, namely, two temperate regions, separated by an equatorial belt, uninhabitable from heat, and two polar regions, considered to be equally unfit for the residence of man, by their excessive cold, the belief in southern lands, to which access was denied solely by the difficulty of traversing the torrid zone, was a natural

* Strabo, lib. ii., p. 65. Ed. Casaub. 1587.

sequence. By a like reasoning, on a system of representation, several ancient poets have spoken of land in the western hemisphere, which has been held, in recent times, to denote an actual knowledge of the existence of the American continent.* The geographers of the middle ages imbibed the same ideas from the study of the ancients; and after Columbus had discovered the western continent, and roused the civilized world to the importance of geographical research, no long period elapsed before the project came to be entertained of seeking the southern continent, which was thought to be a necessary counterpoise to the northern lands. This *Terra Australis incognita* served the same purpose in the south that the North-west Passage did at the other end of the globe, the search for it having led to many notable discoveries, and eventually to our present extensive acquaintance with southern hydrography.

Juan Fernandez is reported to have sailed, in the year 1576, on a west-south-west course, and after a month's voyage, to have arrived at a *tierra ferme*, a pleasant and fertile land, inhabited by highly civilized white people, dressed in woven cloth. The details of this voyage are wanting, and it has been supposed to be altogether apocryphal; but some have conjectured that Fernandez reached the coast of New Zealand. The Spaniards residing in South America, were at this time in expectation of discoveries in the South Pacific, excited by rumours of Alvaro Mendana de Neyra having, in 1567, found the Saloman Islands, which so abounded in silver, that one entire mountain was composed of that precious metal. Mendana was evidently no skilful navigator, able

---

* *See* Select Letters of Columbus, and Early Voyage to Australia. By R. H. Major. Hakl. Soc. pub.

to retrace his former course, and doubtless knew the stories that were current of the richness of the country he had seen, to have little foundation; for he allowed twenty-eight years to elapse without attempting to profit by his supposed discovery. In modern charts, Mendana's discovery of 1567 is recognised in a chain of islands lying to the east of New Guinea, between the parallels of six and twelve degrees of south latitude.

In 1594, Philip II. of Spain having instructed the Viceroy of Peru to encourage new enterprises and settlements, so as to disembarrass the land from many idle gentry, the Marqués de Cañete prepared a naval armament for the settlement of St. Christoval, one of the Saloman Islands, and appointed Alvaro de Mendana to the command, with the title of *Adelantado*. The squadron consisted of four vessels and nearly 400 men, Pedro Fernandez de Quiros embarking as captain and pilot-major in the same ship with Mendana, whose wife also accompanied him. After having sailed about half the distance to the Saloman Islands, the expedition made the island of Madalena, and supposed that this was the land they sought, whereupon there was great rejoicing on account of the shortness of the voyage, and *Te deum laudamus* was sung; but after sailing along the coast of Madalena from one end to the other, the Adelantado acknowledged that it was not the Saloman Islands, but a new discovery. The squadron had, in fact, discovered one of the group of islands now known as *Las Marquesas de Mendo a*. Continuing the voyage on the same parallel of about ten degrees south latitude, the squadron made the island of Santa Cruz, one of the Saloman group, and there a settlement was commenced, the search for San Christoval being abandoned; but the Adelantado soon afterwards died, having, by will, appointed his wife Doña Ysabel Bereto to succeed him. The hostilities of the natives, however, provoked by aggressions of the Spaniards, and the death of some of the leaders, put an end to an enterprise which, Figeroa remarks, was mismanaged in a thousand ways. The governess sailed for Manila, taking with her the corpse of her husband, and having married again, abandoned all thought of re-establishing her government at Santa Cruz.

The pilot-major, however, Pedro Fernandez de Quiros, did not abandon the hope of finding the large land seen by Mendana in 1567; and his arguments appeared to be so plausible to the viceroy, that he sent him to Spain with letters recommending his proposal to the ministry. Being furnished with ships by order of Philip III., Quiros sailed from Callao in December 1605, with the intention of renewing the settlement at Santa Cruz, and then searching for the *Tierra Austral*.* He discovered several small islands, reached another of the Saloman group, named Taumaco, and afterwards, sailing southwards, came to one of the New Hebrides group, which he named AUSTRALIA DEL ESPIRITU SANTO, supposing that it was the great southern continent he was seeking for. He returned to New Spain, and in his report to Philip III. he says—" By all that I have mentioned it appears clearly that there are only two large portions of the earth severed from this of Europe, Africa, and Asia. The first is AMERICA, which Cristoforo Colon discovered; the second and last of the world is that which I have seen, and solicit to people, and completely to discover to your majesty."† The

* Burney, *South Sea Disc.*, (Lond. p. 268-326) quotes the Memorial of Juan Luis Areas. Edin. ed., pp. 17 and 20.

† Peter Bertius in his abridged description of the world, appended to the *Geography of Ortelius*, published in 1661, says that the Antarctic division of

reports made to Quiros by the natives of the Saloman Islands of the vicinity of a great continent, confirmed him in the opinion of the importance of his discovery. An incident of this voyage strongly resembles one that occurred in Frobisher's expedition, related in a former part of this volume. In Taumaco there was a man who had visited the large country named *Pouro*, and brought from thence some arrows, tipped with a metal as white as silver, and in one of the houses of Australia were some black heavy stones, two of which were carried by Quiros to Mexico. One of them was carried to an assayer, but the experiment failed, by the breaking of the crucible ; yet a part remaining, the assayer melted it again, and in it was seen a small point, which expanded under the hammer. He touched it on three stones, and some silversmiths said it was silver-touch ; but some said that the assay should have been made with quicksilver, and others with saltpetre ; yet the assayer affirmed that the metal was good, and two silversmiths said that it was silver.*

Luis Vaez de Torres, being left in the Bay of San Felipe of Australia, remained there for two months, and then sailed to prosecute the original main design of the voyage. Coasting Australia, he soon discovered that it was not the continent he sought ; and after keeping on a south-west course, until he had gone a degree beyond the latitude prescribed by his orders, he turned to the north-west, and coming to New Guinea, coasted its south side for a considerable distance, entered the strait which still bears his name, and had glimpses of Cape York, or the islands lying off the northern promontory

the world, of which a great part was discovered by Peter Ferdinand de Quir, equals Europe and Africa conjoined, and extends through three zones, the frigid temperate, and torrid.

* Burney, l. c. ii., p. 308.

of the real *Terra Australis*. Continuing along the coast of New Guinea, he met Mahometan Malays, and went on to the Moluccas.

In the same year the Dutch Company's yacht Duyfhen coasted New Guinea for above 800 miles, and discovered land in $13\frac{3}{4}°$ S., which must have been part of the coast of *Terra Australis* or New Holland. The point seen was named *Keer veer*, or Turnagain. The subsequent discovery of other portions of *Terra Australis* by Theodoric Hertoge, Pool, Tasman, and Cook, and its more connected survey by Flinders, who proposed the change of name to Australia, which this fifth continent now bears, are not necessary to be mentioned here in detail, as the most southern land of Australia is far remote from the antarctic circle.*

The extent of European knowledge of the high southern latitudes towards the close of the sixteenth century is shewn at once in a mappemonde constructed by Judocus Hondius for a Dutch account of Drake's Voyage Round the World, and reprinted by W. S. W. Vaux, Esq., in his edition of "Drake's World Encompassed."† Thereon an immense *Terra Australis* is depicted, extending from the south pole so as to pass beyond the antarctic circle opposite to the Atlantic, reaching on the Pacific side to the tropic of Capricorn, embracing in its area Australia, and separated from New Guinea merely by Torres Strait, which is not however named.

Captain Cook is the first navigator who is known to have entered within the antarctic circle. His voyage in the years 1772-3-4 and 1775 was undertaken, he tells us, to put an end to all diversity of opinion about the curious and important

---

* The Early Voyages to Terra Australis, Hakluyt Society, by R. H. Major, Esq., 1859, give a full historical detail of various voyages to Australia.

† Hakluyt Society publications, 1854.

question which had long engaged the attention not only of learned men, but of most of the maritime powers of Europe, as to "whether the unexplored part of the southern hemisphere be only an immense mass of water, or contain another continent."* Having ascertained on this long and important voyage that Australia, of which detached portions had been previously discovered by the Spaniards, Portuguese, and English, but chiefly by the Dutch, really possessed the dimensions of a continent, he was not content with that solution of the question, but pushed on to the south and entered the antarctic circle in three separate quarters, namely near the meridian of 40° east longitude, between 100° and 110° west longitude, and between 135° and 148° west, the most southerly point attained by him being 71° 10' of south latitude, on the 107th meridian. On each occasion his further progress southwards was arrested by firm fields of ice. He sums up his doings with the following observations :—" I had now made the circuit of the Southern Ocean in a high latitude, and traversed it in such a manner as to leave not the least room for the possibility of there being *a continent* unless near the pole, and out of the reach of navigation. By twice visiting the tropical sea I had not only settled the situation of some old discoveries, but made there many new ones, and left, I conceive, very little more to be done even in that part. Thus I flatter myself that the intention of the voyage has, in every respect, been fully answered ; the southern hemisphere sufficiently explored, and a final end put to the searching after a southern continent, which has at times engrossed the attention of some of the maritime powers for near two centuries past, and been a favourite theory amongst the geographers of all ages.

* *See* Introduction, p. 9.

"That there may be a continent or large tract of land near the pole I will not deny; on the contrary, *I am of opinion there is;* and it is probable that we have seen a part of it. The excessive cold, the many islands and vast floats of ice all tend to prove that there must be land towards the south; and for my persuasion that the land must lie, or extend furthest to the north opposite the Southern Atlantic and Indian Oceans, I have already assigned some reasons, to which I may add the greater degree of cold experienced by us in these seas than in the Southern Pacific Ocean under the same parallels of latitude. In this last ocean the thermometer seldom fell so low as the freezing point till we were in 60° and upwards; whereas in the others it fell as low in the latitude of 54° S.

"We saw not a river or stream of water on all the coast of Georgia, nor on any of the southern lands. The valleys are covered many fathoms deep with everlasting snow; and at the sea they terminate in icy cliffs of vast height. It is here where the ice islands are formed."*

Cook saw no land to the south of Sandwich Land or Southern Thule, in latitudes 59° and 60°. His description of this group and of the more southern island of Georgia, lying only a short way beyond the parallel of the southern extremity of America, may prepare the reader for the account to be hereafter given of the lands situated within the antarctic circle. Of South Georgia he says :—" The head of the bay, as well as two places on each side, was terminated by perpendicular ice-cliffs of considerable height. Pieces were continually breaking off, and floating out to sea; and a great fall happened while we were in the bay, which made a noise like a cannon. The inner parts of the country were not less savage and hor-

* Cook's Second Voyage Round the World, 1779, ii., p. 239

rible. The wild rocks raised their lofty summits till they were lost in the clouds, and the valleys lay covered with everlasting snow. Not a tree was to be seen, nor a shrub even big enough to make a toothpick. The only vegetation we met with was a coarse strong-bladed grass growing in tufts, wild burnet, and a plant like moss, which sprung from the rocks. The shores swarmed with young seals; there were several flocks of the largest penguins I ever saw, and the oceanic birds were albatrosses, gulls, Port Egmont hens, terns, shags, divers, the new white bird, and a small yellow bird which was most delicious food. All the landbirds consisted of a few small larks, nor did we meet with any quadrupeds; though Mr. Forster observed some dung which he judged to come from a fox or some such animal. The lands, or rather rocks, bordering on the sea-coast were not covered with snow like the inland parts; but all the vegetation we could see on the clear places was the grass above mentioned."* It would appear from this summary sketch that South Georgia lies fully as far within the lower snow-line as Spitzbergen, which is twenty-four degrees nearer to the pole.

On the Sandwich group he did not land, but these islands were erroneously thought by him to be' probably part of a southern continent, as stated in a passage quoted above. He says of them—" I called this land 'Southern Thule,' because it is the most southern land that has ever yet been discovered. It shews a surface of vast height, and is everywhere covered with snow. . . . On the 1st of February we got sight of a new coast, which proved to be a high promontory situated in latitude 58° 27′ S., and from land seen from space to space, was made to conclude that the whole was connected. I was

* Cook, l. c. ii., p. 213.

sorry I could not determine this, but prudence would not permit me to venture near a coast subject to thick fogs, where there was no anchorage, where every port was blocked or filled up with ice, and the whole country, from the summits of the mountains down to the very brink of the cliffs which terminate the coast, covered many fathoms thick with everlasting snow. The cliffs alone were all that were to be seen resembling land. On the 2d, we saw a new land of the appearance of an island of about eight or ten leagues' circuit. It shews a surface of considerable height, whose summit was lost in the clouds, and like all the neighbouring lands, it was covered with a sheet of ice, except a projecting point on the north side, which probably might be two islands. These only were clear of snow, and seemed covered with green turf."*

Captain Cook was mistaken in supposing his Southern Thule to be the nearest land to the antarctic pole, which had been discovered in his time. As early as 1599, Dirk Cherrits, a Dutch navigator, having separated from some ships belonging to Rotterdam, was driven southwards from Staten Island by a tempest, to latitude 64°, when he had sight of a high snowy land, which is now known by the name of South Shetland, and has been traced by the discovery of a broken coast called Graham's Land, to extend within the antarctic circle. Cockburn Island, one of the group, situated in latitude 64° 12′ S., and longitude 59° 49′ W., is the most southern land on which vegetation has been observed, and fortunately we are able to quote the remarks made on it by Dr. Joseph Dalton Hooker, from whose report the following account is abridged.

This island is not above a mile in diameter, but rises 2760

* Cook, l. c. ii., p. 209.
† Sir James C. Ross's voyage to the Southern Seas, ii. p. 335.

feet above the sea-level. Taken as one of the group of islands lying immediately south of Cape Horn, beyond the sixtieth degree of latitude, the number of plants ascertained to inhabit them hardly exceeds twenty-six, and one of these a grass, being the only flowering plant, does not pass the sixty-second degree, furnishing a striking contrast to the forty flowering plants of Spitzbergen or Smith's Sound, situated fifteen or sixteen degrees nearer to the pole. No flowering plant was found on Cockburn Island itself, and only nineteen species of plants in all, belonging to the orders of mosses, algæ, and lichens. Twelve of these plants exist in other parts of the world, some of them being arctic.

The two most conspicuous vegetable productions seen on the approach to Cockburn Island, are the floating sea-weed *Sargassum Jacquinotii*, a plant confined as far as is known to these latitudes, and the common *Lecanora miniata*, a lichen which abounds near the sea on the antarctic islands, as well as in the northern hemisphere, but is nowhere in greater profusion than on Cockburn Island, where it belts the cliffs with yellow. Dr. Hooker thinks that the ammoniacal odours of the Penguin rookeries may promote its growth. The *ulva crispa* grows on the beach near decaying organised substances, and the mosses in the soil which is harboured in the fissures of the rocks, but they are so exceedingly minute, that the closest scrutiny is requisite to detect them. One of the *algæ* was taken from a pool hardly two spans across, and sheltered by a projecting rock that faced the north. Most of the lichens were inconspicuous, and not discoverable without a careful examination of the surface of the rocks. The sea-weeds were all floating, and carried along by a strong current, loaded with masses of ice. Only twelve of the nineteen species are terrestrial.

Vegetation could not be traced above the conspicuous ledge of rocks, with which the whole island is girt, at 1400 feet elevation. The *lichens* ascended the highest. The singular nature of this *flora* must be viewed in connection with the soil and climate; than which perhaps none can be more unfriendly to vegetable life. The form of the island admits of no shelter, its rocks are volcanic, and very hard, sometimes compact, but more frequently vesicular. A steep stony bank descends from the above-mentioned ledge, and to it the plants are almost limited. The slope itself is covered with loose fragments of rock, the debris of the cliff above, further broken up by frost, and ice-bound to a depth which there was no opportunity of ascertaining, for on the day the island was visited, the superficial masses alone were slightly loosened by the sun's rays. Thus, the plants are confined to an almost incessantly frozen locality, and a particularly barren soil, liable to shift at every partial thaw. During nearly the entire year, even during the summer weeks, Cockburn Island was constantly covered with snow. The vegetation of so low a latitude might be supposed to remain torpid, except for a few days in the year, when, if the warmth were genial, and a short period of growing weather took place, the plants would receive an extraordinary stimulus. But far from such being the case, the effect of the sun's rays, when they momentarily appear, is only prejudicial to vegetation. For the black and porous stones quickly part with their moisture, and the *Lecanora* and *Ulva* become so crisp that they crumble into fragments when an attempt is made to remove them. The air was exceedingly dry at the time of Dr. Hooker's visit, and he remarks that such dryness is eminently injurious to all vegetables except *lichens*, which in many cases thrive best under excessive atmospheric changes. The preponderance of the *Lecanora* on this island cannot arise from the exsiccation stimulating its growth, but may be caused by the reaction which arises afterwards, on the rapid condensation of vapour. Fur-seals, penguins, and cormorants innumerable, together with a beautiful white petrel, resort to this island.*

Other discoveries were made subsequent to Cook's voyage in the high southern latitudes, but very few of them within the antarctic circle. The Russian Bellinghausen discovered Petra and Alexander Islands in January 1821, having in his voyage sailed through several degrees into the antarctic area. Weddell, in 1823, advanced three degrees nearer to the pole than Cook had done. Biscoe, in 1831-3, in the brig Tula, discovered Enderly and Kemp Islands, both crossing the antarctic circle between the meridians of 45° and 60° east. In the year 1839, Balleny discovered Sabrina Land; and in 1840, Dumont D'Urville coasted Adelie and Clarie Land between the meridians of 130° and 140°. In the same year, Captain Wilkes of the United States Navy extended these discoveries on the same parallel, rendering it probable that there is a chain of islands just without the antarctic circle, reaching from the 95th to the 150th meridian, often connected by a barrier of ice.

To this period, viz., from the years 1839 to 1843, belongs the remarkable voyage of Sir James Clark Ross, now a Rear-Admiral in the British Navy, who, accompanied by Captain Crozier, carried the Erebus and Terror to a higher southern latitude than any navigators of his own or of any other country have ever attained, and made the only discoveries of extensive land with the area bounded by the antarctic circle.

* Dr. Hooker's, in Sir James C. Ross's voyage to the Southern Seas. ii. p. 340.

The following extracts from Sir James's narrative of his voyage relate to the priority of discovery of the lands seen by Balleny, D'Urville, and Wilkes. "March 1.—At noon we were in latitude 68° 27′ S., longitude 167° 42′ E.; and at 5 P.M. land was seen bearing N. 62° W., appearing like two islands. Although I believe these islands to form a part of the group discovered by Balleny in February 1839, yet it is not improbable they may prove to be the tops of the mountains of a more extensive land. On the 4th of March the land at the distance of thirty or forty miles looked like three distinct islands. It was the same land we had seen on the two previous days, but owing to thick weather, its position can be assigned only approximately as latitude 67° 28′ S., and longitude 165° 30′ E. At noon we had no doubt of the land being that seen by Balleny in the Eliza Scott, whose log states: 'February 9th.—Clear weather, got sights for my chronometer, which gave the ship's longitude 164° 29′ E. At noon observed for the latitude 66° 37′ S., and saw the appearance of land extending from west to about south—got within five miles of it at 8 P.M. February 12.—Tacked at noon, and worked in shore to look for harbour or beach, and at 6 P.M. went on shore in the Sabrina's boat, but found no beach except what was left by the drawback of the sea. Captain Freeman jumped out and got a few stones, but was up to his middle in water. In fact, but for the bare rocks from which the icebergs had broken off, we should scarcely have known it for land, but as we stood in for it we plainly saw smoke arising from the mountain tops. It is evidently volcanic, as the specimens of the cinders prove; the cliffs are perpendicular, and what in all probability would have been valleys or beaches are occupied by solid blocks of ice. We saw no beach or harbour or anything like one.' During the following fortnight as the Eliza Scott sailed to the westward on the parallel of sixty-five, indications of land are noted frequently in her log; and on the 26th of February, when in latitude 64° 40′ S., and longitude 131° 35′ E., and therefore only a few miles to the westward of the high barrier of ice seen by D'Urville on the 30th of January of the following year, and named by him *Cote Clairie*. The Eliza Scott's log states, 'stood for land to the eastward, but at 11 : 30 A.M. made it out to be fog hanging over some icebergs.' Thick weather prevented a further examination of this part of the coast, but there can be no doubt but it was really land that Balleny saw, and it will probably prove to be a continuation of D'Urville's *Terre Adelie* seen in January 1840, and on an islet of which some of his officers landed. Captain Wilkes saw this land, but was then ignorant of the French navigator having seen it a week before him." The last point of land which Balleny saw was seen on the 2d of March 1839, in latitude 65° S., and longitude 122° 44′ E. This was named Sabrina Land by him, and subsequently Tottens high land by Captain Wilkes.

Sir James Ross goes on to remark, "There do not appear to me to be sufficient grounds to justify the assertion that the various patches of land recently discovered by the American, French, and English navigators on the verge of the antarctic circle unite to form a great southern continent. The continuity of the largest of these *Terre Adelie* of M. D'Urville has not been traced more than 300 miles; the others being mostly of inconsiderable extent, of somewhat uncertain determination, and with wide channels between them, would lead rather to the conclusion that they form a chain of islands. But if future navigators should prove those conjectures about a continent to be correct, then the discoveries of Biscoe in the Brig Tula in January 1831, and those of Balleny in 1839, will set at rest all dispute as to which nation the honour justly belongs of the priority of discovery of any such continent between the meridians of 47° and 163° of east longitude, and those of our immortal Cook in the meridian of 107° west, in January 1774 ; for I confidently believe with M. D'Urville, that the enormous mass of ice which bounded his view when at his extreme south latitude, was a range of mountainous land covered with snow." *

* Ross, Southern seas, i., pp. 269-276   D'Umont D'Urville, voyage au Pole Sud, tome ii., p. 7.

CHAPTER II.

DISCOVERY OF VICTORIA LAND.—A.D. 1841-1842.

Sir James Clark Ross's Voyage within the Antarctic Circle—Mount Sabine—Possession Island—Mount Melbourne—Franklin Island—Mount Erebus Volcano—Mount Terror—Barrier of Ice—Balleny Islands.

WE proceed now to give an account of Sir James Ross's great discovery, by extracting such passages from his narrative as comprise the main facts. On January the 10th 1841, the Erebus and Terror steered southwards by compass, heading for the *magnetic pole* as directly as the wind would permit. On the following day, at noon, they had attained the highest latitude reached by Cook (71° 15′ S.), which up to this date Weddell alone had passed. At this time lofty peaks occupied a considerable section of the horizon, judged, when first seen early in the morning, to have been a hundred miles distant, but completely covered by snow, and many on board doubted that it was land. In the evening the ships had approached to the distance of about two leagues from the shore, which was lined with heavy packed ice, but the high surf prevented a landing. This first landfall was named Mount Sabine, and the land generally Victoria. Cape Adare is a remarkable high dark projection of probably volcanic cliffs, and contrasts strongly with the snow-covered coast. Some black rocks rose conspicuously above the white foam of the breakers in the vicinity. Soundings brought up fragments of volcanic stones. In the evening, says Captain Ross, the atmosphere was beautifully clear, "and we had a most enchanting view of the two magnificent ranges of mountains, whose lofty peaks, perfectly covered with eternal snow, rose to elevations varying from 7000 to 10,000 feet above the level of the ocean. The glaciers filled their intervening valleys, which descended from near the mountain summits, projected in many places several miles into the sea, and terminated in lofty perpendicular cliffs. In a few places the rocks broke through the icy covering, by which alone we could be assured that land formed the nucleus of this, to appearance, enormous iceberg."*

The chain of mountains extending to the north-west was named Admiralty Range; and the height of Mount Sabine was found, by means of several measurements, to be rather less than 10,000 feet. Previous to this discovery, the islet seen by Bellinghausen was the most southern known land.

On the 12th, Sir James Ross, Captain Crozier, and several other officers, effected a landing on Possession Island, situated not far from the mainland, in latitude 71° 56′ S., longitude 171° 7′ E. It was found to be accessible on its western side only, and to be composed entirely of igneous rocks. Not the smallest appearance of vegetation was discovered, but inconceivable myriads of penguins covered densely the entire surface of the island, along the ledges of the precipices, and even to the summits of the hills, and by their occupation of the spot for ages had formed a deep bed of guano, the stench of which was insupportable. A strong tide was running between the island and the mainland, but whether it was ebbing or flowing could not be ascertained.

* Ross, l. c. i., p. 185.

Standing southwards, as the wind and weather permitted, on the 15th of January, the sky being beautifully clear, our navigators obtained a fine view of the chain of mountains that they had previously seen stretching away to the southward. They were completely covered with snow, and their summits were, by rough measurement, ascertained to be from 12,000 to upwards of 14,000 feet high. The most conspicuous was named in honour of Sir John Herschel, Bart. The ships kept working to the southward along the coast against a strong breeze, but aided by a strong tide, or rather current, which set to the windward for more than twelve hours. Dark-coloured rocks of small size and curious shapes, one of them perforated, were seen forming a chain of islets.

Sir James Ross mentions an occurrence on this part of the coast which deserves particular notice. Whilst measuring some angles for the survey, an island I had not before noticed appeared, which I was quite sure was not to be seen two or three hours previously. It was above one hundred feet high, and nearly the whole of its summit and eastern side were perfectly free from snow. I was much surprised at the circumstance. On calling the attention of some of the officers to it, one of them remarked that a large berg which had been the object of previous observation, had turned over, unperceived by us. The new surface, covered with earth and stones, was so exactly like an island, that nothing but landing upon it could have satisfied us to the contrary, had not its appearance been so satisfactorily explained; and, moreover, on more careful observation, a slight rolling motion was perceptible.

At noon on the 19th, when off Mount Lubbock, a block of gray granite was dredged up. It had a clean fracture, as if recently broken off from the parent rock. Many other stones of granitic and volcanic structure were obtained from the dredge. A tide and counter-tide ran along the coast, and in calm weather the ships were drifted at the rate of three-quarters of a mile in the hour, near Coulman Island. Mount Melbourne resembles Mount Ætna in general form, but was judged to be much higher than the Sicilian mountain. The land ice here does not rise more than five or six feet above the level of the sea, but blends so imperceptibly with the snowy glaciers that run far into the sea, that it was almost impossible to form a correct notion of the position of the coast line. Along this ledge the ships held their way to the southward, but the strong current setting to the north greatly retarded their progress.

After passing Cape Washington and Mount Melbourne, heavy gales and the intervention of pack-ice compelled the ships to keep out to sea, and for fifty miles the coast line was not seen, owing to its distance.

On the 27th of January, a landing was effected, with much risk, on Franklin Island, which is situated in latitude 76° 8′ S., longitude 168° 12′ E. It is twelve miles long, and is composed wholly of igneous rocks. The northern side presents a line of dark precipitous cliffs, without the smallest perceptible trace of vegetation, not even a lichen or a piece of sea-weed being seen on the rocks. Some broad bands of white and red ochre colours gave a strange appearance to the cliffs. The white petrel and skua gull had nests on the ledges.

From the vicinity of Franklin Island the active volcano of Mount Erebus was discovered emitting flame and smoke in great profusion. Its height was ascertained to be 12,360 feet, and that of Mount Terror, on the east of it, to be 10,884 feet. Running southwards, and inland from them, is the chain of the Parry Mountains. The coast-line was enveloped in an ice-belt, presenting a sea-face of 200 or 300 feet high, perfectly vertical.

At 4 P.M. on the 28th, when the ships had approached nearer, Mount Erebus was observed to emit smoke and flame in unusual quantities, producing a grand spectacle. At each successive jet a volume of dense smoke rose to the height of about 2000 feet above the crater, which, condensing first at its upper part it, descended in mist or snow, and gradually dispersed, to be succeeded, after an interval, by another exhibition of the same kind. Whenever the smoke cleared away, the bright red flame that filled the mouth of the crater was clearly perceptible; and some officers believed they could see streams of lava pouring down its sides, until lost beneath the snow, which, commencing a few hundred feet beneath the crater, continued down to the perpendicular icy cliff that projected several miles into the ocean. Mount Terror was much more free of snow, especially on its eastern side, where were numerous little, conical, crater-like hillocks.

The continuous ice-cliffs were traced from Cape Crozier, which projects from Mount Terror, through thirty-three degrees of longitude to the eastward, and at nearly right angles to the coast-line of Victoria Land. This icy barrier was about 160 feet above the sea, and wherever it was approached within a range of 250 miles from Cape Crozier, it appeared to be without break or indentation, but the depth of the sea in its vicinity convinced Sir James Ross that it was not resting on the ground, though he estimated its thickness to be more than a thousand feet. The highest latitude attained by the ships was 78° 4′ S., or nearly twelve degrees from the pole, and the face of the barrier was a quarter of a degree nearer.

On the 9th of February, the Terror approached within a quarter of a mile of the icy cliffs, and ascertained their elevation at that place to be 150 feet. As seen from the mast heads, their upper surface seemed like an immense plain of frosted silver. Three weeks were spent at this time in examining the barrier, after which Sir James again stood towards Victoria Land.

The coast was examined anew as nearly as the weather and the advance of winter permitted, and the survey extended to the north-west of Mount Sabine, as far as Cape North, which is skirted by a high icy wall similar to that farther to the south. Before quitting the coast, a good view of the land was obtained on the 25th of February, the atmosphere being without haze or cloud, and the land of a spotless white, without the smallest patch of exposed rock.

Standing northwards from the most western part of Victoria Land hitherto seen, Sir James discovered two islands on the 2d of March, which proved to be part of the group seen by Balleny in February 1839, and commented upon in a preceding page.

In the following year, Sir James Ross again examined the icy barrier connected to Victoria Land, seeking to pass round or through it, but was unable to do so. On this occasion, the latitude of the face of the barrier on the meridian of $161\frac{1}{2}°$ west was ascertained to be 78° 11′ S., and the ships were two miles distant from it, being on the highest southern parallel of latitude ever attained. The perseverance, daring, and coolness of the commanding-officer, of the other officers, and of the crews of the Erebus and Terror, who executed this most remarkable voyage, were never surpassed, and have been rarely, if ever, equalled by seamen of any nation; as a perusal of the published narrative will at once convince the reader. Like Cook's first voyage, Ross's southern expedition was designed not so much for the promotion of geographical knowledge as for the making of observations in other branches of science, yet both will ever take positions in the foremost rank of voyages which have contributed to geography.

In 1845, the Pagoda was despatched from the Cape of Good Hope, by orders of the Admiralty, under the command of Lieutenant (now Captain) Moore, for the purpose of observing magnetic phenomena in a quarter of the antarctic seas that had not been visited by Sir James Ross. On the 11th of February, latitude 68° south was attained in longitude $39\tfrac{1}{2}°$ east, and soon afterwards an impenetrable ice-pack stopped further progress in that direction, on which the course was directed towards Enderby Land, which Captain Moore was prevented from sighting by head winds. On the 1st of March the Pagoda crossed the 73d parallel.*

* United Service Magazine. June and July 1850.

CHAPTER III.

ANTARCTIC POLAR AREA.

Desultory Remarks on the Physical Geography of the Antarctic Regions—Predominance of Sea—Its Effect on Climate—Southern Evaporation—Northern Condensation—Tides and Currents—Deep Sea Temperatures—Pressure of the Atmosphere—Mean Atmospheric Temperature in the Antarctic Regions—Wholly within the Snow-Line—Animals—Marine Animals—Microscopic Animals—Whales—Line of Woods—Fossil Trees of Kerguelen Island—Auckland Islands—Trees—Terra del Fuego—Vegetation of Hermite Island—Most Southern Trees.

On comparing the north and south circumpolar maps, a great contrast appears between the arctic and antarctic areas, land predominating in the one, sea in the other. The uncongenial climate of the southern hemisphere* in the high latitudes may be attributed to the comparatively small extent of its land, and especially to the taper-pointed form of its continents in the temperate districts, in strong contrast with the great, almost continuous breadth of the northern continents, extending even across the polar circle. Dovè remarks that "a liquid surface is continually renewed, inasmuch as every depression of temperature causes the water at the surface, which by cooling has

---

* Dovè says, the warmest parallel does not coincide with the equator, but falls in the northern hemisphere, so that the parallel of 10° north is slightly warmer than the equator. Up to 40° south latitude the temperature of the southern hemisphere is lower than that of the northern; this may not be the case in the higher latitudes.—*Distribution of Heat*, p. 15.

become denser and heavier, to sink down and make way for the warmer water which rises from below to replace it; and this goes on until the density of the fluid is the same throughout its entire depth. It is thus that the depths of the sea are deprived of the temperature which they would have if they were as far beneath a solid as they are beneath a liquid surface. The sea is also less warmed by direct radiation from the land, because the process of evaporation, or the conversion of part of the water into vapour, employs heat, which, if the surface were land, would be given off directly by contact to the adjacent air. . . . It is probable that the northern hemisphere may be regarded, comparatively speaking, and to a considerable degree, as the condenser of the great terraqueous steam-engine, and the southern hemisphere as its water-reservoir; that the quantity of rain which falls in the northern hemisphere is therefore considerably greater than that which falls in the southern hemisphere, and that one reason of the higher temperature is, that the large quantity of heat which becomes latent in the southern hemisphere in the formation of aqueous vapour is set free in the northern during great falls of rain and snow.*

Sir James Ross noted many strong currents or tides in the vicinity of Victoria Land, and in other places near islands, but the want of convenient anchoring-places prevented their general direction or limits from being ascertained.

A very remarkable current, coming from southern latitudes, was discovered by Baron Humboldt in 1802, on the Peruvian coast, and since that time described as being 5480 feet in depth, with such a breadth as to form a considerable section of the south polar sea, travelling majestically from south to

* Distribution of Heat, p. 4 and p. 26.

north. Whether there be a counter-current on another meridian, replacing this column in the south polar basin by warmer waters from the equatorial sea, has not been determined. Sir James Ross found that the South Atlantic current, near the Cape of Good Hope, did not interfere with the seas of the high latitudes.

By seven different experiments, Sir James Ross ascertained that between the parallels of 55° S. and 58° 30′ S. there is a belt encircling the earth where the mean temperature of the sea, + 39·5° Fahr., prevails throughout its entire depth, forming a neutral border between two great thermic basins of the ocean. To the north of this circle the sea is warmer than its mean temperature, by reason of the sun's heat which it has absorbed, raising its temperature at various depths in different latitudes. In the equatorial regions this mark of the sun's influence is found at the depth of about 7200 feet, beneath which the ocean maintains its unvarying temperature of 39·5° F., whilst that of the surface is about 78° F.

In these experiments the thermometer was sent down in several cases to the depth of more than 6000 feet, and in most instances the temperature was nearly 39·5°, but in the highest latitude in which an observation of this kind was made, viz., 58° 36′ S., the temperature was 40° at the depth of 3600 feet, and 41° F. at the surface, the intermediate depths being of intermediate temperatures. This was in longitude 104°-105° west, and consequently opposite to the Southern Pacific. A cause for the sea being half a degree hotter in this quarter may be found in the existence of a current setting to the south, as marked in Maury's chart (IX.),* occupying the space between the northerly Peruvian stream and New Zealand, or between

---

* Phys. Geog. of the Sea. 1857.

the 135th and 180th degrees of west longitude. The same chart exhibits currents issuing from the antarctic basin into the South Atlantic and Indian Oceans, the only one indicated as flowing in the opposite direction, and transferring warmer water into the polar basin, being that to the eastward of New Zealand.

In latitude 15° south, longitude 23¼° west, Sir James Ross ascertained the temperature of the sea, at the depth of 7200 feet, to be 39·5° F., while that of the surface was 77° F. And on the north of the equator, near the Cape de Verd Islands, the mean heat of 39·5° was reached at the depth of 7000 feet, the surface being 78·5° F.

The *pressure of the atmosphere* in high southern latitudes was found on this voyage to be constantly inferior to its mean height within the arctic circle. The barometric pressure increased, Sir James tells us, from the equator to the tropic of Capricorn, being greatest in latitude 22¼° south, where the mercury stood at 30·085 inches; it decreased from that parallel to the antarctic circle, where the barometric column stood no higher than 28·92 inches; and in latitude 74° south, it was as low as 28·928 inches. In a corresponding north latitude (74¾° north), at Melville Island, it was 28·870 inches.

The *mean temperature of the atmosphere* within the antarctic circle in the summer months of January, February, and March was + 27·3° F., that of February, which was passed in the highest latitude, being only + 24·3° F., while the temperature of the surface of the sea, for the first two of these months, between the parallels of 66¼° and 78° south, scarcely varied from + 29·2° F. The temperature of the air while the ships were in the polar area never exceeded + 41·5°, and in the two Februarys passed there did not exceed + 35°; only on fourteen days in 1841 was the maximum temperature 32° or above, and in 1842 there were twenty-seven such days of the time occupied in navigating the antarctic area. The whole antarctic area lies within the snow-line, and as has been already stated, no vegetation was detected on any of the cliffs, which were too erect for the snow to lie on them.

There are, therefore, no land quadrupeds. Birds exist there in great numbers, and of kinds which do not frequent the north polar regions. The penguins are perhaps the most abundant and certainly the most curious, for though they resort to the snow-clad ledges of the rocks to breed, they live the lives of fishes. Their short wings are not fitted for flight, but serve as fins to aid them in swimming, while their well-oiled feathers form a dense covering as compact as scales, and not dissimilar in appearance. These birds can swim fast and far under the surface of the sea, and are helpless only on land. Fish must be abundant to supply food to such myriads of penguins, as well as of the multitudes of other oceanic birds which our navigators met with. The few fish that were taken were of previously unknown generic forms.

Even at low mean temperatures organic life is maintained, fourteen species of siliceous shelled polygastrica having been procured from pancake-ice in latitude 75°. Upwards of eighty species of microscopic animals were detected in pancake-ice from the barrier in latitude 78° S., and numbers brought up with the sounding lead from a depth of more than one thousand feet in the same high latitude.

They furnish food to animals of higher organization, and these again to fishes, whales, and seals, the temperature at which sea-water freezes on the surface not being adverse to the well-being of vast numbers of marine animals.

The equatorial limit of the range of the *right whale* in the northern and southern hemispheres, as exhibited in Maury's chart above referred to, and laid down from the experience of whale-fishers, undulates both in the north and south, and indicates perhaps with tolerable correctness the extent to which the cold polar currents intrude on the warmer districts of the ocean. In like manner the lines marking the range of the spermaceti whale have their inflexions towards the poles where the warm currents flow in those directions, and consequently opposite to the curves denoting the wanderings of the right whale. The seals of the antarctic regions differ in species from the northern ones, and probably the *Clio, Limacinæ,* and *Beröes,* observed by Sir James Ross in the south polar seas, and supposed to be the same species on which the whale of the arctic regions feeds, may exhibit specific differences on close examination and comparison with their northern congeners. But some of the lowest forms of vegetable organisms seen near, but outside, the antarctic circle, were, in the opinion of so acute a botanist as Dr. Hooker, in no respect different from northern species as mentioned in a preceding page.

The line of the last woods falls far short of the south polar circle, partly owing to the abbreviation and conical form of the continents in that direction, and the unfitness of small islands for the growth of trees, but chiefly to inferiority of climate.

Kerguelen island, on the 50th parallel and 70th meridian east, no longer nourishes living trees, but beds of coal (tertiary?) exist there associated with basalt and arænacious shale. Highly silicified trunks of trees are found lying under the basalt, some fragments of the wood appearing to the eye to be so recent that their true condition was only discovered by handling them. One fossil tree, seven feet in circumference, was dug out from under the basalt at the height of six hundred feet above the level of the sea.

The Aucklands form a group of which the largest island is about thirty miles long. They are cut by the same parallel of latitude with Kerguelen Island, but reach about a degree nearer to the south pole. None of their heights rise to the level of perpetual snow, and their whole surface is covered with verdure. Like Kerguelen's land, the rocks are mostly volcanic, but few precipices occur. All the shores are skirted by a low forest of *Metrosideros, Dracocephyllum, Panax,* and *Veronica* trees, succeeded by a broad belt of brushwood, above which grassy slopes extend to the summits of the hills. Many ferns grow on these islands.

The islands of Terra del Fuego carry trees furthest towards the south pole, or down to about the 56th parallel south. The Evergreen Beech is the most prevailing tree, and is associated with the antarctic deciduous beech. In North America deciduous beeches (of a different species) have not been traced higher than the state of Wisconsin, or not beyond the 39th parallel of north latitude. Terra del Fuego also supports the winter's bark tree, and many beautiful shrubs of the holly-leaved barbery, fuchsias, and handsome Veronicæ; but as the hills are ascended the woods are replaced by trees so dwarfed as to form an intricate and dense scrub, above which there is a licheniferous moorland; only four flowering plants were traced to the height of 1700 feet. These four belong to the tribes of *Umbelliferæ, Compositæ, Ericeæ,* and *Empetreæ,* and are the outliers of southern flowering plants. Eleven species reached a height of 1500 feet on greenstone. They belong to the genera *Viola, Saxifraga, Escallonia, Azorella, Ourisia, Drapetes, Fagus* (the *antarctica* prostrate, and only three inches long), *Luzula,* and three grasses, *Triodia, Aira,* and *Festuca.* The vegetation of Fuegia includes a considerable number of English plants, though 106 degrees of latitude intervene between them.*

POSTSCRIPT.

As this sheet was passing through the press, a correspondent of the Athenæum (Dec. 29, 1860), residing at Simancas, furnishes information respecting the voyages of John Cabot, which fixes the date of the discovery of Newfoundland to the year 1497, as stated in the text of this compilation (p. 38). The correspondent, who signs himself J. B., states that he sends a short paragraph from a long despatch, written in London by Don Pedro de Ayala, bearing date 25th of July 1498.

"I think your majesties have already heard that the king of England has equipped a fleet in order to discover certain islands and continents, which he was informed that some people from Bristol had found, who manned a few ships for the same purpose last year. I have seen the map which the

* Dr. Hooker, Acc. of Hermite Island, Ross's Voyage, ii. p. 294.

discoverer has made, who is another Genoese like Colon, and who has been in Seville and in Lisbon asking assistance for his discoveries. The people of Bristol have for the last seven years sent out two, three, or four light ships (*caravelas*) in search of the island of Brazil and the Seven Cities, according to the fancy of this Genoese. The king has determined to send out (ships), because the year before they brought certain news that they had found land. His fleet consisted of five vessels, which carried provisions for one year. It is said that one of them, in which went one Friar Buil, has returned to Ireland in great distress, the ship being much damaged (*roto*). The Genoese has continued his voyage. I have seen on a chart the direction which they took, and the distance which they sailed, and I think what they have found, or what they search for, is what your highnesses already possess. It is expected they will be back *(seran venidos)* in the month of September. I write this because the king of England has often spoken to me on this subject, and he thinks that your highnesses will take great interest in it; I think it is not further distant than four hundred leagues. I told him that in my opinion the land was already possessed by your majesties, and though I gave him my reasons, he did not like them. I believe your highnesses are already informed of that matter; and I do not send now the chart or *mapa mundi* which that man has made, and which, according to my opinion, is false, as it gives to understand that (the lands in question) are not the said islands."

In this passage Don Pedro calls Cabot a Genoese, but the evidence of his Venetian origin seems to be unquestionable. We learn from this important letter, that seven years prior to the date of Don Pedro's despatch, the Bristolians, urged by Cabot, had been in the habit of sending out from two to four carvels annually, on voyages of discovery. This makes the efforts of the Cabots coeval with the preparations for Columbus's first expedition ; but there was evidently no notable result until 1497, when the Bristolians brought "certain news that they had found land." The fleet of 1498 was doubtless raised to five vessels by the "three or four small ships of Bristol fraught with sleight and grosse merchandizes," mentioned by Fabian.

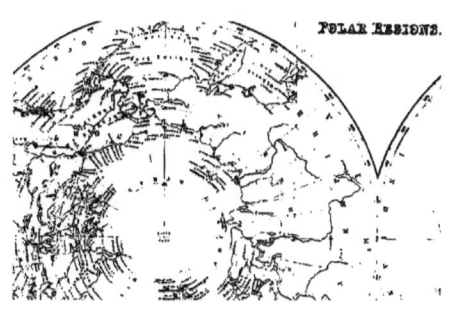

INDEX.

ABENAKI, 298
Acanthocottus quadricornis, 281
Adam of Bremen, 30
Adams, Clement, schoolmaster, 46; his engraving of Sebastian Cabot's card, 38, 47
Adare, Cape, 367
Adelaide Peninsula, 154
Adelie Land, 363
Admiralty, The, ask for opinions on the Franklin Expedition, 159; inlet, 196; scheme of search, 175; range in Victoria, 368
Adrianson, Claes, 72, 73
Advance, The, 200, 224
Aegoland, 55
Aestrymnades, 6
Agnello, Baptista, 82
Ahaknanhelik, 300
Agnese, Baptista, his atlas, 48
Alaseia discovered by Ivanoio, 132
Alaska, Peninsula of, 136, 327
Alba, 22
Albany Fort in Hudson's Bay, 113
Albion, 6
Alca torda, 213
Alcan, 2
Alexander Island, 363
Alexeiew (Fedot) is slain after passing Bering's Strait, 134
Alfred's conversation with Obther, 16
Alfred's long-ships, 15
Alfred's Orosius, 20
Altenfjord, Trees of, 267
Amalchium of Hecatæus, 10
Amazirgh, 12
Amber, where found, noticed in the Odyssy, 3
America, 354
America discovered by Biarni Herjulfrson, 23
American continent, its north point explored, 112
Anadyr, Gulf of, 134, 267, 328
Ancient houses of New Siberia, 329
Anderson (Mr.) descends the Great Fish river, 169, 194
Andreason, 27

Andreef (Sergeant) sees northern land, 181
Andrejew, 329
Andrew-Jackson, Cape, 201
Angekoks, 320, 321
Anghiera (Pietro Martire, De) quoted, 41
Anian, Strait of, 121; referred to Bering's Strait, 94
Animal life in the Antarctic circle, 378
Animals of Smith's Sound, 227
Aniui, Fertile valleys of the, 264; vegetation of its banks, 272
Anjou, Lieut., 222, 230; on Arctic wood hills, 293
Ankudinow Gerassim, 133
Anser albifrons, 278
Anser Hutchinsii, 278
Anskiold, 33
Ante-Columbian period, 14
Antarctic Physical Geography, 374; polar area, 351; polar animals, 378, 379
Antiquitates Americanæ, 29
Aploceros montanus, 277
Approach of an Arctic winter, 260.
Archbishop Eric Wakkendorph, 65.
Arctic America, Tides of, 230; mountains 263; seasons, 256; summer, 256; tern, 213; vegetation uniform, 273; coast-line yet unexplored, 188
Arctica alle, 213
Areas, Juan Luis, 354
Argali or Big-horn, 277, 316
Arimasps, 333
Arkhangel or Archangel, 131
Arkhaya, 332
Aristæus of Proconnesus, 333, 326
Armstrong Point, 232
Armstrong, Dr., on lignite, 291
Arnak-aglærpok, female Shamans, 320
Arngremus, Jonas, 298
Arnold, Bishop of, 27
Artillery Lake, 266
Arzina (or Warzina), 56
Asher, Dr., 45, 90
Atkinson Point, 317
Atlantis, 11
Attal sarazin, 8
Auckland Islands, 380
Auk, Little, 213

386 INDEX.

Austin (Captain and Admiral) sails, 181; surveys Melville Sound, 183; enters Jones' Sound, 184
Australia del Espiritu Santo, 354, 355
Avalanche at Spitzbergen, 207
Avezac, M. de, 37
Avienus (Rufus Festus) consulted Himilco's narrative, 13
Ayala, Don Pedro de, reports the voyage of John Cabot, 381

BACALAOS, or Baccalaos, 38, 47, 49
Bacchalaos, Ysla de, 48
Back (Sir George, Lieutenant, Captain, and Admiral) sails to relieve Captain John Ross, 153; on the Great Fish river, 122; voyage to the frozen strait in the Terror, 118, 156
Bäer, Von, 317, 222, 327
Bæcbord and Steerbord, 17
Baffin's opinion of a north-west passage, 105, 107
Baffin's Bay, 78, 217; currents of, 234
Baffin and Bylot sail round Baffin's Bay, 105; and into Hudson's Strait, 104
Baffin and Hall visit East Greenland, 104
Balleny discovers Sabrina Island, etc., 363, 364, 365, 366; islands, 372
Baillie-Hamilton Island, 283
Baltia of Zenophon of Lampsacus, 10
Banks, Sir Joseph, 146
Banks' Strait, 144; not navigable, 190; tide of, 232
Barclay, Cape, East Greenland, 98
Bardsen, Herjulfr, 22
Bardsen (Ivar) or Boty, 65
Barentzoon or Barentz, William, 64, 65, 66, 67, 77, 221; winters in Novaya Zemlya, 68; dies, 72
Barentz Land, 73
Barlow (Captain George) goes on discovery, 114
Barra Island, 29
Barren ground, 263
Barrow Cape, 126
Barrow Point, 151, 230
Barter Reef, 315
Basques resort to Newfoundland, 50
Bassendine, Woodcocke and Brown, 61
Bastuli, 9
Bathurst Cape, 151
Bathurst Island, 144
Bathurst Inlet, 149
Bay of Biscay, 216
Bay of the Holy Cross, 267
Bay of Mercy, Tides, 232
Bear or Cherie Island, 66, 211, 216
Bear, Norwegian, in America, 277
Bear, Polar, its destructiveness, 163
Bear Lake, Great, 150
Bear Lake River, its tertiary coal, 287
Bear's liver unwholesome, 71
Beauparis, 32
Bedford (Earl of) had Cabot's map, 44
Beecher, Captain, his bottle-chart, 205

Beechy (Lieutenant, Captain, and Admiral), 212, 216, 217; account of Spitzbergen, 56, 142, 206; voyage to Bering's Strait, 150
Beerenberg Island an active volcano, 205
Beghula, a salmonoid fish, 283
Beghula-dessy, 283; woods, 266
Behaim Martin, 34
Beke, Dr., 56, 60, 61, 67, 73
Belcher (Captain Sir Ed.) sails, 186; winters in Northumberland Sound, 187; surveys the Parry Archipelægo, 186; abandons his squadron, 192
Bell Sound, Spitzbergen, 211
Bellinghausen discovers Petra and Alexander Islands, 363, 368
Bellon, the Ichthyologist, 50
Bellot, Lieutenant, French Navy, 160, 185, 197; is drowned, 187
Bellot's Strait, 152, 196, 231
Beluga albicans, 339
Bennet, Stephen, 66
Berber inscription, 11
Bere (Ivar) or Bardsen, 28
Bergos of Pliny, 9
Berens' Point, 315
Bereto, Dona Ysabel, 354
Bering (Captain Vitus), Orthogr. of the name, 127; his discoveries, 134, 135; touches on the American coast, 128
Bering's Island, Kamtschatka, 136
Bering's Strait, 127; fossil ivory of, 295
Bernicla brenta, 213
Bertius, Peter, 354
Bertones, Terra di, 48
Best, Master George, 77, 84
Betula alba, 273
Biarmes, 171
Biarmia or Russian Lapland, 19
Bjarney (Bear Island or Disco.), 26
Bjarni Herjulfrson, 22; discovers America, 23
Biddle's Cabot, 39
Bieloe More or White Sea, 19
Big-horn or American Argali, 277
Billings on currents, 229
Birds of the Frigid Zones all natives, 278
Biscoe, Captain, Antarctic discoverer, 366; discovers Enderly and Kemp's Islands, 363
Bishops Arnold and Endride of Greenland, 27
Bison Americanus, antiquus and latifrons fossil, 296; crassicornis, 298
Black-death epidemic, 27
Bludnaia and affluents of the Chotanga, 139
Bolvanosky nos (Idol Cape), 60
Booth (Sir Felix), Bart., 151
Booth point, 154
Boothia, gulf of, 152; peninsula of, and isthmus, 193
Boty (Ivar), or Bere, or Bardsen, 28
Boulder Island, 315
Boundary of woods, 266

Bourdon, Jean, reported visit to Hudson's Bay, 113
Bows of the Eskimos and Kasuimsky, 308
Brandon, San Island, 48
Bray, De, 160
Brazen weapons dug up in Ireland, 8
Brendan, Saint, 15
Brent goose or geese, 213, 260
Brentford Bay, 152
Brewster's (Sir David) poles of cold, 243
Brigges his mathematicks, 109
Bristol merchants promote discovery, 36
Britain, 216, 217
Britannia Cape, 153
Broer Ruy's Land (Hold with Hope), 98
Brooke Cobham Island, 109
Brunel, Oliver, 73
Byam Martin Channel, 143
Bylot accompanies Sir Thomas Button, 102
Bylot and Baffin sail round Baffin's Bay, 105
Bygd, east and west, 22, 78, 217
Buchan (Commander and Captain), voyage to Spitzbergen, 142
Burbot, The, 283
Burcher's Island in Meta Incognita, 81
Burgermaster, 213
Burrough or Burro, Steuen, master of the Serchthrift, 44, 58; on Samoyed idols, 337
Busa, Jelissei, discovers the Lena, 132
Busse, the Emmanuel of Bridgewater, 87; sunken land of, 88.
Butrigarius, Galeacius, on Cabot, 41
Button (Sir Th.) receives instructions for his voyage from Prince Henry, 102, 103
Button's Bay, Rupert's Land, 102
Button's Islands, Hudson's Strait, 102

CABILLAUD or Baccaalao, 41
Cabo tormentoso, 35
Cabot or Gabotto, John, 37, 40, 381
Cabot, Sebastian (Cabota), 36, 42, 44, 76; dances when aged 88 years, 58
Cæsar's invasion of Britain, 14
Callao, 354
Campeachy woods drifted to Siberia, 221
Canada goose, 278
Canete, Marqués de, 352, 353
Cape Adare, Victoria Land, described, 367
Cape Cod (Kialarnes), 25
Cape Comfort, Southampton Island, 118
Cape Farewell, its names, 21
Cape of God's Mercy, Northumberland Island, 89
Cape Maria, Cumberland Island, 110
Cape North, South Victoria, 372
Cape Walsingham, Cumberland Island, 89.
Carboniferous rocks of Melville Sound, 287

Carcass, the, 214
Carey's Islands, Baffin's Bay, 106
Carthaginean colonies in Spain, 4; large ships of, 4; beads, 8; relics of the trade, 8
Cartris Promontory, 3
Cartwright (the Rev. John), heads a mutiny, 94
Cascathry, one of Hearne's party, 126
Cassiterides (Tin Islands), 6
Castor and Pollux River, 193
Cataio orientale, 43
Cattegat, 4
Cary's Swan's nest, Southampton Island, 102, 109, 117
Celts, 341, 344; migration of, 345
Central heat, 257
Cerealia in the Frigid Zone, 270; in Scandinavia, 268
Cervus alces, 295
Colmogro, 60
Colon Cristiforo (Columbus), 34
Columbus, Letters of, by Major, 33, 352; visits Iceland, 34; in the West Indies, 35; Tratado de los Zinco Zonas habitabiles, 31
Columba, St.; his disciples make voyages to strange lands, 16
Colville River, 315
Colymbus glacialis et septentr., 213, 346
Comfort, Cape, Southampton Island, 104
Compagnie du Nord de Canada, 113
Comparison of height and latitude in respect of temperature, 262
Congecathawachaga, 127, 149
Constinsark, Coasting search or Kostin schar, 73
Contests between the Hudson's Bay and North-west Companies, 146
Cook's examination of the north-west coast of America, 128, 121; Antarctic voyages, 356, 358, 360
Copper Indians, 126
Coppermine River, 151, 231; its woods, 266
Coracle, 4
Coregonus nasutus et sapidus, 282
Cornelius Corneliszoon Nai, 64
Cornwallis Island, 144; circumnavigated, 162
Coronation Gulf, 151
Cortoreale Gaspar, 45, 49
Costa del hues norweste, 47
Coulman Island, Victoria Land, 370
Coulterneb, 213
Countesse of Warwicke's Sound, 80, 81; Island, 84
Croker Mountains (Fata glacialia), 143
Cronian Sea, 9
Crozier, Captain, 156, 363, 371
Chancelor, Richard, 54, 57
Charing Cross, Greenland, 87
Charles' Island, 205; the pinnace, 108
Charlevoix Histoire de N. France, 50; on the Eskimos, 298, 299

INDEX.

Charlotte, Cape, 143
Charter of the Hudson's Bay Co., 112
Chawchinahaw, 123.
Chaya, 339
Chelagski, Chelagskoi or Erri-nos, 133, 230
Cheliuskin, Cape, 63
Cherie, or Bear Island, 66, 98, 211, 241
Cherrits, Dirk, 360
Chesterfield Inlet, or Bowden's, 121, 192
Chidley, Cape, Hudson's Strait, 90, 102
Chotanga, Bay of, 139
Christoval (Saint), Saloman Islands, 353
Chudleigh, Chudley Cape, Labrador, 91
Churchill, or Mississippi River, 102
Chy-calle, 282
Chytræus, Deliciæ Itinerum, 46
Cimbri, 10
Clairie Land, 363
Climate, Determination of, 225, 249; of Siberia, 271; of Spitzbergen, 208
Close season at Spitzbergen, 241
Clowey, or Thlueh Lake, 125
Cluvierus, 32
Coal-fields, 285
Coast, Extent of, explored, 199
Cock-boats (Cymbæ) of Tyrian invention, 4
Cockburn Island, Graham's Land, volcanic, 360, 362; marine animals of, 363
Cockin's Sound, Greenland, 107
Coffin on winds, 244
Cogead Lake, 125
Collinson, Captain, sails for Bering's Strait, 184; passes three winters in the Arctic Sea, 191, 197; in Cambridge and Camden Bays, 191; on currents, 231
Collinson, Cape, 188
Collin's Cape, Spitzbergen, 97
Cumberland Sound or Inlet, Inhabitants of, 321
Cumberland Island, 89; inlet, 320, 322
Cunningham's fiord, 104
Currents of Bering Strait, 227; Davis' Strait, 99, 222; Cape Farewell, 221; Greenland, 96, 234; Labrador, 219; Novaya Zemlya, 220; of the Polar Basin, 219 227; of Siberian coast, 220, 228; of the Spitzbergen seas, 214, 234; of Waigatz and Matoschin-schar, 216; Humboldt's, 375; of Victoria Land setting northwards, 370; of South Atlantic, 376
Cyclops, Hyperborean, 317; dogs, 317

Da Fuca, Strait of, 327
Dance, Sir Nathaniel, 146
Date of arrival of water-fowl, 168
Davis (Master John), his voyages, 88; his hydrographical description of the world, 103; on Labrador, 92
Davis' Strait, 216, 222; traversed by Hudson, 99
Day's voyage, length of, in a row-boat, 23

Dease (Peter Warren), his voyage of discovery, 154; Strait, 154
Death, the black, an epidemic, 27; in the Erebus and Terror, 164, 166
Deer's field, 211
De Haven, Lieutenant, 182
De Laet's novus orbis, 45
Demence, Ysla, 47
Deunis, The bark, 83
Denys (Jean) de Honfleur, 50
Deschnew (Simon) sails through Bering Strait, 133; founds Anadyrskoi, 134
Desire-Provoketh (Akpatok Island), 100
Desolation, Cape, or Torsukatek, 78
Devil's Thumb, Spitzbergen, 205
Devonian rocks, 285
Dier's Cape, Cumberland Island, 89
Digges, Sir Dudley, 99
Digges, Diggs, or Digs, Cape, in Greenland, 99, 101, 106
Dighton rock, of Indian sculpture, 24
Diomed Island, Siberia, 230
Disco or Bjarney, 26; tertiary coal, 287
Disintegration of Spitzbergen rocks, 207
Diver, Great Northern, 213; pigeon, 213; red-throated, 213
Dobbs, Arthur, 114; attacks Middleton, 118; Cape, 118.
Dog-faced nation, 324, 332
Dolgoi ostrov, 65
Dorothea and Trent discovery ships, 216
Dove on temperature, 215, 217
Dovekies, the, 213
Drift of ice at Spitzbergen, 216; on the coast of Iceland, 242
Drift of wood or trees at Iceland, 215; Spitzbergen, 214
Drift of ships out of Lancaster Sound, 222, 233
Drogio of the Zeni, 32
Drontheim, 65
Dumont d'Urville coasts Adelie and Clairie Lands, 363
Dundas Island, 233; Cape, 151
Dunnett, captain of the Prince of Wales whaler, 157
Durfoorth, Cornelius, 54
Dwina River, 56, 60
Dyakow, Fedor, 131

Earliest knowledge of the Polar Regions, 1
Earth, central heat of the, 257
East Bygd, 22, 78; and West Bygds depopulated, 28
East Spitzbergen, 204
Eclipse Sound, Meta Incognita, 196, 231
Eda Island, 29
Edda, Scandinavian, 19
Eden, Richard, 37; decades of, 47
Edgecumbe, Cape and Mount, 136
Egede on Greenland, 323
Eider Duck, 213
Electridas, 9
Elephants, fossil teeth of the, 294

INDEX. 389

Elephas primigenius, 295
El Ghat, 12
Elias, Mount Saint, 128, 136
Eliot or Elyot (Hugh), a discoverer of Newfoundland, 39
Ellis, Thomas, 77
Enara træsk, Lapland, 267
Enderby Islands discovered by Biscoe, 363
English [north-west voyages, 76; of the nineteenth century, 141
Enkhuysen, 64
Enyuin and Enyuk (Eskimos), 299
Engraved stone of Grave Creek tumulus, 11
Engröneland, or Engrovelanda of the Zeni, 31
Equus fossilis, 295
Erebus and Terror winter at Beechey Island, 162; winter off King William's Island, 164; provisioned for three years, 166; numbers of their crews, 166; crews of, leave the ships, 167; fate of, ascertained by Dr. Rae, 193; fate of, made out more fully by Sir Leopold M'Clintock, 198
Erikr Rauthi, colonizes Greenland, 21
Erkight, 324, 332
Erman, Adolph, 321, 333; on Siberia, 4; on temperature of mines, 255
Eschscholtz Bay, 296
Eskimos, or Esquimaux, Chap. on, 298; conduct to strangers, 325; domestic behaviour of, 326; dress of, 306; of East Main, 324; food of the, 312; funeral customs of the, 323; hunting gear, 307; Kashim, 311; Kayacks, 307; of Labrador, 322, have no laws, 324; language, 301; of Meta Incognita, 81; have Mongul features, 302; mourning of the, 323; notions of a future state, 322; notions of the creation, 324; occupations and seasons, 311; physical aspect of, 301; of Point Barrow, 323; range of the, 327, 299; report the deaths of the crews of the Erebus and Terror, 168; resemble the Samoyeds, 302; sculpture of the, 318: sledges of the, 309; of Smith's Sound, 332; snow shoes of the, 310; of social habits, 318; spirits, 320; stature of the, 309; tattoo their faces, 309; temperaments of the, 318; traditions of northern lands, 240; traffic of the, 314; weapons of the, 309; whale-hunts of the, 313; winter dwellings of the, 310
Eslanda of the Zeni, 31
Esquimantzic, an Abenaki word, 298
Estotilanda of the Zeni, 32
Etain, L', (tin) 2
Exeter Sound, Cumberland Island, 89
Eyras, 23

FABIAN, Rob., extract from his chronicle, 40, 43
Fadjevskoi Island, 138

Fair Foreland, Prince Charles' Islands, 97
Fair Haven, Spitzbergen, 207
Fairholme, Lieut., 156; letter from, 157
Fanshaw, Cape, 143
Farewell, Cape, 217; marine currents of, 216
Farväl (Cape Farewell), 21
Feine or Chief, 8
Female (A) of Lapmark, speech of, to Linnæus, 266
Fernandez, Juan, 352
Festus Avienus, on Gades, 7
Finmark, 20, 55
Finn, Magnusen, 27
Finnic Race, 341; runots, 345
Fins, 17, 18, 344
Fishes, Arctic, 280, 281; white, 282
Fish Skins, their use in Eskimo economy, 283
Fitzjames (Commander and Captain), 156
Fletcher, Dr. Giles, on the Samoyeds, 334
Flinders, Captain, 356
Flint weapons of the Eskimos, 308
Floods in Spring, 238
Flushing Point, 71
Fogo, Ysla de, 48
Fonte Da, Straits of, 128
Foraminifera, 222
Forbes on radiation, 253
Forster, John Rheinhold, 359
Forster, Captain, 145
Fortuna, Ysla de La, 48
Fort Confidence, winds of, 245
Four-river Bay, 197
Fossils of Siberia, 138
Foxe, Captain Luke, 108; his furthest, 110, 118
Franklin, Lady, Exertions of in the search, 174; sends out the Albert and Isabel, 184, 188; and the Fox, 194; channel, 198
Franklin (Lieutenant, Captain, and Admiral Sir John), career and character of, 146, 166; his voyage to Spitzbergen, 142; his first overland journey, 145; descends the Coppermine, 146; winters at Fort, 148; his second overland journey and winter at Fort Franklin, 150; his last voyage in the Erebus, 156; instructions to, 158; winters at Beechey Island, 160; sails round Cornwallis Island, 160, 164; descends either Peel Sound or M'Clintock Channel, 165; winters off King William's Island, 161; and dies, 161; Lieutenant Hobson finds a record of the proceedings of, 197, 191
Franklin Island, Victoria Land, described, 370
Fratercula Arctica, 213
Fretum Davis, 101
Friederichstal, 23
Frisians, able seamen, 15

Friesland, or Frizeland, 78; west, 64; or south Greenland, 33; New or East Spitzbergen, 66, 204
Frislanda of the Zeni, 31, 32; of Columbus or Iceland, 34
Frobisher (Sir Martin), 76; salutes the Queen, 78
Frobisher's Straits, 79, 84, 99; and Cumberland Straits, 110
Frozen soil of Jakutsk, Hudson's Bay, and the Mackenzie, 259; trees, 256; Strait of Middleton, 118, 120
Fugle-oe, whale skeleton on, 297
Fulmar Petrel, 213
Furious overfall in Hudson's Strait, 100
Furnace Bombketch, 117
Furthurstranda, 26
Fury Beach, 152
Fury, shipwreck of the, 145
Fury and Hecla, Straits of the, 144, 151, 234; tides thereof, 196, 231

GABOTTO, Giovanni (John Cabot), 36
Gabriel's Island, Meta Incognita, 81
Gades, or Gaddir, founded 1130 B.C., 4
Gale Hanke's Bay, Greenland, 98
Gamo, Vasco de, 35
Gand-vick, or the White Sea, 19
Gardar, Episcopate of, 26, 22
Garry Island, 129; tertiary coal of, 287
Gatehead Island, M'Clintock's Channel, 188
Geese, food of, 279
Geoffrey of Monmouth, Transl. of the Hormista of Orosius, by, 16
Geology, arctic, chapter on, 285
Georgia, South, description of, 358
Getuli and Lybians, destroyers of African Tyrian colonies, 5
Geysers, Iceland, 31
Ghent, John of, 213
Gilbert, Sir Humfrey, his discourse on the North-west passage, 44; his map of Meta Incognita, 80, 82; Sound or Godhaab, West Greenland, 89
Gillam, Capt., 112
Glaciers absent in Arctic America, 239; existing in Melville Bay, Smith's Sound, East Greenland, and Spitzbergen, 200
Glær, or Amber, 3
Glenelg Bay, 188
Glessariæ Islands, 9
Glessum, or Amber, 3
Globigerina, 222
Gneissic rocks, 285
Goat-antelope, 277
Godhaab, 89
Goldner's patent meat corrupted, 162, 163
Goldson on the North-west Passage, 121; mentions Middleton's poverty, 119
Golosme, 32

Golfo quadra, or Gulf of Saint Lawrence, 49
Golzy, 283
Gomara, Fr. Lopez, on Sebastian Cabot, 43
Gomes (Estevan), Voyage of, 50; Terra che discobrio, 48
Goodsir, Dr., 160
Goose, Brent, 213; coast of Lütke, 55; Capes, North and South, 56, 61
Gore, Death of Commander Graham, 164
Gorgossoio-schar, 65
Gothic inroads, 345
Graah, Capt., on the east coast of Greenland, 98, 217
Græsland, 33
Graham's Land, Plants of, 360
Granite, arctic, 285; antarctic, 369
Grave-creek, Ohio, Supposed ancient inscription found in, 11
Great Bank of Newfoundland, Origin of the, 220
Great Bear Lake, 150
Great Fish River, 197, 230, 264; distance from Fort Resolution, Great Slave Lake of the estuary of, 169
Great Land of the Samoyeds, 332
Greeks of Pontus, Trade with the Urals of the, 342
Green (Henry), mutineer, 101
Greenland compared with the barren grounds, 264; Dr. Kane's view of the structure of, 200; mentioned, 66, 217
Greenland, or West Spitzbergen, 204
Greenlanders, The civil state of the, 27; mourning of the, 223
Greipar in Baffin's Bay, 25
Griffin, Lieut., 182
Griffith, Owen, 78
Griffons, Gold-guarding, 333, 334
Grinnell Land of Belcher, Currents of, 234
Grinnell Land of Kane, 201
Grocland, 33
Gröneland, 95
Groiseilliers, or Groiselez and Radisson, 112, 113
Grotius de Origine Gentium Amer, 30
Guillemot, Black, 213
Gulf of Anadyr, 267; of Dantzic yields amber, 3; of Saint Lawrence or Gulf of Markland, 25
Gulf-stream, 219, supposed to reach Norway, Novaya Zemlya, and Cape Tcheliuskin, 220
Gull, The Ivory, 213
Gunbiorn, the son of Ulf Krake, discovers Greenland, 20
Gunningagap, or Baffin's Bay, 26
Gwosden visits America, 128

HAKLUYT, 60; Hakluyt's Island, Greenland, 106

INDEX. 391

Halgoland, Ohther's abode, 55
Hall, Christopher, 77, 79
Hall (James) and Baffin visits Greenland, 104
Halle's Island, Meta Incognita, 80
Hammerfest, Trees of, 268
Hanoteau quoted, 12
Hanrott (Instructions to Hudson, by Prince Henry), 104
Hans Law, 53; Hans Towns, whale-fishers of the, 212
Haswell, Lieutenant, 188
Hatton's headland, Resolution Island, 94
Haughton, Professor, 242; on tides, 234
Hearne (Thomas), first journey, 122; and Elton's quadrant, 126; and Hadley's quadrant, 124; second journey of, 123; third journey of, 125
Heat, distribution of, 250; decrease of towards the pole, and formula, 250, 251; central, 257
Heberstein (Baron), 31, 53; Rerum Musc. Comm., 53
Hecla, 31; cove, 209
Hector (Dr.) on Lignites, 289
Hedenström on fossil wood, 293, 294, 229
Heemskerck, Jacob Van, 66, 74
Heeren, Ideen, etc., quoted, 13
Helluland, it mikla ond it littla, 25
Henlopen, St., 204, 213, 215
Henry, Cape, 103
Henry the Eighth, King, 51
Henry, Prince of Portugal, 35
Herjolfrsnes, 21
Herring, the arctic, 281
Herschel, Cape (King William's Island), 154; Peak, South Victoria, 369
Hertoge, Theodoric, 356
Highest southern navigation, 271
Hind (Professor) on lignite, 290
Hippus, a Tyrian ship-builder, 4
Hobson, Lieutenant, 164, 197
Hogarth Sound, 91
Hold with Hope, Broer Ruysland, 98; East Greenland, 96; Hudson's Strait, 100
Holsteinberg, Greenland, 104, 217
Holy-cross, Bay of, 267
Hondius, map of, 33
Hood (Lieutenant), death of, 149
Hood's River, 149
Hooft-hoek (Angle-head), 71
Hooker, Dr. Jos., on the snow-line, 253; on the plants of Graham's Land, 360
Hooker, W. D., on Norway, 297
Hooper, Lieutenant, on the Tuski, 133
Hope Sanderson, 105
Hope's checked, 109
Hors-hwal, 18
Hormista of Orosius, 16
Hormogenes, Saint, 136
Horn, Cape, 261
Horsburgh, Cape, 287
Horses or ships of Gades, 5
Hotham (Cape), tides of, 233; inlet, 317

Hubart, his hope, 102, 109
Hudson (Henry), 65, 73, 95, 99, 101; the Navigator, by Dr. Asher, 46, 91
Hudson's Bay, 88; Company's Expedition, Dease and Simpson, 153; Dr. Rae, 154; River discovered, 94; Strait, 48, 94
Humboldt on the snow-line, 253; current, 375
Hutchins's Goose, 278
Hvarf (Cape Egede), 21, 22, 23
Hvidsærk (Cape Farewell), 21
Hyperborean Cyclops, 317; Mongolidæ, 331

IBERNIA Phœnicea of Villanuova,' 8
Iberville, M. de, takes York Fort, 113
Icaria of the Zeni, 32
Ice, chapter on, 237; barrier, Parry Archipelago, 240; breaks up, when, 240; bridge, Smith's Sound, 224; bird, 212; bergs, 206; drift, 239; formed in September, lasts over the winter, 241; Haven, Novaya Zemlya, 67; heat required to melt, 239; cliffs, Antarctic, 371, 372
Iceberg mistaken for an island, 369
Iceland, 217; discovered by Naddodr, 15; Greenland and Meta Incognita, 268
Icy Cape, 128
Ierne, 6
Igaliko, 29
Iglu of the Eskimos, 320
Iglut, 310
Iliad, quotations from, 2, 3
Illiseersut, 320
Ilmarinen, 345
Independance, Cape, 202, 223
Indigirka, 342; discovered, 132; fossil wood, 292; Polynia of, 222
Ingaland-mëun, 300
Inglefield (Commander), 184
Ingnersoit, 324
Ingolfr colonizes Iceland, 15
Insula Divi Joanis, 47
Inisfail (Insula sacra), 6
Innuæ, 323
Innuarolit, 324
Inuit (Eskimos), chap. on, 298, 299, 322
Inuk, words derived from, 323
Ireland, 6
Irish discoverers of Iceland, 15
Irminger, Capt., 217
Irving, Dr., 211, 214
Isabnormals, 243
Islanda, 48
Island, Cape, Novaya Zemlya, 71
Isle of Man, 6; Islands of Orkney, 29
Isothærals, of, 43, 45, 266
Isothermals, 209, 261
Ivanoio's discoveries, 132
Ivory, fossil, 294; gull, 213, 260
Ivuktok, 93

JACKSON, Cape, 223

Jakutsk or Yakutsk, frozen soil, 259; lignite of, 292
James, Capt., 110, 111
James, King of Scotland, 43
Jana discovered, 132; and Lena deltas and woods, 267
Jan Mayen's Island (Cherie), 204, 241
Japanese Junks drift across the Pacific, 11
Jefferson, Cape, 201
Jenisei or Yenisei River, 63, 131
Jens Munk, 107
John of Ghent, 213
Johnson, (Richard) on the Samoyeds, 335
Jomard, M., 11, 12; Mon. d'anc. Geogr., 47
Journeys over the ice, 198
Juet (Rob.), mutineer, 101
Jukahirs, Yukagirs or Yukahirs, burn fossil wood, 292; make weapons of fossil ivory, 295
June, temp. in the Parry Archipelago, 260
Jutland or Cartris, 3

KAAFIORD, 249
Kablunak, 300
Kadjak or Kodiak Island, 327; traffic of, 318
Kainulainen or Queen's, 19
Kakertok, 29
Kalalik, 300
Kapiselik, 281
Kamenaya Tundra (Thaw), 258
Kaminsk, 333
Kandalisk, Gulf of, 19
Kane, Dr. Elisha Kent, 199, 202, 222, 223, 224
Kangmali-enyuin, 300
Kanin-nos, 56, 59, 74, 332
Kara Sea, or Karskoie more, 62, 63, 65, 66, 72, 130, 131, 215, 217, 221; gate or Karskoie voratæ, 60
Kasdeer (Tin), 2
Kashim, 311
Kashimiut, 319
Kashimin-wikhak, 319
Kasuimski bow makers, 308
Kassiteros, or Kattiteros (Tin), 2
Kastira (Tin), 2
Keerveer, or Turnagain Point, Australia, 356
Keger in Lapland, 56
Keltikon Kassiteros (Celtic Tin), 6
Kellett, Capt., discovers the Herald Isles, 180
Kemp Island, discovered by Biscoe, 363
Kelsey, Henry, 113
Kennedy, Mr., 152, 185; and Bellot's winter journey, 185
Kennedy's Channel, open water of, 225; tides of, 223; vegetation of, 224
Kennedy, Port, 196
Kerguelen Island, coal beds of, 379; silicified trees of, 379
Khanami, Valley of the, 272

Khan-balik (Pekin), 35
Khasovo, 331
Khaya, 339
Kialernos, or Cape Cod, 25
Kildin in Lapland, 65
Kingitorsoak, or Women's Islands, 26
King Charles's Promontory, 110
King-duck, 213
King James's Foreland, 103; Newland, 202
King, John, stands by Hudson, 101
King William's Island, 160, 152, 197; land, 193, 198; sea, 154
Kirmas, 213
Kittegarëut, 300, 309
Kittewake, 213
Kjolen firs, 267
Klaproth's Asia Polyglotta, 328
Knight, Master John, voyage to Labrador, 95; accompanies Barlow and Vaughan on discovery, 114, 115
Kodiak, or Kadjak Island, 327
Koedesniks, 338
Kola, River of, 19, 59, 74
Kolgoi, or Kolguev Island, 59, 332
Kolmogorui, or Colmogro, 60
Kolyma River, 222; polynia, 222; seasons, 238
Kolyutschin tribes, 327
Kongueserokit, 324
Kongskuggsiö, or Speculum regale, 27, 29
Kotelnoi Island, 229, 230; ancient erection on, 229
Kotzebue (Lieut. and Adm.), 137, 296
Kroksfiorthr, 26
Krusenstern, P. Von, 339
Kublai-khan, 35
Kundsha, 283
Kuskutchewak, 319
Kwichpak, woods of, 267
Kyerdlek (Cape Farewell), 21

LABRADOR, 48, 212; Cape, 80; current, 219
Lagmand, or Justiciary, 22
Lagopus albus, 212
Laet, map of Newfoundland by De, 49
Lake Kennedy, Cumberland Island, 110
Lancaster (Sir James), Sound named by Baffin, 106; entered by Sir John Ross, 143; sailed through by Parry, 143
Land of Desolation, West Greenland, 88
Landnama bok, 23, 29, 30
Langenès, 65
Lapland, 56, 74; bows, 346; burrows, 345; coasts, 265; marriages, 346; metallurgists, 345; pulea or sledge, 346; snow-shoes or skeuts, 346; line of woods, 267; witches, 320; boats, 340; rein-deer, 348
Laplanders or Lapps, chap. on, 331, 344; dress of the, 346; figures of the, 346; funereal customs of the, 248; nomades, 346

INDEX. 393

Laptew, Demetrius, surveys Bear Island, 137, 140
Laptew or Leptew, Lieut. Chariton explores Samoyeda and part of Siberia, 139, 140
Larus glaucus, 213
Laughing goose, 278
Laurentian range of rocks, 285
Leaf beds, 288
Leems, Knud, 349
Leerboord or Larboard, 17
Leifde Bay, 202
Leifr, Erikson, baptized, 24; winters in Vinland, 24
Leifrs buthr, 24
Lena River, 132
Lenok, 283
Leslie on the snow-line, 252
Lewes (Sir Cornwall), quoted, 13
Ley or Lee, Dr. Edward, 39, 51
Liakhow, the merchant, discoveries of, 137; on currents, 229
Liakhow Islands, 134, 138, 329; wood huts of, 293; snow-line of the, 254
Liassic basin, 287
Libyan idiom, 12
Ligneous deposits in the arctic basin, 289
Lignite of Banks's Island, 290; on the flanks of the Rocky Mountains, 289; at Jakutsk, 292; in Siberia, 291; Tertiary, 285; of Vancouver's Island at Nanino, 289
Linnæus on the Lapland wastes, 265
Lincolnshire amber, 3
Linschoten, 65
Linzacha, or Jakuts, 131
Lisle (De) touches on the American coast, 128
List of Searching Expeditions, 172
Litho-slaves, 344
Little fish-hill, 125
Little table-island, 213
Lobassy, 329
Lodia or Lodji, 73, 131
Lofoden Isles, 19
Log-huts, Ancient, 343
Lok, Michael, 76; manuscript of, 81
London coast, Greenland, 93
Loom or Lom (Colymbus), 213, 346
Lord Mayor's Bay, Boothia, 152
Lord Mulgrave (Phipps, Capt.), 211
Lord Weston's Portland, Cumberland Island, 110
Low Island, Spitzbergen, 205, 211, 214, 215
Lubbock (Mount), in South Victoria, 369
Lumley, Lord, 108
Lumley's (or Lumlie's) Inlet, 60, 100, 108
Lütke, Admiral, on the currents of Novaya Zemlya, 220
Lütke's land, 63; north extreme, 73
Lutwidge, Capt., 214

MacClintock (Lieutenant and Captain, Sir Leopold), 182, 195, 196, 222; on drift ice, 241; on the glacier in Melville Bay, 200; on tides, 23; his long sledge-journey round the Parry Archipelago, 184; visits Ireland's Eye, 186
MacClinteck's Channel, 188
MacClure (Commander Capt. Sir Robt.), winters in Prince of Wales' Strait, and in Mercy Bay, 187; makes the passage from Bering's Strait to Baffin's Bay, 189; and abandons his ship, 189
M'Cormick, Dr., 212
Mackenzie, Sir Alex., 129; his accuracy, 129
Mackenzie River, 231, 238; tertiary coal of the, 287
Madalena Island, 353
Magdalena Bay, Spitzbergen, 206
Mageröe, Ligneous plants of, 268
Magna Britannia, Hudson's Strait, 100
Magnetic Pole, North, 152, 197
Major's notes on Russia, 20
Maldonado (Laurent Ferrer), his pretended voyage, 94
Maloy broun, 204, 209
Malte-Brun on currents, 220; on lignites, 291
Matvyeca ostrog, 65
Mammoths, 295; in Siberia, 294
Mangaseia, Town of, 131
Manila, or Manilla, 354
Mansfield, or Mansel Island, 103
Mappemonde du Sebastian Cabot, 37, 39, 45
Marble, White, or Brook-Cobham Island, 114
Marco Polo, 35
Margaret of Denmark, 28
Marine animals 211, 279
Markham, Lieutenant, 199
Markland, or Nova Scotia, 25
Markland's Gulf, or Gulf of St. Lawrence, 25
Marmora, or Sea of Kara, 221
Marriage among the Samoyeds, 388
Marsh, Antony, 62
Martens on Spitzbergen, 210, 221
Martire, Pietro (d'Anghiera), 41
Massacre of Eskimos, 127
Massilia, or Marseilles founded, 9
Mastodon, 296
Mathew's Land or Matpheoue, Novaya Zemlya, 63, 65
Matiuschkin, M., 328
Matochkin-schar, 63; currents of, 216
Matonabbe takes Hearne under his protection, 124
Matuschan-yar, see Matochkin-schar, 63
Maupertuis, 349
Maury, Lieutenant, 219, 222; his wind-chart, 244
Mediterranean water-shed compared with that of Kara, 215
Medusharsky, 73
Meech on solar intensity, 253
Melbourne, Mount, Victoria Land, 370

Melville Island, 143, 209, 232; plants of, 269; strait not navigable, 190
Memel and Dantzic yield amber, 3
Men with two faces, etc., 326
Mendana, Alvaro de, 352, 353, 354
Mendoça, Las Marquesas, 353
Meres de glaces in Greenland, 200
Mermaid, The bark, 93
Meta Incognita, 82, 241; Best's account of, 85
Meyer's formula for decrease of temperature, 251
Mew, 213
Michael, The, 78
Midianites knew the use of tin, 2
Middle ice, 94
Middleton's voyage to Repulse Bay, 117; his vindication, 119; captain of the Shark, 120
Mill Island, 104
Mississippi, or Churchill River, 102
Missions, Blat, 91, 322
Mistaken Straits of Frobisher or Hudson's Strait, 83, 85
Moffen Island, Spitzbergen, 214
Mollusks, Elevated remains of, 296
Molvas, Porto di, 49
Molyneux's globe, 90
Mongolidæ, Hyperborean, 331; Yugrian, 331
Monte de Lions, 32
Montreal Island, 194
Moonshine, The Bark, 88
Moore's (Captain) Southern voyage, 373
Morgiouet's harbour, 59
Morimarusa of the Cimbri, 10
Morton, Mr. William, 201, 202, 223, 224
Müsur, 25
Mount Erebus, an active volcano, 370, 371; Melbourne, Victoria Land, 370
Mount of God's mercie, 95
Mount Raleigh, Cumberland Island, 89
Mount Saint Elias, 128
Mount Warwick, Frobisher's Strait, 81, 83
Mount Sabine, Victoria Land, 367, 372; plants of, 368
Mount Terror, Victoria Land, 371
Mozine River, 332
Muckla jokel (Cape Farewell), 21
Mulgrave, Lord, 97, 127, 215
Müller, Max, 345
Munckenes, Vinterbavn, 107, 109
Munk (Admiral), 104; Cove, 107
Muscovy, 67
Muscovy Company, 61, 77, 57
Musk ox, its range, 276; in North Greenland, 227; Fossil, 295
Mussel harbour, Spitzbergen, 210
Mutnaia, River, 131

Naddodr discovers Iceland, 15
Naggœuktormëut, 300
Nai, Corneliszoon, 64, 65
Namollos, 328

Nanino, Lignite of, 290
Nantuckel Island seen by Biarni Herjulfrson, 23
Nassau or Pet's Strait, 62, 65; Cape, 73
Navy Board Inlet, 196
Ne ultra, 117
Nelluangmeun, 300
Nelma, 283
Nelson River, 102; entered by Foxe, 110
Nerigon of Pliny (Norway), 9
Nerrim-Innua, 324
Nevado, Rio de, 49
New Denmark, 107
New Friesland, 204
Newfoundland, 216, 221
New Guinea, 353, 355
New Hebrides, 354
Newland, King James', or West Spitzbergen, 97, 98, 202
New Siberia, 138, 229, 230; Wood-hills of, 293
New Wales, Ruperts' Land, 102; of Foxe, 110
Neyra, Alvaro Mendana de, 352
Noatak, Woods of the, 267
Nootka Sound, 327
Nordvich, Bay of, 137
Norman contests with the Kainulaineu, 19; pegs of lead, 2
North Bay of Baffin, 118
North Cape, 19, 216, 230
North Devon, 144
North-east Cape, 130; voyages of the Dutch, 64
North Foreland, Meta Incognita, 80
North Georgian Islands, 233
North Somerset, 144, 197
North Star, 93
Northursætæ (summer haunts), 25, 26
North-west Foxe, 108
North-west passage, 76; not navigable, 170
North-west voyages, Early, 76; of the 19th century, 141
Norton Sound, 267
Norway, 18; trees of, 267; bear, 277
Norwegians, 350
Nova Francia, 32
Nova Scotia, or Markland, 25
Novaya Zemlya, 31, 55, 60, 61, 65, 67, 209, 234
Novogorod merchants, 333
Nicholas, St., Harbour of, 57, 59
Nikolai, Cape, 152
Nijnei Kolymsk founded, 132
Nunatangmëun, 300, 315, 316
Nuwuk, 300
Nuvujarschiut (Cumberland Sound), 92
Nye Hernhut, 89
Nyenecb, 331
Nynei, or Nijnei Kolymsk, founded, 132

Ob, Obi, or Obba, The River, 65, 60, 131, 308, 332, 333, 334
Obi, Gulf of, 131

INDEX.

Obdorian Samoites, 334
Obdorsk, 273, 333; vegetation of, 272
Observation on the line of search by Great Fish River, 177
Ogle, Point, 194
Ohther, Octher, or Audher, 349; voyage round the North Cape, 16; Reisebericht, 18.
Olaf on currents, 220
Olaf, King Triggveson, 24
Olaus Magnus, de Hist. Gent. Sept., 30
Olekma, 132
Olenü-nos, 63
Omalik, 324, 343
Omenarsorsoak (Cape Farewell), 21
Ommaney, Commander and Captain, 182; finds traces of Franklin, 183
Omoki, 342
Oonalashka, 327
Ooze, composition of the sea, 222
Orange Islands (of Novaya Zemlya), 65, 71
Organisms at the bottom of the sea, 221
Ornithology of the Arctic Sea, 278
Orosius, Alfred's, 20
Ortelius, Theatr. Orbis of, 30
Osborn (Lieut. and Capt.), 184, 186
Oudney, Dr, 12.
Ounartok, hot springs of, 31
Oussa, river, 238
Ovibos moschatus, 276, 295

PAGODA, brig, Capt. Moore, 373
Pagophila eburnea, 213
Parliamentary reward for the discovery of the North-west Passage voted in 1745, 119
Parmenides, 351
Parrot, Diving, 213
Parry (Lieut., Capt., and Admiral Sir E.), 212, 214, 217; accompanies Sir John Ross round Baffin's Bay, 142; sails through Lancaster Sound, 143, 144; examines the coast from Repulse Bay northwards, 118, 144; lands the Fury's stores, 145; his boat voyage towards the North Pole, 145, 211; on tides, 232; cape, 151; mount, 201; mountains, 371
Pasquiligi, Pietro, 50
Pässida or Tundra, 131; river, 131
Pässida, Tonguse village on the, 140
Pauli (Dr.), Kœnig, Alfred of, quoted, 18
Peace of Utrecht, 113
Peckham, Sir George on the N. W. Passage, 44
Peel's Strait, 196, 231
Pekin or Khan-balik, 35
Pemican, 159; quantity of taken in the overland expedition, 176
Penguins of Victoria Land, 368, 378
Penny, Capt., 182; finds Franklin's first wintering place, 183; surveys Wellington Sound, 183; in Hogarth Sound, 91
Permäkow (Jakow) reports the existence of the Liakhow Islands, 137

Permia, 19; Permiäki or Permians, 18, 19
Pernambuco, Drift-wood from, 221
Pert (Sir Th.) and Cabot, 42
Peschschori (subterranean dwellings), 333
Peshew Lake, 266
Pestilence of the black death, 27, 28
Pet (Arthur) and Jackman, 61
Pet's Strait or Yugorsky schar, 65
Petchora, River and Country of, 59, 63, 131, 238, 332, 339; bay of, 62, 73
Peter's pence paid in walrus tusks, 24
Petra Island, Antarctic Sea, 363
Petrel, the Fulmar, 213
Peyrouse (La), his voyage to Hudson's Bay, 84
Phenian miners, 8
Philip II. of Spain, 352, 353
Philip III. of Spain, 354
Philo of Byblus, 13
Phinn or Giant, 8, 344
Phipps, Earl Mulgrave, 127, 205, 209
Phocœan Greeks, 6
Phœnicians, the first Atlantic navigators, 1; supposed of one or more of their ships to America, 10; miners and metallurgists, 7; pig of lead, 2; inscription at Malta, 7
Physical geography, 203
Piasina, 63
Pigeon-diver, 213
Pinus cembra, 273
Pistol Bay or Rankin's Inlet, 120
Planisphere of Sebastian Cabot, 45
Plants of Spitzbergen, 210
Playse or Pleyce, writer of Hudson's journal, 95
Plectrophanes invalis, 212
Pleuronectes glacialis and scaber, 281
Pliny, 326; on metals of India, 2; on plumbum album, 2
Point Desire, 71; lake, edge of woods, 266; Peregrine, 110; Turnagain, 149
Polar bears, 163; Polynia, 222; tides, 190
Pole, magnetic, 197; (North) thermal days at the, 253
Poles of cold, 243
Polo, Marco, 35
Polygastrica inhabiting the depths of the sea, 378
Polynia of Kennedy's channel, 222; polar, 222; Siberian, 222, 229
Pomarine Skua, 213
Pond's Bay, 196; Strait, 234
Pool, 356
Pope Nicholas the Fifth, 28
Porsanger Fjord, 268
Port or larboard, 17
Porter, Cape, 193
Possession Island, Victoria Land, 368
Potherie (M. Bacqueville de la), his history of North America, 113
Pouro, 355
Preserved meat tins found on Beechey Island, 162
Press, on the search for Franklin, 185

INDEX.

Prickett (Abacuk) relates Hudson's fate, 101; accompanies Sir T. Button, 102
Prince Charles Island, Spitzbergen, 97
Prince Henry's instructions to Sir Thomas Button, 103
Prince Patrick's Island, 232, 241
Prince of Wales' Land or Island, 197; Strait, currents of, 232
Prince William's Sound, 136, 327
Princess Royal Islands, wood drifted to, 232
Prior's Sound, 80
Procellaria glacialis, 213
Promischlenniki or fur-hunters, 131
Proteus, Eskimo, 322
Provisions corrupted, 57, 166
Ptarmigan, 212
Puffin, 231
Pullen (Lieutenant and Commander), 181
Purchas, his pilgrims, 62
Pustoserk, 330
Pytheas, a Massilian navigator, 9

QUALÜEN, dead trees of, 297
Quäns or Queens, Finnick queens, 350, 349
Qwænvick or Kainulainen (Queenland), 19
Queen's Cape, 80
Queen Ann's Cape, Greenland, 217; Foreland, 103
Queen's Foreland, Meta Incognita, 83
Queene Elizabeth's Foreland, 80
Quiros (Pedro Fernandez de), also De Quir, 353, 354, 355
Quoich River,

RACE (or Ras), Cape, 49
Radiation from snow, 254
Radisson, M., 113
Rae (Dr. John), voyage to Repulse Bay, 154; surveys the south coast of Wollaston Island, 178; and the Gulf of Boothia, 180; also the isthmus of Boothia, 192; Quoich River, Chesterfield Inlet, 192; learns the death of the crews of the Erebus and Terror, 193; receives the Admiralty reward of £10,000,
Rafn, Antiquitates Americanæ, 29
Rain in winter at Spitzbergen, 246
Raleigh, mount on Cumberland Island, 92
Ramusio, 31; imperfectly acquainted with the voyages of Sebastian Cabot, 45; Viaggi de, 33; discorso sopra los Bacchalaos, 43; his map of Newfoundland, 49
Rane (or Rein)-deer, 18
Rankin's Inlet, Douglas Bay, Corbet's Inlet, or Pistol's Bay, 120
Ras (Raso), Cabo de, 48
Rask, Professor, 345
Ratcliffe, 62
Rathsher, The, 213

Ratisson and Groiseleiz, 112
Rawlinson's Herodotus, 7
Razor-bill, 213
Record of the fate of the Erebus and Terror, 161
Red Indians, 300
Redpole, 212
Redshob (Dr.), quoted, 5
Regent's Inlet, 143, 151, 196, 231
Rein-deer or Rennthiere, 274, 275, 276; fleetness of the harnessed Lappish, 349; migrations of the, 275
Rensselaer Bay, 200
Refraction at Ice-haven, 69
Repulse Bay, 117, 120; winds of, 247
Resolute, drift of the, 192
Resolution Island, Hudson's Strait, 89, 90
Reynolds Point, 188
Rhodostethea Rossii, 213
Richards, Commissioner, 186
Richardson (Dr. Sir John), 150, 177; Point, 153
Rijp, Jan Corneliszoon, 66
Ringed plover, 212
Rio Nevado, 48, 83; en Golfo de Castello, 48; or Hudson's Strait, 49
Riphæan or Ural Mountains, 342
Rissa tridactyla, 213
Rivers break, up at 36° or 39° F., 237; freeze again, 238
Rockball, 216
Rocks, series of arctic, 285
Rocky Mountains, north end of, 267
Roe's (Sir Thomas) Welcome, 117
Romanzoff, Count, 137
Rondelet, 50
Rosmuister, 63
Ross (Captain and Admiral Sir John), 151, 182, 211; sails round Baffin's Bay, 142; on the Arctic Highlanders, 105; on Cumberland Island, 92; voyage to Boothia, 151, 152; his letter on the Franklin expedition, 158
Ross (Lieutenant, Captain, and Admiral Sir James Clark), 217, 221; surveys Boothia, 152; sails in the Enterprise, 179; surveys Peel's Sound, 180; is caught in the pack and drifted into Baffin's Bay, 180; on atmospheric pressure, 377; discovers Southern Victoria Land, 367; on priority of antarctic discovery, 364; his remarkable antarctic voyage, 363; on southern temperature, 366, 367
Ross's Gull, 213; Islet, 145, 201, 213
Rotge, 213
Route of the Davis' Strait whalers, 241
Runic stone of Kingitorsöak, 27
Runots, Finnish, 345
Rupert's Land, 113
Russia Company, 57
Russian exploration of Siberia, 130; sailors' residence on Maloy Broun, 210

INDEX.   397

SABINE'S Gull, 213; Mount, Victoria Land, 367
Saborne, 345
Sabrina Island discovered by Balleny, 363, 364
Sacheuse, an Eskimo interpreter, 106
Sacred promontory of Wrangell, 63
Saint James, Island of, 60
Saint John, 38; Island of, 47
Saint Lawrence Island in Bering's Straits, 316
Salisburie's foreland, 100; his island, 103
Salix arctica, 202, 224; uva-ursi, 202
Salmo leucichthys, 283; salar, 283
Saloman Islands, 352, 353
Sama, 321
Same or Swampy, 345
Samogedi, 332
Samoyeds, chapter on, 331, 359; area of, 341; belief in a Supreme Being, 338; dress, 341; eaters of raw flesh and fish, 339, 340; ethnological relations of the, 341; idols, 60; of the Jenisei, 131; land, 62; physiognomy, 340; stature of the, 340; the physical powers of, 339; wealth, 339
Sana or Sanak, 321
Sanchoniathon, 13
Sanderson's Hope, 93, 202
Sandpiper, the purple, 212
Sandwich, Earl of, 127
Sandwich Islands or Southern Thule, 359; land, 358
San Felipe Bay, in Australia del Espiritu Santo, 355
Santa Cruz, Saloman Islands, 353
Santarem, Visconte de, Essais sur Cosmographie, 52
Saunders, J., Master, R.N., 180
Scandia of Pliny (Scandinavia), 9
Schalarov on currents, 229; his map referred to, 230
Scilly, 6
Sclave migrations, 345
Scolvo, Scalva, Scolvus, or Sclolvus, 33
Scoresby, Dr., 212, 217, 241, 242; on East Greenland, 98; on Spitzbergen vegetation, 254; on winds, 248
Scotland, 216
Scroggs, Master John, 114
Scythic language, 345
Sea-horse Point, Southampton Island, 104; beasts, 279
Sea-ice, 239
Sealing district, 241
Seals, 279
Search, Admiralty scheme for, 175; by the Great Fish River, 177
Searching expeditions, list of the, 172; overland, 175
Seasons for navigation, 241; on the Parry Islands, 260
Secular elevation of Spitzbergen, 215; of Siberia, 230
Sedentary Tchukche, 328
Seeds travelling on the ice, 275
Senjün or Seynam, in Norway, 55
Serchthrift pinnace, 58
Settle, Dionise, 77
Seven Islands, Spitzbergen, 98, 201
Seynam or Senjün, 55
Shaminism, 320
Shamans, the dress and practices of the, 321, 325, 335, 338; Samoyed, 335
Shefferus (John) on Lapland, 19
Shells, Elevated, 296
Ships of Tarshish, 6
Siberia, Russian exploration of, 130
Siberian coast, 217; lignite, 291; sturgeons, 284; tundren, 261, 333
Sidonians, 2
Sieveroi vostochnoi-nos, Cape Cheliuskin, or the North-east Cape, 63, 130, 332; never sailed round, 140; snow-line of, 255; Polynia of, 220
Silla, or the atmosphere, 324
Sillagiksertok, 324
Silurian rocks, 285, 286
Simpson (Mr. Thos.), his discoveries, 151, 154; on the Eskimos, 299; Peninsula, 192; Strait, 154
Sinus codanus, or the Baltic, 9
Siragnezi, 331
Sir Thomas Roe's Welcome, Island, and Strait, 109, 201, 225
Sir Thomas Smith's Sound, 106
Skäl, 298
Skeletons found, 168
Skrællingar or Skrællings, Chap. on, 28; of Meta Incognita, 81, 298
Skrige, 344
Skrithifinnia, or Finnish Lapland, 19
Skrithifinni, 344
Skua, 213
Slata baba, 334
Slaty lands, Labrador and Newfoundland, 25
Sledge Island,
Smeerenberg, or Oily Hill, 25; coffins deposited in the harbour of, 212
Smith, Commodore, 120
Smith's St. Columba quoted, 16
Smith's Sound, 106, 201, 225, 199
Snorre Thorfinnson, born in Vinland, 25
Snow-bird, 212
Snow-houses, 310
Snow-line, 252; on the Arctic circle, 254; Meech on the, 253
Soiots of the Samoyed nation, 341
Somateria mollissima, 213
Sorüe in Norway, 217
South Georgia, birds of, 359; snow-line, 359
South Shetland, 360
Southampton, Cape, 102; Island, 118
Southern outlets of Melville Sound, 179
Southern Thule, 358, 359
South Goose-cape, 61
Spitzbergen, 66, 201, 217, 259; current from, to Kennedy's Channel, 225;

drift ice of, 218; East, 204; geology of, 204; Hudson's visit to, 97, 98; physical geography of, 203; quadrupeds of, 210; vegetation of, 210; origin of the name of, 205; West, 66.
Spring birds, 239
Squier, Mr., 12
Staats Iland, 204
Staduchin winters on the Kolyma, 132
Stán or Stannum, Tin, 2
Statenhuk (Cape Farewell), 21
Staten Island, South Sea, 360
Steereboord, Steorbord, or Starboard, 17
Stenodus Mackenzii, 283
Stercorarius parasiticus, 213
Sterna arctica, 213
Stewart, Captain, 182
Stock-fish or Bacalaos, 50
Stone-tools, 344
Stoney Mountains, 126
Strabo, 351, 355
Straumfiorthr or Buzzard's Bay, 25
Strugannia, 340
Strunt-jager, 213
Sturgeons not found in Arctic America, 280; Siberian, 284
Subterranean dwellings of Timansk, 333
Summer, progress of, 261; in America, 259
Sunken land of Busse, 88
Sun's rays, action of the, 255
Sunshine, The bark,
Supperguksoak, 322
Supreme Being, belief of the Samoyeds respecting the, 388
Sutherland (Dr.), on tides, 233
Svatoi (accursed), 134
Swane of Ter Veere, Nai, master of the, 64
Swethland or Sweden, 19
Syevernuy, Guisnuy muis (Goose Cape), 55
Szkolni, or John of Kolnus, or Scolvo, 33

Tabin of Pliny
Table of mean, summer, and annual temperatures, 261; Island, Spitzbergen, 215
Tadebes, 338
Taimur Bay, tides of, 221; Cape, 63, 130; attempts to sail round the Cape, 138; Lake, 140; River, 139
Takak, 321.
Tamaco, Taumaco, or Taomaco, Saloman Islands, 354, 355
Tarandu's fossilis, 295
Tarshish or Tartessus, 5, 6; meaning of ships of, 7
Tartaria, 67
Tasman, 356
Tartarinow, 230
Taxites, 288
Tchéndoma, 132
Tchúr, 282

Tchirikow, Lieut., 135, 136; touches on the American coast, 128
Tchugatch Bay, 327
Tchukche, Tchutchi, or Tchuktchi, 131, 328, 343; first heard of, 132; mode of bartering with the, 132; rein-deer of the, 133; sedentary, 328; traders, 316
Tchukotchi or Tchukotski Cape, 328, 343
Temperature, Chapter on, 249; of antarctic districts, 376; atmospheric, 377; influence on vegetation of, 249; of June in the Parry Archipelago, 260; of Lapland and Great Bear Lake compared, 262; in deep mines, 254, 255; table of mean, summer, and annual, 261
Teneriffe, 216
Tenn or Tin, 2
Teploi weter, in Arctic America, 246
Terfynnes or Terfins, 18
Tern, Arctic, 213
Terra Australis, 356; incognita, 352; de Baccalaos, 82; Cortorealis, 32; map of T. Cortorealis, 50; nova, 32; nuova, 49; primum visa, 88
Terre Adelie, 26
Tertiary lignite, 285
Teutons, 341
Thaw, depth in the Kamenaya tundra of the summer, 258; descent into the earth in various localities of the, 257
Thermal days in the frigid zone, 253; rays heat the earth to the depth of 100 feet, 256
Thermic anomaly, 243
Thlueh or Clowey (Trout Lake), 125
Thomas Williams' Island, meta incognita, 81
Thorne, Mr. Robert, 39; declaration of the Indies of, 50
Thornæsting, 21
Thorwaldson of Icelandic descent, 27
Thousand Isles, Spitzbergen, 204
Thule of the Romans, 14; described by Pytheas, 9; voyage from Nerigon to, 9; of Columbus or Iceland, 34; southern, 358, 359
Tides of Arctic America, 230; Bay of Mercy, 232; and currents of Kennedy's Channel, 223; in Fury and Hecla Strait, 196; of 156 feet, 34; Polar, 190; Dr. Sutherland's register of the, 233; of Spitzbergen, 221; of Victoria Land, 375
Timæus on Raunonia, 10
Timansk, 333
Tierra Austral, 354; Fermè, 352; trees of Tierra del Fuego, 380
Tilia, 288
Tin, first notice of, 1; Islands, 6
Tinné or Northern Red Indians, 122, 325, 327
Todd, Dr., quoted from Natural History Review, 16

INDEX. 399

Torfæus' Grœnlandia quoted, 65; Vinland of, 30
Torneå, Gulf of, 350
Tornait, 92
Torngak, or Eskimo familiar, 320, 321, 324
Torngarsuk, an Eskimo deity, 320, 321, 324
Torres, Luis Vaez, 355; Strait, 356
Torsukatek, or Desolation part of West Greenland, 78
Totnes Rode, Cumberland Island, 89
Touareg, 12
Trafalgar, Franklin present in the action of, 146
Traffic of the Eskimos and Tchukche,
Treaty between Erik of Norway and Henry VII. of England, 28
Trees, elevation to which they ascend, 268; freeze, 256; their line of termination in Lapland and Norway, 267; their drift on the Spitzbergen coast, 214; of Tierra del Fuego, Kerguelen's Land, and the Aucklands, 380
Treurenberg Bay, 145, 211; coffins deposited in, 122
Tringa hypoleuca and maritima, 212
Trouts, Arctic, 283
Troye, three Hudson's Bay forts taken by M. Le Chevalier de, 113
Trumpet Island, Meta Incognita, 81
Tsinagh, or Berber writing, 12
Tundren, 264
Tungalhtuk, 320
Tunnersoit, 324
Tunudiakbek, Cumberland Island, 111
Turks, 341
Turnagain Point, 149
Turnchauka River, 131
Two-faced nation, 317
Tyrians, 2; reputed to have invented ship-building, 4
Tytler, Frazer, on discovery in North America, 50

Ubygds, 22
Umiak, 309
Ungava Bay, 100
Urals, Ural Mountains, 19, 332, 333
Uria brunnichii, grylle and troille, 213
Ursus arctos, 277
Urville, Dumont d', 364; on Cote Clairie, 365
Utica founded, 4
Utkuhikalik or Utkuhikaling-mæut, 300, 306, 317.
Utkuhikalik-kok, 193
Utlulik Point, 198
Ut ultra of Button, 102; seen by Foxe, 109

Vaccinium uliginosum, 223, 288
Vaigatz or Waygatz (Pet's Strait), 60, 62, 63
Vaigatz or Henlopen Strait, 66

Vancouver's Island, 328
Varanger fiord, 19
Vasco de Gama, 35
Vaughan (Captain David), 114
Veer, Gerrit de, 66, 69
Vegetation of the Aniui Valleys, 272; of Arctic Siberia, 272; of the barren grounds, 263; of the Mackenzie Valley, 269; of Rensselåer Harbour, 226; of the Obdorsk M., 272; of Smith's Sound, 225; of Spitzbergen, 210; Scandinavia, 268; chapter on, 263
Victoria Harbour, 151; Land, 197
Victoria (South), chapter on, 367; glaciers and mountains of, 368; tides of, 368, 369, 370
Victory, The, 151
Vigdisa's monument, 29
Vogel-hoek, Spitzbergen, 97
Vogul country, 333
Vostochnoi sieveroi-nos, 234
Voyages, Dutch, to N.E., 64; English to N.E., 53; of Hudson's Bay Company, 114; to N.W. (Rundall), 94

Wager Inlet, 122; river, 117
Waggon-way, Spitzbergen, 206
Waigatz or Waygatz Island and Strait, Novaya Zemlya, 221, 332; currents of, 216
Wakash Indians, 327
Wakkendorph, Archbishop Erik, 65
Walden Islands, Spitzbergen, 215, 113
Walle, John de, 64
Wallich (Dr.), on marine animals living at great depths, 221
Walker (Dr.), on temperature of the earth, 257; on the freezing point of sea-water, 223
Walker, Cape, 144, 158, 175, 186
Wallace, James, his account of the Orkneys quoted, 29
Walrus, 279
Walsingham, Cape, Cumberland Island, 92, 94
Warbeck, Perkin, 43
Wardhuus in Finmark, 55, 57, 60
Warmow, Moravian brother, 91, 322
Warwick, Cape, 109; Foreland, 90; Island, 94
Warwicke, Earl of, 76
Water-fowl, arrival of, 168
Warzina in Finmark, 56
Washington, Cape, Victoria Land, 370
Watery surfaces contrasted with expanses of land, 374
Weide Bay, 200
Weide Jans Water, 204
Wellington Channel, 160, 200; entered by the American expedition, 182; currents of, 233
Wellington Sound or Channel, 143
West Bygd, 22
West England, Greenland, 87, 88
West Spitzbergen, 203

Westray Island, 29
Weymouth, Captain George, 94
Whales, 279
Whalebone, 316
Whale-fisher's bight, 241
Whale fishery, 280; origin of the Spitzbergen, 211
Whale Island, 129; Point, 102; Sound, 106
White Dolphins (Beluga), 316
White-fish, 282
White Sea, 19, 56, 63, 74
Wilkes, Captain, 363, 364; sees Terre Adelie, 365
Willes (Richard) on the North-west Passage, 44
Wilhelmina, drift of the bark, 216
William Torr, whaler, drift of the wreck of, 216
Willoughby, Sir Hugh, 54, 55; 's Land, 55
Winds, chapter on, 244; Maury's chart of, 244; currents affected by, 228; at Fort Confidence, 245; Repulse Bay, 247; in Siberia, 245; at Spitzbergen, 248; warm winter, 246
Winter harbour, Melville Island, 143
Witches Land, 204
Wolf, 316
Wollaston Land, 151
Wolstenholme, Sir John, 99, 106; 's ultima vale of Foxe, 110
Wolverine, 316
Women's Islands, Greenland, 93, 105
Wood, Captain James, 111
Wood, alluvial deposits of at Iakutsk, 292; fossil, in Siberia, 291; hills of the Liakhow Islands, 293

Woods on the Ob, 62; terminations of, 266, 267; antarctic line of, 379
Worthington (Will.), keeper of Cabot's writings, 45
Worsenholme, Cape, 101
Wrangell (Baron), 222, 317; on Nomades, 342; on the movements of reindeer, 275; on the Siberian coast, 134; on Siberian winds, 230, 244
Wynniatt, Lieutenant, 188; Wynniatt's farthest, 197.

XEMA SABINI, 213

YASSAK (Tribute), 131
Yell, a kind of fir, 62
Yenisei or Jenisei, 63
Yeniseians or Jeniseians, 341
Ygorsky-schar, 62
Young, Captain Allen, 194, 197
York, Cape, Australia, 355
Ysbrantzoon, Brant, 64, 65
Ysla de los aves (Bird Island), 48
Yugorsky-schar (Pet's Strait), 65
Yugrians, 341, 345
Yugrian Mongolidæ, 331; marriage, 345; skulls, 345; stock, 344
Yukahirs, Yukagirs, or Jukahirs, 333, 341, 343
Yukon River, 266, 296
Yuzhnuy Guisiney muis, 55

ZALFI, 48
Zeelandt, 64
Zeni, map of the, 32; supposed voyage of, 30
Zoology, chapter on, 274

www.ingramcontent.com/pod-product-compliance
Lightning Source LLC
Chambersburg PA
CBHW022112290426
44112CB00008B/640